Lecture Notes in Earth Sciences

42

Editors:
S. Bhattacharji, Brooklyn
G. M. Friedman, Brooklyn and Troy
H. J. Neugebauer, Bonn
A. Seilacher, Tuebingen

Antonio Cendrero   Gerd Lüttig
Fredrik Chr. Wolff (Eds.)

Planning the Use
of the Earth's Surface

Springer-Verlag Berlin Heidelberg GmbH

Editors

Antonio Cendrero
DCITTYM, Division of Earth Sciences, Faculty of Science
University of Cantabria
Av. de los Castros, S/N, 39005 Santander, Spain

Gerd Lüttig
Lehrstuhl für Angewandte Geologie, Institut für Geologie und Mineralogie
Universität Erlangen-Nürnberg
Schloßgarten 5, W-8520 Erlangen, FRG

Fredrik Chr. Wolff
Geological Survey of Norway
P. O. Box 3006, Lade, N-7002 Trondheim, Norway

"For all Lecture Notes in Earth Sciences published till now please see final page of the book"

ISBN 978-3-540-55353-3   ISBN 978-3-540-47031-1 (eBook)
DOI 10.1007/978-3-540-47031-1

This work is subject to copyright. All rights are reserved, whether the whole or part of the material is concerned, specifically the rights of translation, reprinting, re-use of illustrations, recitation, broadcasting, reproduction on microfilms or in any other way, and storage in data banks. Duplication of this publication or parts thereof is permitted only under the provisions of the German Copyright Law of September 9, 1965, in its current version, and permission for use must always be obtained from Springer-Verlag Berlin Heidelberg GmbH.
Violations are liable for prosecution under the German Copyright Law.

© Springer-Verlag Berlin Heidelberg 1992
Originally published by Springer-Verlag Berlin Heidelberg New York in 1992

Typesetting: Camera ready by author

32/3140-543210 - Printed on acid-free paper

Preface

The papers contained in the present volume were prepared from the contributions presented during an international Advanced Workshop held in Santander, Cantabria, Spain between 1-5 November 1989.

The workshop was a joint activity of the Working Group on Geology and Land-Use Planning (program "Geology and Environment", UNESCO), the Commission on Applied Quaternary Research (INQUA), the Sub-Commission on Maps of Environmental Geology (Commission of the Geological Map of the World) and the Grupo Español de Geología Ambiental y Ordenación del Territorio (Spanish Association for Environmental Geology and Land-Use Planning).

The aims of the meeting were to discuss a series of topics in which the four participating scientific bodies share an interest, to analyze the existing problems and trends and to identify certain lines along which work and/or actions will be particularly necessary in the near future. It was expected that the discussions and the conclusions of the meeting would provide useful guidelines for earth scientists working on environmental problems and for other professionals and officials who deal with environmental analysis, planning and management, either on a scientific basis or in a decision-making capacity.

The topics of interest in this field are many and varied, but the obvious limitations of a workshop make it impossible to touch all of them. Therefore, a selection had to be made, which is reflected in the contents. The editors are well aware of the fact that certain relevant subjects are not included in the volume, but the harsh realities do not always coincide with the desires or, indeed, the maximum expectations of the organizers of a meeting.

The workshop and the preparation of the contributions presented here were possible thanks to the sponsorship of the organizing scientific groups, as well as the following institutions: Consejería de Ecología Medio Ambiente y Ordenación del Territorio, Diputación Regional de Cantabria (Department of Ecology, Environment and Land-Use Planning, Regional Autonomous Government of Cantabria); Dirección General de Medio Ambiente, Ministerio de Obras Públicas y Urbanismo (General Directorate for the Environment, Ministry of Public Works and Urban Planning, Central Government of Spain); Dirección General de Investigación Científica y Técnica, Ministerio de Educación y Ciencia (General Directorate for Scientific and Technical Research, Ministry of Education and Science, Central Government of Spain); University of Cantabria; Caja de Ahorros de Santander y Cantabria (Savings Bank of Santander and Cantabria). The organization was carried out by the Instituto Regional de Medio Ambiente de la Diputación Regional de Cantabria (Environmental institute, Regional Autonomous Government of Cantabria) and by the DCITTYM, División de Ciencias de la Tierra, Universidad de Cantabria (Division of Earth Science, Department of Earth Materials Science and Engineering, University of Cantabria). The cooperation of these institutions, as well as the work and the contributions of the participants, were essential for the success of the meeting and are gratefully acknowledged. Dr. U. Mattig was the technical editor of the volume and her contribution was essential to the preparation of the final manuscript.

It is hoped that the results of the scientific presentations and the discussions, reflected in the contents of this volume, will provide some insights into certain problems of great relevance for the sound use of our environment and will constitute a useful guide for practitioners and authorities concerned with environmental planning and management.

Santander, Erlangen, Trondheim   A. Cendrero, G. Lüttig and F.Ch. Wolff
September 1990   Editors

Table of Contents

Chapter 3: Specific Geoenvironmental Topics.

Chapter 4: Organisational and Institutional Topics.

Chapter 5: Synthesis and Conclusions.

Chapter 1:

Planning the Use of the Earth's Surface: an Overview

Planning the use of the Earth's Surface: an Overview

A. CENDRERO *

It is a well known fact that the growth of human population and its increased technical capacities have brought about an important qualitative change in the processes that affect the surface of our planet. For the first time in the history of the Earth the actions of the human species have an influence on the global environment which is comparable to that of many large-scale natural processes. The effects of those actions are felt in the composition and physical conditions of the atmosphere, the hydrosphere and the soil layer, on the rates of transfer of solid matter from one point of the crust to another and on the disappearence of many plant and animal species.

These global changes have been a source of great concern for some years (see, for instance, FYFE, 1981; CHAFEE, 1986; MALONE & ROEDERER, 1986; EMBLETON, 1989), but it is only recently that they have become one of the important concerns of international bodies at the highest political level (WORLD COMMISSION ON ENVIRONMENT AND DEVELOPMENT, 1987).

The many and varied problems which affect the environment of mankind are in most cases of an interdisciplinary nature. But not only do they cut across traditional disciplinary boundaries, they also cut across national borders, having more often than not an international dimension.

The strategies which have been proposed or implemented for the analysis and

* Author's address: Prof. A. CENDRERO, DCITTYM, Division of Earth Sciences, University of Cantabria, Santander, Spain.

solution of such problems, both at national and international levels, do in general mention the need of an interdisciplinary approach. Nevertheless, they often fail to recognise the fact that environmental problems stem from the use of the surface of our planet by human beings and that, therefore, a proper understanding of those problems and the finding of adequate solutions for them cannot be achieved without due consideration of the constitution and dynamics of the earth's surface (ARCHER, LÜTTIG & SNEZHKO, 1987).

In the author's opinion, this is due to two main factors. First, the number of earth scientists is in most countries relatively small as compared to other specialists. Second, traditionally earth scientists have been involved mostly in the exploitation of mineral and energy resources and in the study of bedrock geology, devoting much less interest to surficial geology and environmental problems. The result has been that, in general, the level of geological input into environmental programs does not reflect the potential contribution of earth science to the solution of many problems derived from the interaction between humans and their environment (BARAHONA et al., 1987).

Nevertheless, since the first textbook on environmental geology appeared twenty years ago (FLAWN, 1970), we have been able to witness a steady increase in the interest of geologists towards the subject, interest which runs parallel to the growing demand for geologists trained in the analysis and solution of environmental problems. In recent years, this interest has been shown by the publication of several international volumes dealing with environmental-geological topics (ARNDT & LÜTTIG, 1987; WOLFF, 1987; ARCHER, LÜTTIG & SNEZHKO, 1987; SINGH & TIWARI, 1988; KOZLOVSKY & SYTCHEV, 1988; CENDRERO, 1989 a; WOLFF & CENDRERO, 1990) and by the organisation of graduate training courses in many countries and by various international bodies (CENDRERO, 1989b). The analysis of these and other publications shows very clearly that there is a general agreement on the need to take mainly preventive measures in order to avoid environmental problems, rather than solving them once they have appeared. If preventive strategies are to be implemented, planning is of the foremost importance.

The commissions and working groups concerned with environmental-geological matters within UNESCO, INQUA, CGMW and GEGAOT, considered that the time was ripe to hold an international workshop to examine a series of issues related to the application of earth science for environmental planning, discuss approaches and methods and reach some conclusions about actions to be taken in the near future, be it in the field of research and development or in the field of planning, management and decision-making.

In this workshop, special attention was devoted to what constitutes one of the main instruments in environmental planning: the production of maps as a basis for spatial or land-use planning. The basis of any preventive approach in the use of the earth's surface is the identification of areas or sites which are particularly vulnerable to human influence, or areas which are not suitable for certain uses, either because they may suffer environmental damage or because they may represent a hazard forhuman lifes and/or works.

A comprehensive analysis of geoscientific maps for land-use planning prepared in a variety of countries is presented by MATTIG. In her analysis, she identifies a series of problems in relation to the elaboration of these maps, such as:

a)   the difficulty to make comparisons between maps made in different countries for different purposes or to prepare small-scale synthesis maps at the international level, due to the variety of criteria and methodologies in map-making. The need for standardisation is pointed out.

b)   the problem of generalisation to pass from a variety of specialised thematic maps to a synthesis map, which should be the basic document for the planner and most potential users.

c)   the need to translate earth science data and assessments into statements which are meaningful for planning and understandable by the non-specialist. This implies that such statements must have a prognostic value, indicating what kind of behaviour is to be expected from the earth's surface if certain actions are to be implemented.

MATTIG suggests certain tasks which should be undertaken in the near future:

a)   increased cooperation with specialists in ecology, regional planning or economists,

b)   the participation of geologists in planning teams, throughout the planning process,

c)   a permanent up-dating and supplementation of data used for the preparation of maps, in order to be able to supply pertinent and good-quality assessments,

d)   the introduction of training programs of an interdisciplinary character, aimed at providing geologists with an understanding of the methods and needs of spatial planning and, viceversa, planners with a basic understanding of the nature of the earth's surface. The need of cooperation at the international level, for the testing of models, the adoption of standards and the setting up of training programs, is stressed.

The complexities of preparing earth science maps that meet the planners' requirements are discussed by MARKER. He points out that, although sound planning requires good information on the area to be planned, the connection between the planners' main concerns (industry, housing, retailing, infrastructure, health, education, recreation, tourism, etc.) and information on the 'ground' is not always obvious and may sometimes be neglected. There is an essential difference in approach between the earth scientists and the planner; the latter is driven by needs and aspirations, whereas the former starts by considering the nature and properties of the terrain. The author contends that, to be effective, a study aimed at providing the basis for planning should:

a)   address the right issues;

b)   be presented in a manner which is conductive to easy application;

c)   be linked to a formal procedure for its use;

d)   be properly drawn to the attention of the potential audience.

In his conclusions, MARKER indicates that there is a need of cooperation between planners and earth scientists in the specification of studies, the course of the work and in the designing of results; this means that the research to be carried out must be carefully targetted on the planning issues. Other points in his conclusions refer to the need to present results clearly and concisely, to disseminate the results of the studies and to establish a legal or advisory mechanism to incorporate such studies into the planning process. Systematic evaluation of studies, in order to assess their usefulness and to identify necessary changes is essential. The benefits of earth science studies for planning can be greatly increased through the use of computerised data bases and digital cartography, which facilitate the permanent up-dating of maps and the obtention of different map products.

Among his suggestions for actions to be undertaken in the near future he points:

a) improving public understanding of earth science issues related to planning, through the preparation of information booklets, by direct appearences and presentations to public audiences and by publications in planning journals.

b) making comparisons of the various ways in which: results are taken into account in different planning processes; planners are involved in the work; dissemination and evaluation of studies are carried out.

c) analyse the potential applications of GIS and interactive digital cartography in order to judge future trends.

d) preparation of an advisory manual on earth science mapping, covering material on planning for earth scientists as well as advice on the conduct of work, presentation and dissemination of results.

The important issue of the application of computer-assisted methods to the elaboration and display of geoscientific maps is discussed by GABERT. The author describes the three basic possibilities that exist for the preparation of such maps:

a) map construction directly from geological field data;

b) digitization of existing maps;

c) conversion of remotely sensed data in raster or vector maps.

Different methods are explained and examples of their application presented. He concludes that knowledge-based techniques will play an increasing role for the comprehensive evaluation of large databases stored in information systems and that the integration of digital data from terrain models, geoscientific maps and satellite images will in the future replace two-dimensional maps by three-dimensional representations. Tasks which will have an increasing importance in the near future in the field of cartography for planning are: design of geoinformation systems, construction of digital maps from databases, and integration of interdisciplinary data for the compilation of thematic and synthesis maps.

One important aspect of geoscientific cartography applied to environmental planning is the incorporation of natural hazards into environmental impact assessment, which is the topic developed by PANIZZA, based on the consideration of geomorphological hazards. He introduces the concepts of 'geomorphological environment' as the conjunction of 'geomorphological resources' and 'geomorphological hazards' and 'anthropical element' as represented by 'human activity' or a situation of 'area vulnerability'. In turn, he defines 'impact' as the result of the interaction between 'geomorphological resources' and 'human activity', and 'risk' as the interaction between 'area vulnerability' and 'geomorphological hazards'.

This author proposes a series of methodological steps in the application of this concepts to hazards and environmental impact assessment. The combination of data on the geomorphological history and dynamics of a region with others on the natural context and human activities will enable the definition of areas of geomorphological instability and the preparation of maps of 'geomorphological hazards'. These, combined with data on area vulnerability (which includes the socio-economic context) can in turn be used to obtain 'geomorphological risk' maps. On the other hand, certain geomorphological elements, when combined with a certain social organisation, constitute 'geomorphological resources' which can be negatively affected by human actions. According to the concepts presented by PANIZZA, the incorporation of geomorphology into environmental impact assessments can be made through the preparation, first, of maps of geomorphological hazards and risks and of geomorphological resources, and then combining these with the characteristics of the proposed project, to obtain maps of 'induced geomorphological risk' and of 'negative effects on resources'.

A particular kind of planning problems are the ones related to urban areas. The growing interest on urban geology amongst earth scientists and city planners, as a result of explosive urban development in the last two decades is mentioned by DE MULDER. He points out the specific problems that urban geology faces, such as difficulties for data acquisition because of the scarcity of observations points and the restrictions for systematic site investigation, sudden lateral changes in sub-surface beds due to human modifications, or problems related to the establishment of reliable, relevant boundary lines. Among his recommendations, the author includes: the need to set up a board of specialists and representatives of the munincipality to control the progress and monitor the quality of any urban geological mapping; to organise automated municipal databases for the storage, processing and display of urban geological data, with the help of GIS, so that up-dated, plotter produced maps can be supplied according to users' needs; to establish and define map units on the basis of their lithological constitution and their geotechnical and geohydrological properties; to develop methods to improve the reliability of lines on (large scale) maps and the clarity of map presentations.

Landsliding or, more widely, mass movements are probably the most widespread processes that can affect different types of human structures. The assessment of slope instability and the preparation of hazard and risk maps for planning are one of the most frequent tasks in environmental geological work. COROMINAS reviews the state of the art in this field, describing the main instability indicators and the existing methods for the identification and zoning of potentially unstable areas, as well as for the prediction of failure. This author indicates a series of topics on which research could be particularly promising. For the recognition of slope instability, this author recommends more detailed sedimentological studies on the fabric of deposits from mudflows, solifluction or rock avalanches as comapred to other colluvial or glacial deposits; improvement of the application of remote sensing techniques to the reconnaisance and follow-up of the evolution of an area after extreme events, such as intense rainfall, in order to establish the behaviour of slopes. In relation to the localisation of potentially unstable areas, COROMINAS poses several questions that need to be answered, such as: may all ancient landslides be reactivated? What kind of reconnaisance studies should be undertaken to be sure of the actual dimensions of an unstable slope? Is brittle failure possible without warning or some premonitoring signs always occur? How to predict the kind of movement which will take place and its possible transformations? What kind of influence on slope stability will have the uplifting of mountain areas, or the effect of weathering or cohesion decay? Efforts should also be made to improve predicting capabilities regarding the volume and reach of movements as well as the probability and moment of failure. In this respect, the application of dating techniques to obtain long and detailed series of landslide activity may be particularly interesting.

A very enlightening discussion of the application of geological concepts and models to the solution of specific planning problems is presented by JOHNSON, with a case study about assessment of groundwater pollution hazards in a glaciated Appalachian region. The concept of depositional systems, applied to the mapping and modelling of Quaternary glacial deposits, is used to predict the subsurface facies distribution and, therefore, the conditions leading to groundwater pollution hazard. The author concludes that a well-founded conceptual framework enables the development of predictive models that will inform about fluid behaviour in the subsurface. This sort of approach does not necessarily need costly investments and can be of use to many small communities relying mainly on groundwater for their supply.

Finally, in a paper reviewing the application of geoenvironmental maps to the solution of a variety of non-geological problems, CENDRERO et al. show how map units defined on the basis of geological factors can provide a good framework for the incorporation of other types of information and for expressing different terrain qualities relevant for planning and management, such as landscape quality, value for conservation, land-use suitability, hazard and risk, etc. These authors conclude that morphodynamic or geoenvironmental units can be used nor only for terrain evaluation, but also for the establishment of specific management plans and for the estimation of costs.

An other group of contributions are centered on different, more specific environmental topics, some directly related to planning problems and others addressing more general environmental issues.

The paper presented by MOLDAN contains a general discussion of the contribution of biogeochemistry to the understanding of environmental processes and changes and a brief consideration of some global problems. The author points out that the technology for the acquisition of high-quality data on a great variety of environmental variables from the different biogeochemical reservoirs (the continents, the sea, the atmosphere and the biota, and their subdivisions) is already available; the difficulties for an improved understanding of the processes taking place come from the need to obtain large numbers of reliable samples, and from the limited capability to make overall creative interpretations. From the available data, it appears that human contribution to or influence on the main biogeochemical fluxes is already very considerable, and probably increasing. Thus, according to the author, on the earth's surface the flow of matter due to human activities is three to five times higher than the one due to sedimentary processes. Another example of the degree of influence on biogeochemical cycles is the acquisition of net primary production by direct or indirect human activities, which amounts to more than 26 % globally, and as much as 39 % of terrestrial NPP. This means that only 61 % is left for all other terrestrial consumer species. When it comes to the atmosphere, the present rate of CO_2 emissions due to fossil fuel combustion represents 10 % of the carbon dioxide used by plants during photosynthesis. The rate of increase in this component and other gases producing greenhouse effect indicates that a global warming of 3.5 C is to be expected within the next 50 years. But the transformation of the nature of the atmosphere is also shown by the fact that the present rate of emission of most trace metals from human sources is nearly always greater than the one from natural sources.

MOLDAN suggests that the study of small catchments represents a promising methodological approach that will enable the acquisition of improved knowledge on the functioning of different biogeochemical processes and cycles. This methodology allows a better control of the different factors and makes experimentation also possible, thus facilitating the analysis of the "metabolism" of the systems, with or without human interference, and providing reliable, quantitative information about environmental impacts.

The environmental significance of geochemical information is very clearly illustrated by the relationships between human health and trace element distribution, discussed by THORNTON. The author reviews the work carried out to date for the preparation of geochemical atlases and for the establishment of correlations or causal relationships between the abundance of some trace elements in soils, river sediments, water, plants or airborne dust and the incidence of certain diseases. So far, the data available, mainly from the U.K., U.S.A., Scandinavia, India and China, show a clear cause-effect relationship only for iodine and goitre, fluorine and fluorosis or dental caries, and selenium and certain selenium-responsive diseases in China. Less clear is the relationship between cardiovascular disease and water hardness. The problems of establishing relationships, even at an empirical level, are related mainly to the differences in sampling criteria and strategies for mortality and morbility and those for trace element contents.

In his conclusions and recommendations THORNTON mentions the need to continue with the systematic elaboration of multielement geochemical maps and atlases, because they provide very useful information to identify potentially toxic or "problem" areas. He points out that sampling and analytical strategies must be developed to characterise the chemistry of the urban environment, as most people live in urban areas, and their contact with the natural environment is indirect and probably less relevant for their health. Research is also needed to define the processes which determine the incidence of certain elements in human health, in order to go beyond the present (mostly empirical) correlations to the establishment of causal relationships.

One important aspect that must be taken into consideration when planning the use of the environment is the availability of resources and the problems related with their exploitation. humans have traditionally extracted most of the mineral and energy resources they use from land areas, but in recent decades a growing proportion of those resources have come from the oceans. For obvious reasons, the part of the ocean most intensely affected by human activities is the continental shelf, especially the inner continental shelf, and this is also the area with the highest possibilities for the exploitation of a great variety of resources. The mining potential of the inner continental shelf is discussed by CHAR-LIER, who reviews the different types of substances that can be extracted, dissolved, suspended, from living matter, from placers or from consolidated deposits. The environmental impacts derived from mining activities on the shelf are also briefly described. This author concludes that the mineral wealth of the continental shelf is condsiderable and that it is likely that in the near future they will represent an important contribution to the world economy. The recent upsurge of interest in ocean mining has prompted the development of a series of new technologies and has increased greatly the extent and intensity of environmental impacts. Efforts will be needed in the near future to asses with detail the existing reserves of different substances on the inner continental shelf (especially such low unitary value resources as sand and gravel), and to develop new technologies for the reduction of environmental impacts derived from mining.

The contribution by LÜTTIG reviews the present trends in the use, exploration, exploitation and processing of industrial minerals and rocks. He points out that only three of these materials, sand and gravel, hard and dimension stone, and limestone and dolomite, account for nearly 60 % of the mass of energy and raw materials mined in the world. When it comes to value -- not counting the fossil fuels -- sand and gravel rank first, followed by iron ore and gold and, closely, by limestone and dolomite and hard and dimension stone. The exploitation of these and other surface materials is producing increasingly important impacts, and new technologies are being developed for improved exploitation, dressing and re-use. The need to improve exploration methods, including geoenvironmental criteria at the planning level, as well as for further research efforts to develop environmentally sound extraction and handling methods is indicated. These tasks will require the training of a new type of specialists.

One important consequence of the exploitation of mineral resources is the environmental degradation suffered by mining areas. The correction of environmental impacts due to mining activities, on land, is the subject of the contribution by GALLEGO and VADIL-LO, who review the current situation with regards to reclamation of areas affected by mining with particular attention to the situation in Spain. The authors consider the different types of impacts and the different possibilities of intervention: reclamation, restoration and rehabilitation. They stress the importance of follow-up programs in any action aimed at recovering degraded mining areas, in order to monitor the evolution of the zone. The costs of reclamation in spain represent an incidence of about 0.5 % of final costs.

Two courses of action are proposed for the correction of the existing problems. In the case of abandoned operations, plans should be set up for the inventory and classification of existing typologies. An initial diagnosis ought to follow, in order to identify the main impacts, to define possible actions to be undertaken and estimate their costs. The reclamation plans would normally come under the jurisdiction of munincipal or regional administrations, except for very large operations. To carry out these plans, it is recommended that public organisations be created, with funding from taxes on the resource sold. Tax incentives should also be given to companies that undertake the reclamation of old exploitations within their mining concessions. In general, priority ought to be given to the re-use of degraded land. The type of treatment and use to be applied to each exploitation should be included in the master plans for the zone. In the case of new exploitations, preventive studies are recommended for the inclusion of mining activities within general land-use plans. The studies ought to include maps showing zones with different types of mining potential, the main environmental units to the expected impacts; conflict maps should be prepared and, finally, guidelines should be defined for land-use in each area, and for the preventive and corrective measures to be implemented to mitigate impacts.

The restoration of mining sites is closely linked with the problems of earth science conservation, a subject which, despite early actions going back to the middle of the nineteenth century, has received little attention as compared to other types of conservation, as explained by GONGGRIJP. His paper reviews the main threats to geological and geomorphological landscapes and features and stresses the importance of preserving sites that represent irreplaceble scientific and educational resources. The main aims of earth science conservation are summarized by the author as: documentation of geological history throughout all parts of the geological column; protection and conservation of rare, valuable or threatened sites; protection of a sufficiently comprehensive and durable suite of localities to cover the needs of research and education. The preservation of such sites requires a compromise with other demands on land resources; therefore, a series of criteria are defined for the classification and ranking of sites. A brief survey of actions undertaken in several countries shows, according to the author, that although interesting national efforts have been made, there is a strong need of international cooperation and action. The initial steps in that direction have been taken by a recently created European Working Group on Earth Science Conservation. The aims of this group are: analysis of existing legislation, exchange of information, promotion of conservation through the dissemination of information, execution of common projects. The first projects to be undertaken, which are also guidelines for action, include: preparation of an information paper on earth science conservation illustrated with examples from different countries; publication of a manual on earth science conservation methods and procedures, which will help to achieve standardisation; inventory of Europe's most important sites ("Eurogeosites"), as a means to support and promote national programs, to inform national and international organisations and to gain public recognition and understanding. The development of a common set of standards and criteria and the systematic inventory and classification of sites appear as the most urgent tasks.

When trying to consider planning of the use of the Earth's surface, it is not sufficient to be "problem oriented", we must also be "regionally oriented". the specific nature of many environmental problems and the possible solutions for them, are determined not only by the conditions of the natural environment, but also by the socioeconomic context (WORLD COMMISSION ON ENVIRONMENT AND DEVELOPMENT, 1987). Thus, the type and intensity of such problems are very different in the developed and developing countries (ARNDT & LÜTTIG, 1987), and so should be the strategies to tackle them. An insight of the problems existing in developing countries is contained in the papers by PENHA and HERMELIN.

The case of Brazil, presented by PENHA, is representative of a very large country with a great diversity in the physical, ethnical and cultural characteristics, as well as in the degree of development. Wide areas with very low population density -- like the Amazonia, with 3 inhabitants/km -- show an intense degree of environmental degradation, mostly due to forest burning and to Hg pollution derived from gold mining. On the other hand, dense urban-industrial agglomerations, like the Rio de Janeiro-Sao Paulo region, with 40 % of unplanned, uncontrolled development, resulting in the occupation include a marked increase of runoff and erosion, with subsequent silting up of river channels and coastal wetlands, flooding of lowlands, and intensification of slope instability problems. Large losses of life and property have occured in recent years as a result of those processes. According to PENHA, the cause of this type of occupation is mostly socioeconomic and political, and technical solutions by themselves will not solve the problem. He concludes that the solution of the problems described requires careful preventive planning, based on an appropriate knowledge of the physical character of the land, and efforts in the application of existing legislation. He points out, as a most urgent task, the acquisition of basic data on the constitution of, and processes active in, the main areas of urban development, with the elaboration of geoscientific maps that could serve as a basis for planning.

Much the same situation is described by HERMELIN, who claims that many of the problems related to environmental geology in Latin America in general, and in Colombia in particular, are due to the occupation (often ilegal) of unsafe areas due to the explosive increase of population, particularly in urban areas, and to the fact that formerly safe areas have become hazardous as an indirect effect of human activities. Among this activities, the destruction of the vegetation cover for various purposes, with its concomitant increase in erosion rates, silting up of river courses and increased flood and slope instability hazards, is probably the most extensive. The problem is, according to HERMELIN, to a great extent of a social and political nature and socio-political structures often make it difficult to apply technical planning solutions. After reviewing some of the actions taken at the national and provincial levels the author points out several lines of action for the immediate future. These include training aspects; on the one hand, of specialists in environmental geology and natural hazards, through the establishment of postgraduate programs and the creation of one or several specialised 'graduate schools in Latin America and, on the other hand, of other professionals concerned with land planning and management, through the inclusions of basic courses on environmental earth science. According to HERMELIN there is a great need to set up geological research programs focused on Quaternary -- especially Holocene -- processes and geological hazards, to establish monitoring and alert systems, to carry out a zonification of urban areas in terms of hazards and other environmental factors and to integrate the resulting scientific knowledge in a set of legal norms and rules to regulate land-use at the national, provincial and minincipal level. The author claims that geologists must make an effort to communicate with the authorities and the general public, in order to achieve a proper understanding of the existing problems and possible solutions.

A final group of papers discuss some general aspects related with the incorporation of ideas and concepts on environmental geology into the structures that constitute the "real world", the framework within which actions have to be integrated to obtain practical results. This is particularly important in the field of environmental geology, which, by its very nature, is a problem-oriented discipline, rather than a theoretical one.

Very enlightening in this context is the paper by AURENHEIMER on the role of environmental geology within a regional administration. The author, a Ph.D. in Geology by training, with several years of experience in the administration of environmental matters in one of the main autonomous regions of Spain, as Director of the Environmental Protection Agency, points out that geoenvironmental concepts, methods and instruments can be very useful in very diverse fields of environmental management, but that environmental geology is only a means to solve certain problems and should not become an end in itself. Administrators and earth scientists should, in each particular region or situation, ask themselves where, when and how environmental geological knowledge can help in finding a solution to existing environmental problems, instead of trying to define how environmental geology should be managed. The potential contribution from earth science is greater in regions where environmental problems are related mainly to land occupation and/or degradation, natural hazards, water and soil conservation and protection of natural areas, as in the case of most Mediterranean regions.

The author comments that one of the main difficulties for incorporating geological expertise into environmental planning and management is the dispersion of jurisdictions between different governmental departments of a regional administration, and between the munincipal, regional and central administrations. In his opinion, master plans are the most suitable instrument for the integration of geological considerations into environmental policies. He suggests that E.I.A. should become mandatory for such plans, as this would reduce the impacts due to ill-planned land occupation.

In Europe and in other parts of the world, many people feel that an easy and reliable way to predict future social and scientific trends is to observe present and recent trends in the U.S. Although this view probably represents a simplification of reality, there is no doubt an element of truth in it. The well known influence of the publication of the National Environmental Protection Act in 1970, is a good example of the application of such principle to environmental matters. In this context, the review of present practice in the U.S., presented by LUNDGREN, is very interesting. Environmental geological issues seem to be dealt with, to a great extent, by state agencies. The work which is being carried out is concentrated mainly on coastal zone problems, hazards, hydrogeology (including waste disposal) and general mapping and GIS. It seems that the demand for specialists in these fields, particularly hydrogeology, is increasing.

With regards to environmental geological mapping, LUNDGREN emphasizes the need to prepare maps which combine geological and non-geological information, bearing in mind that these maps will be used as tools for the solution of environmental problems by people with different types of training and as a means to communicate with a broad audience. The importance of the public-information and communication role that environmental geologists are increasingly playing is also stressed by the author.

The observed practice leads LUNDGREN to make some recommendations with regards to training of future environmental geologists. Geologists working on environmen-

tal problems should understand that such problems are never purely technical nor manageable by science alone. Therefore, they should incorporate in their training the study of
basic of economic and social issues, the learning of communication skills and with other
skills related with the many non-geological tasks geologists often have to face when
dealing with hazards or other environmental issues. The consideration of ethical issues and
the recognition that personal involvement in such issues is perhaps not only inevitable but
desirable, should also be part of the training. Finally, thinking about the future, that is, the
likely evolution of land and water systems in their interactions with human systems should
be included in any training scheme.

The general application of data, procedures, instruments, etc. into everyday practice
and activities requires the development and acceptance of common standards. This issue
is discussed by STENESTAD, who points out that, besides scientific research, environmental geologists should develop tools for the solution of environmental problems. Some
of the tools would be monitoring, networking and cost-benefit analyses of environmental
solutions. The adoption of common standards can assure that relevant, high-quality data
are available and facilitate the exchange of data and knowledge, both at regional and international levels. Common standards would be useful both for the users - mainly international organisations and agencies establishing data networks for the inventory, analysis
and monitoring of the environment - and the producers, who more and more will find that
their databanks will become part of major networks, in general through the use of Geographical Information System. STENESTAD indicates that standardisation will have to be
accelerated on all levels of the "data chain", including environmental geology data,
methods of analysis and formats for the exchange of data. The standardisation of map
formats, contents and legends receives particular attention by the author. Due to the
complexities of the task, if common standards are to be established and accepted they
should be "minimum standards".

Among the tasks to be undertaken in the near future STENESTAD suggests the
development of minimum criteria for all links of the environmental geology data chain,
the development of standard environmental geological maps and map sets, the systematic

introduction of economic aspects in relation to environmental geology, and the concentration of efforts on hazard and impact assessments. Efforts should also be placed on the development of thesaurus's, catalogues of existing standards, indexing systems, codes and standards. To have professional standards accepted it is crucial to obtain a concensus, that should involve both producers and users. Because of the costs and difficulty of the task, there should be strong incentives for map and data producers to adopt existing standards, and the use of home-made systems should be discouraged.

If one of the main aims of the application of earth science expertise to environmental problems is to help solving them, it is obvious that people must be trained to do so. The paper by CENDRERO presents a series of guidelines for the organisation of graduate training courses in this field. The author points out that, on the whole, earth science graduates lack sufficient knowledge on topics related to planning and management, and often even on certain earth science subjects of particular relevance with regards to the environment, such as surficial processes, geomorphology, soils etc. A proposal for the organisation of a course is presented, including environmental earth science topics and others from related disciplines, aimed at providing a sound background for people interested in the professional aspects, as well as for those intending to go into research. The course proposed has a multidisciplinary, plurilingual and international structure.

It is hoped that the brief comments presented above will provide the reader with a general view and digest of the contents of this volume. The author has just touched upon some of the points which appear especially relevant to him, but his selection and emphasis will probably not agree with many readers' criteria or interests, and certainly do not reflect the many interesting problems and issues discussed by the different contributors.

When trying to plan for the correct use of the Earth's surface, there are many issues to be considered and many tasks to be performed, both in the field of basic research and in the implementation of specific solutions to existing problems, using already available knowledge. Obviously, not all of those, perhaps not even the majority of them, are included in this volume; but it is certain that the reading of its contents will provide ideas on methods and procedures, suggestions for approaches to problems, guidelines for action, identification of lines of work to be undertaken, etc. Hopefully, the work to be carried out in the next few years by earth scientists, in cooperation with other groups, will provide answers to some of the questions presented here and will result in an advance towards the solution of the problems described.

References

[ARCHER, A.A., LÜTTIG, G.W. & SNEZKO, I.I.] (1987): Man's dependence on the earth. - 216 p., Stuttgart (Schweizerbart).

[ARNDT, P. & LÜTTIG, G.] (1987): Mineral resources extraction. environmental protection and land-use planning in the industrial and developing countries. - 337 p., Stuttgart (Schweizerbart).

BARAHONA, E., CALVO, S., MAGARIÑOS, A., MAURE, B. & VILALLONGA, I. (1987): El medio ambiente en las organizaciones internacionales. - 126 p., Madrid (MOPU).

[CENDRERO, A.] (1989 a): Special issue on land-use problems, planning and management in the coastal zone. - Ocean and Shoreline Managem. **12** (5-6): 365-586, Barking/Essex.

CENDRERO, A. (1989 b): Environmental problems in the teaching of Earth Sciences at the graduate and postgraduate level. - In: [SUSANNE, C., HENS, L. & DEVUYST, D.]: Integration of environmental education into general university teaching in Europe: 95-108, Brussels (UNESCO-UNEP).

CHAFEE, J.H. (1986): Our global environment: the next challenge. - In: [TITUS, J.G.]: Effects of changes in stratospheric ozone and global climate, 1: 59-62, Nairobi etc.

EMBLETON, C. (1989): Natural hazards and global change. - I.T.C. J. **1989**-3/4: 169-178.

FLAWN, P. (1970): Environmental geology. Conservation, land-use planning and resource management. - 313 p., New York (Harper & Row).

FYFE, W.S. (1981): The environmental crisis: quantifying geosphere interactions. - Science **213**, 4503: 105-110, New York.

[KOZLOVSKY, E.A. & SYTCHEV, K.I.] (1988): Geology and the environment, 1: Water management and the environment. - 178 p., Paris-Nairobi (UNESCO-UNEP).

[MALONE, T.F. & ROEDERER, J.G.] (1986): Global Change. - 512 p., Cambridge (CSU Press-Cambridge University).

[SINGH, S. & TIWARI, R.C.] (1988): Geomorphology and environment. - 764 p., Allahabad, India (Allahabad Geomoph. Soc.).

[WOLFF, F.Ch.] (1987): Geology for environmental planning. - Spec. Publ., geol. Survey Norway 2, 121 p., Trondheim.

[WOLFF, F.Ch. & CENDRERO, A.] (1990): Special issue on Geology and the environment. - Engin. Geol., Amsterdam (in press).

WORLD COMMISSION ON ENVIRONMENT AND DEVELOPMENT (1987): Our common future. - 400 p., Oxford + New York (Oxford University Press).

Chapter 2:

Geoenvironmental Mapping and Assessment

Methods and Approaches of Environmental Geology Mapping: Meeting the Planner's Requirement

BRIAN R. MARKER [*]

Abstract

Environmental policies are influenced by a wide range of interests. Sensible planning of land use and decisions on planning applications depend on sound information. Much of this concerns economic, social and environmental issues. Whilst the physical and chemical characteristics of ground materials are relevant to many land uses, there is often a limited appreciation of this amongst planners. Collaboration between earth scientists and planners is needed to ensure that the right information is sought and provided.

Applied earth science mapping is one route towards providing the information. Such studies need to address the right issues and to present the results in the right manner. The results should be concentrated on the information essential for land-use planning. They need to be presented in a clear, concise and easily used form. Relevant basic concepts should be carefully explained and illustrated. Maps are most useful if they show factors directly related to planning policies. These will normally show economically significant earth resources and constraints to development. Maps showing the suitability of land for various uses are particularly valuable. Examples are given for sand and gravel resources and for mined ground.

[*] Author's address: Dr. B.R. MARKER, Minerals and Land Reclamation Division, Room C16/15, Department of the Environment, 2 Marsham Street, London SW1P 3EB, U.K.

Important steps in map preparation are selection of the scale and simplification of the key so that it can be used readily by non-geologists. The map scale should be linked to the level of detail required by the land-use planning system. The classification of factors on the map needs careful consideration and should be as straightforward as possible. Maps consisting of small complicated areas of ornament should be avoided.

If a map is very complicated it is likely that either the scale or classification needs amendment.

Results are only useful if they are applied. Dissemination of the results is important. Evaluation of the research and its application is needed for improvement of future mapping exercises. Results are most likely to be used effectively when there is a formal or advisory mechanism for incorporating them in the planning process. Long-term benefits accrue from data-bases if they can be maintained and updated. Direct digital cartography has potential for overcoming the problem of maps becoming rapidly outdated.

Public responses to the applied earth sciences need to be improved so that the significance of applied earth science mapping can be better appreciated.

Five topics are suggested as needing discussion and action.

Introduction

Environmental policies are made and implemented at many levels of society. Everyone concerned with the development and use of land, or with conservation, contributes, directly or indirectly, to policy formulation. The contributions may be through their actions, consumer and investment decisions, membership of special interest groups, and as electors. Central government usually sets a legal and advisory framework for balancing the various demands for development and for environmental protection, but the local authorities are, more commonly, the principal policy making bodies within the national framework. The balancing of interests is achieved through land-use planning, control of development in the course of carrying out development or during the subsequent land use.

The details of the system used vary greatly from country to country and it is not possible to review these in a short paper. The main points are, however, that in all countries a wide range of individuals and organizations, of varying backgrounds and interests are affected by these matters and that a system of policy statements advice and regulations is used to promote and limit the various activities. Information is needed to develop sensible policies and to allow rational decisions on specific development proposals. This includes material on social, economic, and environmental aspects of areas of interest and of individual planning applications. The physical and chemical characteristics of the ground may also be relevant although this is not always apprechiated by planners and developers.

Planners, whilst having a direct interest in natural resources through topics such as agriculture, forestry, fisheries and mineral resources (including soils and water), are largely taken up with issues concerning industry, housing, retailing, infrastructure, health, education, recreation, tourism and so on. The links between these topics and information on the ground are far less obvious, and may sometimes be neglected.

To take a hypothetical example, the planning aims in a particular area might be to promote new development whilst improving the urban environment and safeguarding wildlife and recreational amenities. Policies to achieve this might then have the objectives:

--   to secure reclamation of derelict land in the urban area,

--   to encourage redevelopment in the urban area,

--   to make provision for new housing on the fringes of the town, where this does not affect wildlife and recreational interests adversely,

-- to safeguard nature reserves, areas of attractive landscape, historic buildings and archaeological sites,

-- to constrain unsuitable development in a water catchment area,

-- to identify a new site suitable for long term waste disposal requirements,

-- to make sufficient provision for extraction of sand and gravel, over the next ten years, in environmentally acceptable locations.

The purpose of giving this list is to demonstrate that the perspective from which the planner views the issues is different from that taken by most earth scientists. The planner is driven by needs and aspirations in the area whilst the earth scientist commences by considering the state and properties of the ground before defining what it might be useful for. The planner must meet the objectives even though there may be problems and costs along the way. Of course this might not always be possible. If, for example, no economic sand and gravel resources remain in the area these will have to be sought elsewhere. Alternatively, all of the sand and gravel may be under wetlands of high conservation value or under potential housing land. The planner will then be faced with difficult choices. The earth scientist can provide considerable assistance by defining the resource areas. If we now take a contrasting example, the planner is unlikely to think, in the first instance, of a relationship between earth science information and policies for recreation. However our hypothetical area might contain areas of hills much used by walkers. These might have been the sites of ancient lead mines and be dotted with old shafts and adits. These could be a risk to the public using the ground and need to be taken into account when designating routes and amenity areas. The earth scientist can, again, provide assistance but, in this case, the planner might not ask for it.

Thus, in some cases the planner can give a brief to the earth scientist on what information is needed but in others the earth scientist needs to examine the policies and advise the planner on what information is required. The exercise needs to be a collaborative one. In recent studies carried out in Britain, this has been done by consulting planners on specifications of geological studies and involving them in steering committees convened to manage the research. In some cases planners have been included in the research team.

Such cooperation requires planners who are willing to explore what the earth sciences have to offer and earth scientists who are able to appreciate the way in which the planning system works. An open-minded approach is essential if the benefits of the work are to be fully realized.

Whilst those involved in the study may be fully attuned to the issues, the diverse audience which will ultimately use the results may not be. The way in which the results are presented and explained is crucial, therefore, both to fit them into the planning process and to ensure that sufficient notice is taken of them.

The remainder of this paper is concerned with the design of studies, presentation of results, and communication of them to meet these needs.

The Study

The success of any practical study depends on the way in which it is planned and executed, but it is equally important that the results should be properly applied. It is a waste of time and money if results, no matter how good, remain unopened in a shelf. The results are most likely to be used if they:
-- address the right issues,
-- are presented in a manner which is conductive to easy application,
-- are linked to a formal procedure for their use,
-- are properly drawn to the attention of the potential audience.

It is important, therefore, to take all of these matters into account when considering the best approaches to earth science mapping.

At the most basic level, such studies are concerned with the collection, collation, interpretation and presentation of earth science information in a form which can be readily used by non-specialists for the specific purposes of land-use planning, development control and environmental protection.

Collection and collation of information

The balance of data collection varies from area to area. It depends on the state of existing knowledge of the area including the amount of available information in, for example, site investigation reports, old mine plans, mineral borehols, records of past flooding, and so on. Some studies on areas which are well documented may require no further field investigations or only limited field revision. At the other extreme are poorly known areas which may require detailed mapping with drilling and testing in order to produce a useful result. Geologists, as scientists, quite properly prefer to have the fullest possible information on which to base their interpretations. However studies of this sort are usually carried out on limited budgets. It is important to carefully define what activities are needed to give a sound basis for the planning purposes rather than the scientific ideal. Some ranges of costs for recent British studies are given in Table 1. The design of the study needs to be flexible rather than a standard process applied blindly to all areas.

Equally the themes to be addressed need to be aimed at the specific planning requirements. Since these also vary from area to area it follows that economically planned research will concentrate on different issues in various areas. Of course, during the course of the work unforseen needs may be identified. The programme needs to be able to accomodate these. The implication is that a continuous dialogue between the study team and the ultimate users of the results throughout the study is prudent.

The assembly of information, whether from a primary survey or from existing sources, is an important task. The compilation needs to be in a form which gives flexibility of recovery for the interpretation of the material and its compilation in maps and tables. There is an obvious advantage if the data are organised into a form suitable for entry into a computer data-base. Such data bases range in scale from large national systems (CLAYTON et al. 1987) to small systems suitable for personal computers (FREEMAN FOX Ltd. 1988). The latter can be designed to be compatible with systems operated by local authorities and can form a valuable local source of reference at the end of a study.

Presentation of results

It has already been suggested that results should be presented in a form which is suitable for non-specialists and is directed towards the land use planning issues. The results will normally consist of maps and a descriptive report.

All results aimed at non-specialists should have:
-- a minimum of technical terminology; that which is essential needs to defined in the text or in a glossary,
-- concise presentation keeping to the information which is essential for the planning purpose and putting any necessary detail in tables or appendices,
-- concepts well explained and clarified using diagrams, including basic text book explanations where necessary.

The text should be self-contained and should not rely on the user seeking out references, unless these are very easily and widely available.

Apart from containing any necessary background information, the report needs to cover what work was done, how and when it was carried out and by whom, the results, their implications and limitations what additional advice might be needed, where it can be obtained and what types of expertise are needed by those who give the advice.

It is convenient for the reader of the report if explanatory diagrams and tables are set beside the relevant text. If the report accompanies one or more large maps it becomes difficult to handle both together. It is then useful to include summary maps at smaller scales in the text so that the points made can be quickly related to the maps and will guide the reader to the relevant areas of the main map sheets.

Sets of applied earth science maps, environmental geology maps, engineering geology maps or whatever they are termed, are very variable in their content. Commonly the output consists largely or entirely of maps showing single factors judged relevant to planning by the geologist. These might include, for example, sheets showing the extent of sand and gravel, thickness of peat deposits, depth to rockhead, recorded locations of landslides, the extent of mined ground and so on. It is quite true that these are relevant to planning and development decisions. However non-specialists are likely to be at a loss how to apply them. This is because such maps are not interpreted and presented to the stage where they are clearly identifiable with planning policies. They are essential working steps towards the interpretative and summary maps which will fulfill the needs of the user.

These maps which are more specifically targetted on the audience indicate factors such as:

-- the extent of resources which might be economically significant over a given period,
-- the extent of areas which are, or might be, subject to geological hazards or, if possible the areas likely to be at risk,
-- vulnerability of areas, for example the vulnerability of soil to erosion,
-- suitability of areas for various land uses, for example as a foundation medium for heavy structures,
-- the types of site investigations needed in support of development proposals,
-- the relative costs of development consequent on ground conditions.

At the broadest level, all of these matters can be covered by two maps showing, respectively, resources for development or constraints to development.

To take one example, mined ground is very widespread in England. Most has been mined over the last 200 years but some goes back to Roman and even Iron Age times. A map showing the distribution of mined ground and of recorded subsidence incidents is of obvious value if construction is contemplated in a particular mined area. However the map might be somewhat misleading to a non-specialist. For several reasons:

-- The existence of mined ground does not necessarily mean that there is a problem, the workings may have subsided long ago.

-- The records of past mining are very unlikely to be comprehensive; in England there were no clear commitments to lodge mine plans in a public archive until 1872 or, for some small mines, as late as 1911. This was after the peak period for mining in many of the areas.

-- The accuracy of many early records is suspect and it may be very difficult to transfer the information from old maps to a modern map-base.

-- The gross distribution of mined ground may not be related to subsidence potential which is influenced by the ground conditions, age, type and depth of mining, behaviour of groundwater and other factors.

There are obvious difficulties in preparing a more useful type of map. An attempt can be made, however, if there is sufficient knowledge of mining practices in the area and of past subsidence events. A recent study of a coal mining area in part of West Yorkshire, for example, collected considerable material on the extent and depths of workings (GILES 1988). Most subsidences in the area were caused by workings at depths of 30 meters or less below rockhead. It was possible, therefore, to define areas on the basis of this depth (Table 2) and to make allowance, also, for areas where recent open-cast extraction of coal would have removed any shallow mine workings. The result gives better guidance on areas where problems might be experienced in the future. This can only be done when the data on mined ground is fairly comprehensive. In another study, this time of part of the City of Bristol, extraction of coal was known to go back to mediaeval times. These early workings were from shallow adits driven into the outcrop and from small bell pits sunk into the seam at shallow depths. Most of the workings were, therefore, close to the outcrop on steeply dipping seams but extended back further when the dip was shallow. Whilst it was not possible to indicate where workings actually exist, a map could be constructed to show belts of land in which they might occur by drawing envelopes related to the outcrop position and dip of the individual seams (HUMPHREYS et al. 1987).

A second group of examples concerns maps of mineral resources. Whilst a map may show the extent, say, of sand and gravel in an area this will not help the planner a great deal.

It is more useful if the map takes into account the thickness of the deposit and of any overburden, and some measure of its quality. Several countries have programmes to produce maps this type (see, for example, THURELL 1981 and NEEB 1987). Such programmes normally apply a set of standard criteria for assessing the resource. One such set, used by the British Geological Survey in over 120 studies, is shown in Table 3 (an example of its application is CLARKE 1981). The standard approach is very useful in comparing various gravel-bearing areas throughout the country. In recent years, the Department of the Environment, which commissions the work, has recognised that these criteria, whilst useful, describe a long-term resource. Consultations with the industry have shown that in specific areas the economic criteria for the sort of period which is of interest to planners, say 10 to 20 years, needs to be more tightly defined. In current studies, therefore, two sets of criteria are used: the original set for comparibility and a set defined in terms of local conditions. An example is shown in Table 3. This allows the planner to consider shorter term areas of search for working and to have the longer term information to help decisions on safeguarding resources from other forms of development (an example of a study of this type is CRIMES et al. 1989).

Recent British studies also review planning constraints on land-use in the resource areas so that a wider range of relevant information is at hand. This has been done on studies elsewhere for some time, for example in France (DORIDOT & RÉSENDE 1977). It is not possible in the space available to describe more of the numerous resource and constraints maps which have been produced in many parts of the world. A wider range of these are reviewed in McCALL & MARKER (1989). The important point is that, with thought, maps can be made more directly useful and that if the relevance is direct the non-specialist is more likely to recognise the value of the results.

As with reports, maps need to be clear, concise and presented in straight-forward terms. The key should be as simple as possible. This is the reason that many of the excellent engineering geology maps produced over the last 20 years are not ideal for planners. They often have complex legends including considerable detail on the ground conditions and summaries of geotechnical data which non-specialists are likely to find daunting (see, for example, GOSTELOW & BROWNE 1986).

Clarity depends partly on scale. The scales of maps needed for various levels of consideration of land use have been reviewed by LÜTTIG (1987). The scale is also influenced by the detail of the available data. In general, it is important that the areas shown on planning maps should not be too small and complicated. If they are, it is likely that the wrong scale has been selected or that the wrong classification has been used in the key. There can be problems in trying to reduce the numbers of categories in a complex set of information. One approach which has been developed is to assign values to various sets of parameters and to combine these, usually with some appropriate level of weighting, to give scores to different parts of the map. Examples of this approach are the work of CENDRERO et al. (1976) in Spain or the habitability maps of parts of Morocco (HAFDI 1987). It is important to remember that non-specialists often have difficulty in reading the vertical dimension into a map. Numerous studies contain thickness or depth maps which attempt to convey this. A limited number have attempted alternative approaches. A map of foundation conditions in the vicinity of Marseilles, for example, used superimposed coloured ornaments to indicate the properties of 3 m thick slices of ground. The resulting map (IAEG 1976) is, however, visually complicated. More recently, the problem has been addressed in a study of part of the Clyde Valley in Scotland (BROWNE & McMILLAN 1987). This divided the drift sequence above a specific till horizon into coarse and fine components. Areas containing only coarse of only fine deposits and areas of outcropping bedrock were mapped. The remainder of the study area was then classified according to the combinations of coarse and fine units present. The result was a visually clear map. This now needs to be tested by users of results.

In many cases studies have to fall back on the use of three-dimensional diagrams such as the excellent block representations which accompanied a study of the Franconia area, West Virginia (FROELICH et al. 1978).

However well the results are presented, problems may attend the publication of applied maps. The most notable of these are the possible blighting effects of maps showing hazards and risks. These may lead to a loss of property values, withdrawal of insurance cover, or decline in investment. These are serious matters but ones which cannot be sidestepped. It is incumbent on those preparing such maps to be as accurate as possible. However most planning maps are indicative rather than definitive. The style of representation of lines on the map is often used to denote uncertainly. Some examples of the pixel-like style is a study of landslide potential in the Rhondda Valley, south Wales (SIR WILLIAM HALCROW & PARTNERS 1988) in which a series of values was attached to each square in the study area and these were combined to give a general measure of landslide potential. An example of the use of scatters of symbols was a housing suitability map forming part of the Franconian study (FROELICH et al. 1978).

The emphasis on simplification for non-specialists can lead to a slim, straightforward report and summary maps which are then spoken of in derogatory terms by earth scientists. That is not a helpful attitude, since it can cause non-specialists to lose confidence in what are, in fact, valuable aids to planning. It needs to be clearly appreciated by earth scientists that clear communication is important, will benefit earth scientist and planner alike, and should be given much care and attention. It is not a second rate task to be rushed off after "the interesting part of the work has been completed".

Reaching the audience

The results of any study only meet the objectives of the work when they are used. Given the wide range of individuals and interests involved in land use, this can only be when the results have been published. However the work should not end with publication since there are other important tasks which need to be carried out by the earth scientist. The first of these concerns dissemination of the results. In England, the approach taken by the Department of the Environment is to issue press notices to announce results, to organize seminars and presentations to selected audiences, to contribute papers to journals, and to distribute leaflets. Display material is taken to conferences. It must be emphasised that these efforts should not be aimed at other earth scientists. They need to be focussed on the potential users of the results. This may not be easy. If, for example, a seminar is organized on a geological study for planners it is usually only those who already appreciate the relevance who attend. Those who do not, and are thus the most important group to reach, are those who have no such appreciation. This is a difficult problem to overcome. A possible tactic is to subordinate the earth science aspect to a more general planning topic which might draw in a wider audience.

It should not be lost sight of that planners are busy experts in their own field. It is useful, therefore, to preface every study by a brief executive summary of the contents. Planners often find this useful when they need to present a concise summary of relevant points to elected members, or to other committees.

Application of the Results

Drawing attention to results is only half of the battle. These are most likely to be used if there is a clear mechanism for taking them into account in the planning process. It is no accident that environmental geology coverage is furthest advanced in systems where the maps have legal force (for example in the FRG; WILLE 1987) and in centralized systems such as that of Czechoslovakia (MATULA 1971 and 1979).

In a system which is based more on guidance, such as that in England, a less direct approach has to be taken. On land instability, for example, it has been necessary to prepare planning guidelines (DEPARTMENT OF THE ENVIRONMENT, in preparation) which outline the principal types of land instability, the appropriate responses within the planning and development system, and the sources of information which are available. Such government advice has to be noted by planners because the Secretary of State for the Environment has a role in approving certain planning documents prepared by the local authorities and in deciding appeals against refusals of planning permissions by the authorities.

However much depends on the local authorities receiving adequate information with planning applications, thus developers as well as planners need to take note of results. A way of drawing attention to such matters is the preparation of a guidance leaflet which can be given to the prospective developer before an application is lodged. Such a leaflet was prepeared, for example, as part of the out put from the study of landslide potential in the Rhondda Valley (RHONDDA BOROUGH COUNCIL, not dated). In addition the BOROUGH COUNCIL has prepared a policy statement for inclusion in the local plan for the area which indicated that applicants will need to satisfy the authority of the stability of sites within areas indicated as having significant landslide potential. Approaches such of these translate the results from research to application and are an essential part of an environmental geology initiative.

Lasting Benefits

In this day and age, it is becoming increasingly important to evaluate activities to measure their success and the value in relation to the expenditure. It is important to assess if research has been well executed, whether dissemination has been effective, and whether results are being properly applied. This helps in assessment of improvements which may be needed in future work. It is also important to establish the benefits which come from carrying out the research but, to date, comparatively few attempts to do this have been published. Close attention needs to be paid to work on this by DE MULDER (1988) and AYALA et al. (1986).

One clear source of lasting benefits is the data base which may be compiled as the basis for a study. If it is organized into an easily accessible system, it is a long term reference source for scientists and non-scientists alike. However this benefit can be greatly increased if there is an arrangement for maintaining and updating the collection. The ideal is to have a computer data-base linked to a graphics system. This can be used as part of a geographical information system which are increasingly used by local authorities. It can also form the basis of a system for continually updating maps as more information becomes available.

With that in mind, the British Geological Survey and the Department of the Environment jointly funded a study of the Southampton area which attempted to produce thematic maps directly from a data base using an ORACLE data management system and an INTERGRAPH work station (LOUDON & MENNIM 1987; EDWARDS 1987). The exercise was only partly successful but sufficient experience was gained to take the techniques forward on a current study of the Wrexham area of north Wales, with very promising results. This work uses a large and expensive system. The Department of the Environment therefore commissioned the Soil Survey and Land Research Centre to undertake a parallel study of soils in part of the Southampton area on a smaller computer. This has successfully produced maps showing erosion risk, land suitability for various uses, and other interpretative themes (HODGSON & WHITFIELD 1989). Such developments in a number of countries are leading the way towards a time when the age-old problem of outdated maps can be overcome without full resurveys.

The Non-Geological Dimension

At a number of points this paper has referred to earth scientists rather than geologists. Land-use planning has wide information requirements and needs to use the talents of a broad range of specialists including hydrologists, geomorphologists and climatologists as well as geologists. The issue goes wider than this. Because planning involves the whole scope of land use and conservation, biological aspects and the human heritage may need to be considered alongside the physical and chemical characteristics of the ground. This is likely to become increasingly the case as more attention is paid to environmental assessments of major development proposals (LEE & WOOD 1978). Some multidisciplinary studies have already been undertaken, especially in Spain (for example CENDRERO et al. 1976; COMMUNIDAD DE MADRID 1986) and the FRG (BECKER-PLATEN et al. 1987). It is a short step from environmental geology mapping in the broad sense to environmental mapping in general.

Improving the Response

The low-key reception which is often given to environmental geology results is partly a consequence of training. WORTH (1987) has pointed out that, in Great Britain, very few courses for planners include any earth science. In addition, the earth sciences play a comparatively small part in school curricula. The media give some coverage to the earth sciences but this tends to concentrate on large-scale theoretical aspects, such as plate tectonics and end-Cretaceous extinctions, or on the human effects of major disasters. Applied earth sciences are rarely mentioned. Conversely, very few earth scientists have any introduction to planning issues. It is not surprising, therefore, that the two disciplines do not collaborate as much as they might.

Over the last 20 or so years, numerous texts have been published on earth science relevant to planning (for example, LEGGETT 1973; UTGARD 1978), but these tend to be sizeable texts and do not get the circulation which they merit. A few attempts have been made to reach a wider audience with slimmer, well presented, volumes; notably ROBINSON & SPIEKER (1978), and two excellent booklets from the Netherlands (LABORATORIUM VOOR GRONDMECHANICY 1984; RIJKS GEOLOGISCHE DIENST et al. 1986). Understandably, all three of these publications concentrate on examples from the country of origin. In addition they are targetted on the topics of environmental geology mapping and engineering geology rather than the broad range of benefits which come from the earth sciences.

This broader issue of improving public appreciation of applied earth science is perhaps more important than promoting the use of applied earth science maps which are, after all, only one method of illustrating results. There is a need, therefore, to prepare more booklets drawing attention to these matters and for earth scientists to take every possible opportunity to speak to the public on them. As it is, scientists commonly spend most of the time communicating with each other rather than with the wider potential audience.

Conclusions

The principal conclusions are that

-- land use planning involves a wide range of environmental issues and a broad spectrum of society, much of which is little acquainted with the earth sciences.
-- Planners approach problems of land use from a different perspective to earth scientists thus it may not be obvious to them what geological studies may offer for policy and decision making.
-- Cooperation between planners and earth scientists is needed in specification of

studies, in the course of the work, and, especially, in designing results.

-- The research needs to be carefully targetted on the planning issues.

-- Results need to be presented clearly and concisely: close attention should be paid to the ways in which results are to be used thus maps which address the planning issues are preferable to those which only show factual material.

-- Dissemination of the results is an essential part of the exercise. The methods need to be carefully thought out and visual and other aids prepared during the study.

-- Results are most likely to be applied if there is a formal legal or advisory mechanism for taking them into account in the planning process. Liaison between earth scientists and administrators is needed to achieve this.

-- Evaluation of the study and of the application of the results helps to demonstrate effectiveness and indicates where changes are needed in future studies.

-- The benefits of studies can be extended if data-base are properly organized and are subsequently maintained and updated. Direct digital cartography may overcome the problem of maps becoming quickly outdated.

-- There is a need to consider when topics from outside the earth sciences should be included in the research in order to give a more general planning base.

-- Initiatives are needed to improve public understanding of what the earth sciences have to contribute to decision making are needed.

These conclusions lead to five suggestions of matters which could be usefully discussed when considering the way forward for applied earth science mapping. These are:

a)   that public understanding of the applied earth sciences needs to be improved by the preparation of cheap, well illustrated booklets. To have the greatest effect, these should concentrate primarily on examples from the country of origin,

b)   that earth scientists should take every opportunity to speak to the public about these matters and to contribute articles to popular and planning journals,

c)   that comparisons of the various ways in which

-    results are taken into account in different planning processes,

-    planners are involved in the work,

- dissemination of results and evaluation of studies are carried out, would be useful,

d)   that discussions of the potential applications of geographical information systems and of interactive digital cartography would assist in judging future trends,

e)   that an advisory manual on applied earth science mapping is needed but this needs to cover material on planning for earth scientists as well as advice on the conduct of work, presentation of results, and the dissemination of results. Given the fact that different planning systems exist in the various countries of the European Community there would be an advantage for potential contractors if this is done before 1992.

Acknowledgement

© Crown Copyright, 1989.

This paper is reproduced by kind permission of Her Britannic Majesty's Stationery Office. The views expressed in this paper are those of the author alone and do not necessarily correspond to the views of the Department of the Environment.

References

AYALA, F. (1986): El impacto socio-economico de los riesgos geologicos in Espana.- 421 pp., Madrid (IGME).

BECKER-PLATEN, J.D., DORN, M. & LOOK, E.R. (1987): An introduction to the legend of the geoscientific map of the natural environments potential (GMNEP) of Lower Saxony and Bremen.- In: [ARNDT, P. & LÜTTIG, G.]: Mineral resources extraction, environmental protection and land-use planning in the industrial and developing countries: 119-126, Stuttgart (Schweizerbart).

BROWNE, M.A.E. & McMILLAN, A.A. (1987): Geology for land use planning; drift deposits of the Clyde valley.- Brit. geol. surv. Res. Rep., 96 pp., 2 maps, Edinburgh.

CENDRERO, A. et al. (1976): A technique for the definition of environmental geologic units and for evaluating their environmental value.- Landscape Planning **3**: 35-66, Amsterdam.

CLARKE, M.R. (1981): The sand and gravel resources of the country north of Bourne mouth, Dorset.- Inst. geol. Sci., Mineral Assessment Rep. **51**: 1-128, 1 map, London (HMSO).

CLAYTON, A.R. et al. (1987): Geological data bank pilot study.- Brit. geol. Surv. Res. Rep. **ICSO/87/1**:1-23, Keyworth.

[COMMISSION ON ENGINEERING GEOLOGY MAPS, IAEG] (1976): Engineering geological maps: a guide to their preparation.- Earth Sci. Ser. **15**: 1-79, Paris (UNESCO).

[COMMUNIDAD DE MADRID, CONSERJERIA DE AGRICULTURA Y GRANADE RIA] (1986): Mapa fisiografica de Madrid.- Escala 1:200 000, 42 pp., 1 map, Madrid.

CRIMES, T.P. et al. (1989): Assessment of sand and gravel resources to the east of Royal Leamington Spa.- 1 + 2, 101 pp., 7 maps, Liverpool (Univ. Liverpool, Departm. Earth Sci.).

[DEPARTMENT OF THE ENVIRONMENT] (1989): Planning policy guidelines: Development on Unstable Land.- London (1990 in preparation).

DORIDOT, M. & RÉSENDE, S. (1977): Étude des gisements de materiaux alluvionaires pour la préparation d'un SDAU.- Bull. Liaison Laborat. Centr. Ponts et Chauseès, spec. Rep. **IV**; Granulats ressources et prospection des gisements: 141-151, Paris.

EDWARDS, R.A. et al. (1987): Applied geological mapping: Southampton area.- Brit. geol. Surv. Res. Rep. **ICSO/87/2**: 1-69, 11 maps, Exeter.

[FREEMAN FOX Ltd.] (1988): Methods of compilation, storage and retrieval of data on disused mine openings and workings.- 132 pp., London (HMSO).

FROEHLICH, A.J. et al. (1978): Franconia area, Fairfax County, virginia.- In: [ROBIN SON, E.D. & SPIEKER, A.M.]: Nature to be commanded.- U.S. geol. Surv. prof. Pap. **950**: 68-89, Washington.

GILES, J.R.A. (1988): Geology and land-use planning: Morley-Rothwell-Castleford.- Brit. geol. Surv. techn. Rep. **WA/88/33**: 1-48, Keyworth.

GOSTELOW, T.P. & BROWNE, M.A.E. (1986): Engineering geology of the Upper Forth Estuary.- Brit. geol. Surv. Rep. **16 (8)**: 1-56, 8 maps, London (HMSO).

HAFDI, A. (1987): Approach of a methodology for drawing up a habilitability map. In: [ARNDT, P. & LÜTTIG, G.W.]: Mineral resources extraction, environmental protection and land-use planning in the industrial and developing countries: 271-278. Stuttgart (Schweizerbart).

HODGSON, J.M. & WHITFIELD, W.A.D. (1989 in preparation): Applied soil mapping in the Southampton area.- Special Survey 16: VI + 127, Silsoe (Soil Survey and Land Research Centre).

[HUMPHREYS HOWARD & Partners] (1987): Environmental geology study of the Bristol area.- 1-3: 3 vols, 99 pp., 15 maps, Leatherhead (Howard Humphreys etc.).

[LABORATORIUM VOOR GRONDMECHANICY & RIJKS GEOLOGISCHE DIENST] (1984): Ingenieurs-geologische Kaarten van Nederland.- 20 pp., Delft and Haarlem.

LEE, N. & WOOD, C. (1978): Environmental impact assessment of projects in EEC countries.- J. engin. technol. Man. 6: 57-71, Amsterdam.

LEGGETT, R.F. (1973): Cities and Geology.- 624 pp., New York (Mc Graw Hill).

LOUDON, T.V. & MENNIM, K.C. (1987): Mapping techniques using computer storage and presentation for applied geological mapping of the Southampton area.- Brit. geol. Surv., Res. Rep. ICSO/87/3: IV + 21 pp. Edinburgh.

LÜTTIG, G.W. (1987): Large scale maps for detailed environmental planning. Geology for Environmental Planning.- Norg. geol. Unders., spec. Publ. 2: 71-76, Trondheim.

MATULA, M. (1971): Engineering geologic mapping and evaluation in urban planning.- In: [NICHOLS, D.R. & CAMPBELL, C.C.]: Environmental planning and geology: 144-153 - Menlo Park (USGS & Dep. Housing and Urban Developm.).

[McCALL, G.J.H. & MARKER, B.R.] (1989): Earth science mapping for planning, development and conservation.- 268 pp., London (Graham & Trotman).

MULDER, E.F.J. DE (1988): Thematic applied Quaternary maps: a profitable investment or expensive wallpaper? - In: [MULDER, E.F.J. DE & HAGEMAN, B.P.]: Applied Quaternary Geology: 105-117, 3 Fig., Rotterdam (Balkema).

NEEB, P.R. (1987): Development of a Norwegian national ADP-based file for sand, gravel and hard rock-aggregate.- Norg. geol. Unders., spec. Publ. 2: 61-67, Trondheim.

[RHONDDA BOROUGH COUNCIL] (1989): Development on unstable land: guidance note for developers. Leaflet (Rhondda Borough Council).

[RIJKS GEOLOGISCHE DIENST, STICHTING VOOR BODEMKARTERING & DIENST GRONDWATERVERKENNING TNO.] (1986): Subsoil uncovered.- 36 pp., Haarlem, Wageningen and Delft.

ROBINSON, G.S. & SPIEKER, A.M. (1978): Nature to be commanded - earth science maps applied to land and water management.- U.S. geol. Surv. prof. Pap. **950**: 1-96, Washington DC.

[SIR WILLIAM HALCROW & Partners] (1988) Rhondda landslip potential assessment - planning guidelines.- III + 61 pp., Cardiff (Sir William Halcrow and Partners).

THURRELL, R.G. (1981): Quarry resources and reserves: the identification of bulk mineral reserves.- Quarry Management & Products **8**: 181-192, Nottingham.

UTGARD, R.O. et al. (1978): Geology of the urban environment.- 363 pp., Minneapolis (Burgess Publishing).

WILLE, V. (1987): Examples of maps for the planning process. - In: [ARNDT, P. & LÜTTIG, G.W.]: Mineral resources extraction, environmental protection and land-use planning in the industrial and developing countries: 271-278. Stuttgart (Schweizerbart).

WORTH, D.H. (1987): Planning for engineering geologists.- In: [CULSHAW, M.G. et al.]: Planning and engineering geology. Geol. Soc. London, engin.Geol. spec. Publ. **4**: 39-46, London.

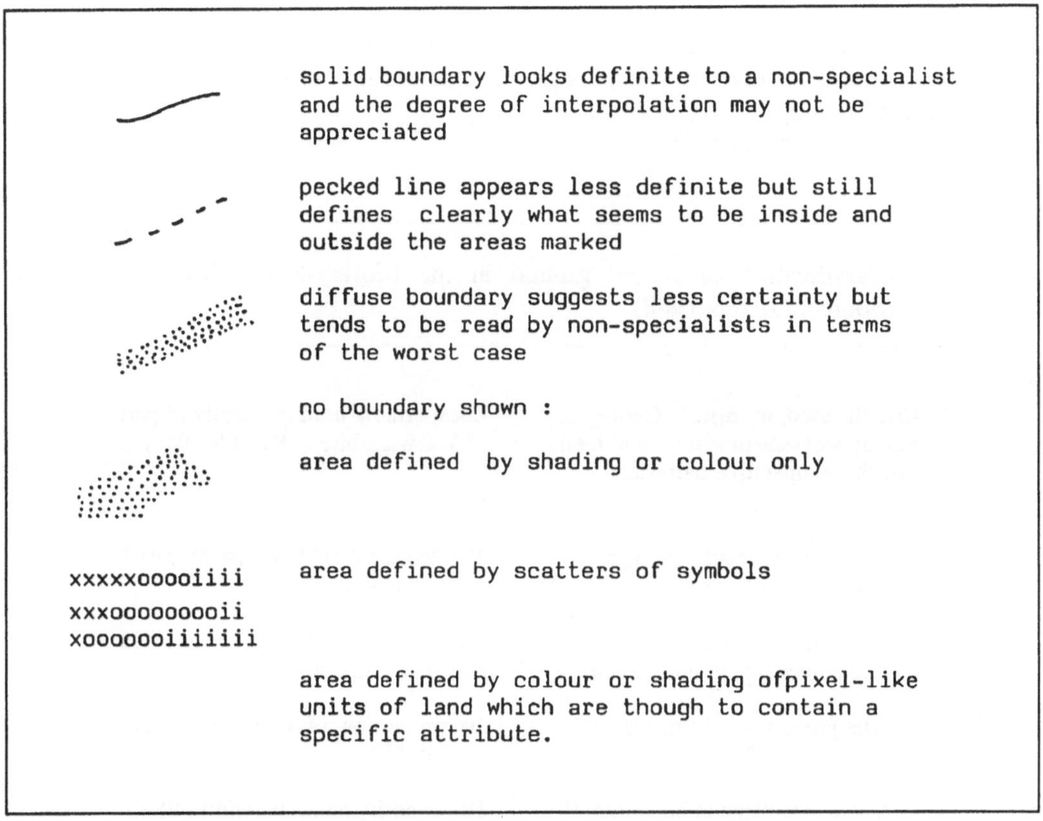

solid boundary looks definite to a non-specialist
and the degree of interpolation may not be
appreciated

pecked line appears less definite but still
defines clearly what seems to be inside and
outside the areas marked

diffuse boundary suggests less certainty but
tends to be read by non-specialists in terms
of the worst case

no boundary shown :

area defined by shading or colour only

xxxxxooooiiii   area defined by scatters of symbols
xxxooooooooii
xooooooiiiiiii

area defined by colour or shading ofpixel-like
units of land which are though to contain a
specific attribute.

Figure 1:   Selected examples of lines and symbols used to define areas on maps with
             varying degrees of certainty.

Currency	Full geological field mapping	Check mapping or no field mapping
U.K.	770-2460	200-1260
GFR	2400-7600	620-3900
France	7900-25300	2060-13000
Netherlands	2700-8600	700-4400
Spain	147100-470000	38200-240700
ECU	1140-3650	300-1900

Note: The figure are for recent studies in the UK. The figures are given in several
currencies for convenience. The high values are for fairly complicated coalfield areas with
very large amounts of existing data. Low values are for geologically straightforward areas
with limited amounts of existing data.

Table 1:   Costs per km of recent applied geological mapping studies

Workings known or inferred at 0-30 m. depth below rockhead.
Workings known or inferred at depths greater than 30 m. below rockhead.
No record of mining.
Worked out open-cast sites.

Table 2:   Classification of mined ground in the Morley-Rothwell-Castleford area.
(Source: GILES 1988)

Criteria used in British Geological Survey assessment studies and retained for comparative purposes	Local criteria used in a study of part of Warwickshire (CRIMES 1989)
1. The deposit should average at least 1 m. in thickness	The deposit should average at least 1 m. in thickness
2. The ratio of overburden to sand and gravel should not exceed 3:1	The ratio of overburden to sand and gravel should not exceed 2 : 1
3. The proportion of fines should not exceed 40 %	the proportion of fines should not exceed 20 %
4. The deposit should be within 25 m. of the surface	The deposit should be within 25 m. of the surface

Table 3:   Criteria used to define resources in sand and gravel resource studies

Geoscientific Maps for Land-Use Planning
-- a Review

ULRIKE MATTIG *

Abstract

This paper gives a brief overview of the most important models and suitable drafts for an evaluation of the natural environment's potential in different countries and explains the development, the state of the art and the main tendencies in the field of geoscientific maps.

The basic prerequisites and ideas for the production of these maps will be introduced by five central questions. Their philosophy and system is described by means of a selected example (Geoscientific Map of the Natural Environment's Potential/FRG). This is followed by a discussion of the specific problems and main trends (e.g. computer application).

After a general overview of lines of approach for further research and development, a definition and a recommendation for the compilation of geoscientific maps, relevant to planning, is given.

The contribution of the geoscientist's complex and expert knowledge of regional planning and spatial management can only be made available by providing a comprehensive, clearly arranged cartographic presentation, which has to be simple, distinct, adaptable and capable of development.

* **Author's address:** Dr. U. MATTIG, Chair of Applied Geology, Institute for Geology and Mineralogy, Schloßgarten 5, D-8520 Erlangen, FRG.

Geological information and planning

The increasing world population and the development of human technology, with its increased demand for space and raw materials, has lead to uncontrolled urban spreading and extended consumption of non-renewable, non-increasable and, particularly, unmovable natural resources such as soil, groundwater, minerals and landscapes. In spite of knowing the "limits of growth ", irreplaceable geogenetic potential is wasted every day by misplanning or lack of planning all over the world. This leads to considerable economic loss, pollution and hazards (ARCHER et al. 1987).

These problems have strengthened the public awareness for the limitation of resources and the vulnerability of the environment, caused by the massive human interference in the natural cycles and the anthropogenic and geogenic hazards, incidental to that interference. The complex body of knowledge of geosciences is essential for a proper and responsible use of the natural environment's potential. This means, firstly, repair and correction of already existing damages and hazards, and, secondly, foresighted, optimal planning for better management of resources by using new approaches and technologies (LÜTTIG 1977, BGR 1983, KÖNIGSSON 1987).

The concept of the "natural environment's potential" (**NEP**) can be very useful for planning (VON DANIELS & LÜTTIG 1982, LÜTTIG 1983, AUST & BECKER-PLATEN 1985, WOLFF 1986, 1987b). The NEP includes all natural materials and processes which affect or can affect the human population and its environment, including positive and negative changes of these potentials, caused by human actions. The natural environment's potential is often unsatisfactorily included in considerations concerning planning, although an optimal use of the environment at different levels of planning is only possible after an interpretation of geoscientific information. This is caused by the fact that there is a lack of information, or the information is not presented in a form, directly understandable by the planner. The reaction is often one of disregard of geological input, as geological maps have a great density of information which is not useful for the non-specialist.

Because geoscientists have a basic knowledge about the past and present behaviour of the earth's surface, their active participation in planning is of utmost importance. This can be achieved by geological expert support, advice and, above all, by influencing the administration concerning local and regional planning measures, whenever environmental geology can be relevant. A **map** is a suitable medium for the storage and presentation of information and is familiar to both earth scientists and planners. The complex geological information should be presented in different thematic maps reflecting selected features and qualities. These maps, however, have to be adapted to the specific needs and requirements of the planner, in order to ensure utilization and application in decision-making. Geological "know how" has to be prepared and translated for planning purposes.

International overview of thematic geoscientific maps

Thematic cartography is a graphical way of presenting fundamental geographic-geological information which is the basis for all planning. The need for such information is steadily increasing in all fields. In our cosmopolitical age, which is strongly influenced by the massmedia, verbal information is increasingly replaced by image, film and graphics. Therefore the map is an obvious means of information and communication in daily life and not at all simply an accessory element among other things because it provides statements that cannot be presented verbally (MEINE 1979).

Since the beginning of the 1960's geoscientific thematic maps of the natural environment's potential utilization have been developed in North America (US

DEPARTMENT OF HOUSING AND URBAN DEVELOPMENT & US DEPARTMENT OF THE INTERIOR 1971) and since the early 1960's also in Europe (PASEK & RYBAR 1961). A comprehensive overview of the current international state of the art is given by GSL (1985), ARNDT & LÜTTIG (1987), WOLFF (1987b) and MC CALL & MARKER (1989). In the last decades their number has increased explosively. While the fast and comprehensive stock-taking of huge areas in extensive North America quickly led to general cartographic results, problem-oriented investigations with analytical methods based on comprehensive geographical and geological knowledge were carried out in smaller Europe. Today thematic maps cover 60 - 80% of the total map production, and their number is steadily increasing (MEINE 1971). The reasons are the progressive expansion, methodological deepening and technical differentiation of sciences. Especially the geosciences strive for practical application of geoscientific knowledge in the field of spatial and regional planning.

Since the beginning of the 1970's **West-Germany** has intensified its geoscientific activities in presenting maps relevant to planning (LÜTTIG 1972, 1978a, 1978b). First the Geological Survey of Lower Saxony developed a "Geoscientific Set of Maps of the Natural Environment's Potential" (GMNEP) at the scale of 1:200 000 (LÜTTIG & PFEIFFER 1974, LÜTTIG 1979, BECKER-PLATEN & LÜTTIG 1980, VON DANIELS & LÜTTIG 1982, LOOK 1984). Based on fundamental data from the geological map and the soil habitat map, the following 11 thematic maps of the natural potential were produced (BECKER-PLATEN 1983, BECKER-PLATEN et al. 1987):

* soil habitat map -- susceptibility to desiccation
* soil habitat map -- agricultural productivity
* geotechnical properties
* groundwater -- basic information
* groundwater -- utilization
* near-surface resources -- mineral deposit occurences
* near-surface resources -- reserve areas for mineral resources
* deep-lying resources -- ores, coal, industrial minerals
* deep-lying resources -- salt and potash

* deep-lying resources -- oil and natural gas
* sites worth protecting for their geoscientific value

Combining the most important, geoscientifically-founded claims for utilization, a synthesis-map of "recommendations for land-use from a geoscientific point of view" was finally developed. It contains selected, important information from the thematic base maps.

Thereby the claims and recommendations for utilization are made clear to the planner. If required, they can be supplied with precise proposals in order to facilitate the weighing up with other non-geoscientifically based proposals for utilization. Furthermore, there is no limitation for the drafting of more thematic maps for special topics, relevant to planning (e.g. susceptibility to earthquakes, suitability for refuse dumps etc.).

There are similar thematic maps at different scales in the other federal states of the FRG and in many other European countries:

The idea of the Geoscientific Map of the Natural Environment's Potential (GMNEP) of Lower Saxony was adopted by **Austria** in 1978. On the initiative of the national government selected areas were tested in project studies and mapped to the scale of 1:50 000 (GATTINGER 1980, HADITSCH 1980, ARBEITER-CZERNY 1983, GRÄF 1986, ÖROK 1988). At the same time, an example of evaluation of current domestic and foreign models for the production of maps of the natural environment's potential was carried out (HÖNIG 1984, WALTERS & WIGAND 1987). Special emphasis was laid on specific questions, such as the protection of near-surface raw materials.

In 1985, the Geological Survey of **Czechoslovakia** started the project "Geological and Derived Maps of the Natural Potential and Raw Materials, scale 1:50 000", which is part of the national environment programme. For each sheet 10 different thematic maps are produced, the final synthesis-map indirectly contains recommendations for the planning authorities, by emphasizing important positive and negative parameters (VORACEK 1984,

MOLDAN 1987, MOLDAN et al. 1987). The complete edition, for the whole country, should be finished within the next 10 to 12 years.

In the middle of the 1980's the **Dutch** Geological Survey, together with the Soil Science Institute and the TNO-DGU Institute for Applied Geosciences, presented 15 maps concerning different geoscientific topics at the scales of 1:50 000 to 1:250000 (LABORATORIUM VOOR GRONDMECHANICA DELFT & RIJKS GEOLOGISCHE DIENST 1984, RIJKS GEOLOGISCHE DIENST 1986). Special attention is paid to geotechnical, hydrological and geochemical problems, because in The Netherlands more than 99% of the surface consists of unconsolidated quarternary deposits. The priorities lie in the field of planning densely populated areas (urban geology) and coastal areas (HAGEMAN 1984, DE MULDER 1987).

In **Spain** the concept of the Geoscientific Map System was started in 1975 (CENDRERO 1975, CENDRERO et al. 1976). Based on a set of descriptive thematic maps of different natural potentials, maps relevant to planning were produced by help of automatic data processing ADP (digitalizing, storage, evaluation of models). Starting in the province of Cantabria, which has been mapped to the scale of 1:50 000 (Santander Map System), work was extended to the provinces of Almeria and Valencia to the scale of 1:200 000 (ABAD et al. 1982, INSTITUTO GEOLOGICO Y MINERO DE ESPANA/IBERGESA 1982a, 1982b, CENDRERO et al. 1986, CENDRERO & DIAZ DE TERAN 1987). At the same time the topics were modified and adapted to current problems (GONZALEZ-LASTRA et al. 1980). The pilot-study "Map of Natural Hazards", carried out in the province of Vizcaya to the scale of 1:5 000 (INGEMISA 1985, CENDRERO et al. 1987), is an example for the combination of different scientific basic data to a useful contribution.

Another example is the pilot-project of Glenrothes town/Scotland of the **British** Geological Survey (NICKLESS 1982, FLOYD et al. 1983, MONRO & HULL 1986, 1987, FORSTER & CULSHAW 1987). Based on 18 thematic maps, a set of 4 additional maps was presented, which were summed up in 5 geopotential maps. Later project studies have been carried out in Scotland, England and Wales; especially problems of urban geology and physical and chemical hazards are discussed (WILSON & SMITH 1985, WADGE 1987).

At the end of the 1970's the "Geoscientific Map of the Natural Environmental Potential Utilization" to the scale of 1:250 000 was developed in **Norway** (WOLFF 1977, 1987a). It gives the planner a comprehensive survey of the actual utilization and offers, to some extent, recommendations, but excludes all conflicting proposals. In the last few years, the spectrum of
geoscientific thematic maps has extended, according to the changing needs and requirements of society (HESJEDAL 1983). Particularly problems of near-surface raw material protection, geochemistry and geomedicine are of topical interest (BØLVIKEN 1986, GARNÅSJORDET et al. 1981, OTTESEN & BØLVIKEN 1987). The promising project "Geotope Maps 1990" (Fig. 1), which produces both large- and small-scale maps on the basis of digital techniques has, for the time being, stopped because of lack of resources.

Besides the classical geologic maps, the **Swedish** Geological Survey produces a set of thematic maps, concerning different topics, especially industrial minerals and environmental geology (heavy metal-, radon-strain) (MINELL 1987, ÅCKERBLOM 1987).

In addition, other European countries have also produced thematic geoscientific maps for land-use planning:

* France (JOURNAUX 1975, ARNOULD et al. 1979, GODEFROY & HUMBERT 1983, HUMBERT & VOGT 1983)
* Italy (MERLA, MERLO & OLIVIERI 1976)
* Switzerland, Liechtenstein (ALLEMANN 1977, HUBER 1979)
* GDR (HAASE & SCHLÜTER 1980, KRAUSE 1988)
* Poland (SILIWONCZUK 1969, KOZLOWSKI 1973, LOZINSKA-STEPIEN 1979)
* Hungary (RADO & PAPP-VARY 1976)

*   Portugal (MONTEIRO 1987, ROMARIZ et al. 1989)
*   USSR (MELNIKOV 1979, ZHAMOIDA et al. 1984, KAVOUN 1983, CHERNEGOV 1987)

Particularly in the **USA** and **Canada**, numerous publications exist as regards "environmental maps", "maps of environmental geology", "planning maps" or "land-use maps". However, they are comparable only to a certain extent with the European drafts, concerning contents and system (BROWN et al. 1971, FISHER et al. 1973, LARSEN 1973, TURNER & COFFMANN 1973, MC CARTHY & MATHESON 1974, BERGSTRÖM et al. 1976, KOCKELMANN 1976, SPANGLE et al. 1976, GRIGGS & GILCHRIST 1977, FROELICH et al. 1980, KEMPTON 1981, BROWN & KOCKELMANN 1983, BRABB 1984, ALGER & BRABB 1985, LUNDGREN 1986). Starting from engineering geological maps, the USA are far advanced in environmental geological mapping (MARKER & MC CALL 1989).

In **Australia** "environmental maps" cover a multitude of activities, especially in the field of climate, raw materials, water, soil and vegetation. They are presented in series of thematic maps at different scales (HOFMANN 1976, JACOBSEN 1977, HENDERSON 1980, CHRISTIAN 1982).

At the end of the 1970's the Geological Survey of **India** started two environmental geological research projects. A set of thematic maps was produced in cooperation with several organizations concerned with the pilot project "Operation Anantapur" and "Hyderabad" (RAJU 1975, 1987, RAJU et al. 1979, GSI 1980). Eight further projects were in preparation (MURTHY 1982).

In other non-European countries also, there are examples for the development of the production of geoscientific thematic maps, relevant to planning:

*   Israel (AMIR 1984)
*   Morocco (HAFDI 1987)

*   Togo (ALLAGLO et al. 1987)
*   Nigeria (EGE, GRIFFITS & OVERSTREET 1985)
*   Japan (GSJ 1979)
*   China (MINGDE 1987)
*   Thailand (JAPAKASETR 1987)
*   Indonesia (PATTY & WONGSOSENTONO 1987)
*   Columbia (HERMELIN 1984, BUENAVENTURA 1987, BUSTAMANTE & HERMELIN 1988)
*   Argentina (SAYAGO 1982)

According to the different geological, geographical, social, economical, political and spatial differences, the maps differ more or less in conception and contents. Although they are not completely comparable, comprehensive drafts of small-scale thematic maps can be prepared.

From 1979 to 1986, the Geological Surveys of Finland, Sweden, Norway and Greenland, in cooperation, produced a cartographic synthesis of geoscientific information of Northern Fennoscandia. The **Nordkalotte Project** consists of different thematic maps for various topics (geophysics, geology, petrography, mineralogy, stratigraphy, geochemistry and geology of mineral deposits) at the scale of 1:1 000 000 (KAUTSKY 1986). The results, which had been gathered by interdisciplinary working groups, were fed into data banks. They form a complex and uniform base, not only for geoscientific activities. Since 1987 the project has been carried out in the adjacent southern strip (Mid-Norden Project).

A pilot project "**Pilot-Strip**" of the Subcommission for Environmental Geological Maps (SC-MEG) of the CGMW was launched in 1988. It embraces the north-western European continent. The Geological Surveys of Lower Saxony, The Netherlands, Belgium, Great Britain, Ireland and the BRGM are cooperating in developing a set of 10 thematic maps. A part of the strip has already been completed in the scale of 1:500 000.

The production of Environmental Geology Maps or Geoscientific Maps of the Natural

Environment's Potential is supported by various national and international groups (European Community, Subcommission of Maps on Environmental Geology within the International Union of Geological Sciences, United Nations Educational, Scientific and Cultural Organisation, International Quarternary Union; LÜTTIG 1975, 1987b) in selective projects (e.g. "Geology and Environment", UNEP/UNESCO) (DOTTIN 1987). At the 28th International Geological Congress in Washington in 1989 the initiative for setting up a new IUGS-Commission on Geology and Environmental Planning has been put forward.

The ideas of the GMNEP and the environmental maps had their origin in the highly industrialized and densely populated countries, responding to the need for careful planning of the remaining natural potential and to avoid irreparable damages. By using this type of map, the planning could be carried out on the basis of a better knowledge of the areas of concern. In this way many planning mistakes could be avoided and the losses diminished. This is also the case for developing countries, in which many of the mistakes commited in industrialized countries could be avoided. The aim is a better management of the natural resources on the basis of appropriate methods of presentation and evaluation of geoscientific data (LÜTTIG 1987c).

Five crucial questions

Quality of life is determined among other things by the nature and quality of necessary facilities and their spatial distribution. Important parameters are: easy accessibility, sufficient "protective space" to ensure the undisturbed coexistence of different utilizations, protection and utilization of natural and historical advantages of site as well as protection of the basic natural elements of life and resources. Land-use planning aims at solving land-utilization problems through a balanced distribution of activities and a coordination of the different demands for space (WILLE 1987).

The main purpose of thematic maps is not to set priorities for the utilization of geogenic potential, but to make an intelligent contribution to environmental planning by presenting basic facts and interpreting them for the planner. Based on the information, given in the map, the planner assigns a certain utilization to a selected area. This assignment, ideally, should be made in a dialogue between the planner and the geoscientist responsible for producing the map. However, after an evaluation, well founded proposals for utilization can be refused due to overlapping of conflicting claims. When establishing priorities for use the general world situation with regards to resources and environmental problems must be considered, as well as the local one. In this case detailed investigations might be necessary.

The prerequisites and fundamentals for the production of geoscientific maps for planning, can be summarized in five crucial interrelated questions:

* **Why** geoscientific maps?
* For **what purpose**?
* **Who** needs geoscientific maps?
* **What** should these maps contain?
* **How** to produce geoscientific maps?

The first question has already been answered above. The approach to the problem and the setting of the task (question 2) primarily concern the **scale** and **topics** to be dealt with, which should enable an interpretation and incorporation into programmes of spatial planning. The **scale** has to be adapted to the specific plan (Fig. 2). The plan also determines the thematic contents of the maps (single-, multi-purpose maps). Large-scale maps (1:50 000 and larger) are used for semidetailed and detailed planning on community level; small-scale maps are used for general planning at province, state or national level. Obviously, the character and the density of the information depends on the scale of the maps.

The third question, the **target group**, implies that the production of geoscientific thematic maps should not be regarded as an end in itself. The ultimate users of GMNEP-maps are usually not geoscientists. Therefore, a simple, clearly understandable presentation of maps and legends is essential. Technical terms should be avoided in explanations restricted to those features which are relevant to planning purposes.

Following the organization of the Geoscientific Map of the Natural Environment's Potential in five levels, the gradual construction and arrangement of geoscientific information for planning (questions 4 and 5) can be explained in the following manner (Fig. 3; LÜTTIG 1987a):

Level 1 contains geoscientific and other base maps (e.g. bedrock geology, surficial geology, pedological, geomorphological, slope angle, hydrological and other maps).

Level 2 combines elements of the base maps, relevant to planning, in derived maps (maps of natural hazards, engineering geological maps, maps of mineral deposits).

Level 3 includes maps of "land-use claims" from the point of view of different disciplines concerned with the earth's surface (raw material classification maps, groundwater potential maps, agricultural yield potential maps, nature conservation maps). In addition to fundamental geoscientific data, non-geological (mostly economical) factors are evaluated, in order to define the priorities for use which are "claimed" for different uses (mineral extraction, groundwater protection or extraction, agriculture, conservation).

In **Level 4** a conflict map is developed by superimposing all claims on maps of level 3. It shows conflicts and facilitates the establishing of priorities between competing uses. This intermediate map product should serve as a scientific "aide-memoire" and does not have to be published.

Level 5 includes the final "land-use priority map from a geoscientific point of view". Precise and reasonable spatial and factual proposals for utilization are given, which provide the foundation for the next step in the planning process. The preparation of this map taxes the geoscientist severly, because the different data have to be linked to the problems relevant for planning. The map should contain one proposal or a combination of recommendations, which should be put into distinct and simple phrases. Their basis and limits should be clear to the planner. Also a "protective" statement should be included (in this and in the other maps) indicating that detailed investigations might be needed to solve specific problems.

The construction as a set enables the planner to fall back on the individual stages if necessary. Close cooperation between the geoscientist and the planner should improve the understanding of the meaning and limitations of the map, thus helping their acceptance and actual incorporation into the planning process. Updating the maps is very important if they are to remain as useful instruments in the future.

The use of this kind of map set can help considerably in the management of the environment and its resources and in the reduction of damages due to natural processes or to human actions.

These questions and their answers give rise to some specific **problems**:

*   The quality of thematic maps depends primarily on the basic information and the topographical base maps. These have to be of sufficient accuracy (based on modern techniques such as aerial photography, satellite imagery, spectral-colour photography) and have to be easily accessible (MATSUDA 1980, ALBERTZ 1986, BARDINET et al. 1988).
*   New techniques of graphical representation and interpretation create newer and newer form of maps (HAAG et al. 1987). There is a wealth of publications with a great variety in content, methods of presentation, colour and symbols, thus complicating common interpretations. A comprehensive evaluation and translation of geoscientific data into maps for supranational planning is very difficult, to some extent impossible. The development of a standard cartographic system of characters and rules for drafting small-scale thematic

maps on an international level, which are produced by final generalization and combination of large-scale potential maps (BOSSE et al. 1982) is an important future task (STENESTAD, this volume).

* The quantity of "indispensable" thematic maps which are to be analyzed, complicates the final composition of the mosaic pieces into an integrated whole. The final product must, therefore, include one single "synthesis map", from which one can refer back to more detailed thematic maps, if necessary. The basic prerequisite is generalization that is not only a simplifying omission, but rather the emphasis on the essential, dominant and typical. The main task is not to produce maps, but to **design** maps, because their make-up finally decides their practical use and acceptance.

* Spatial typifying requires factual typifying. This is a mental process which is both complex and submitted to subjectivity. A complete synthesis, as well as the demand for an absolute accuracy, is an unattainable ideal. The impossibility to reach this ideal within the limitations of time, manpower and financial resources inherent to planning, hinders many geoscientists who prepare thematic maps.

* The difficulty (and, therefore, the reluctancy to undertake the task) is often intensified by legal problems due to the interpretation of maps (e.g. questions of copyright and systematics with analytical maps, accuracy and limits of interpretations, way and degree of certainty of phrasing).

* Technical-cartographic problems for the presentation of specific circumstances complicate the readability and, therefore, the applicability (e.g. overlapping, positive/negative figure). The representation of geoscientific parameters relevant to planning, means a translation of three-dimensional problems into a two-dimensional map and this is not always easy to solve in such a way that the picture is clear for the non-specialist (LESER 1984).

* The time-problem is vital concerning the translation of geoscientific statements for planning purposes. On the one hand, the prognostic statements, which are difficult to make, have to be valid for a time span of about 20 years; on the

other hand, the frequent modification of complex plans requires the adaptation and up-dating of the geoscientific maps.

Up to recent times, this is especially difficult because most maps, drawn by a variety of experts from different fields, were "hand made" and printed. The increasing acceptance of **ADP** concerning areal planning is bringing a new **trend** into geoscientific cartography (VINKEN & VOSS 1983).

Geoscientific data are stored in different ways. The ADP enables the storage, processing and controll of data from different sources in computer language. In general a mixture of various data types accumulate. The storage with the help of ADP enables a better selection, classification, interpretation and combination of data, thus making an adequate consideration of areal heterogenity possible. The present-day international standard of data acquisition, storage and processing is unclear. Often the programmes are not published, but used only internally in different departments. Different systems are used, particularly concerning the collection of environmental data in landscape planning or ecological mapping. Most **geographic information systems** (GIS) meet the requirements for drafting computerized geoscientific maps. The difficulties in computerized processing of geoscientific data and the limits of their presentation in thematic maps, relevant to spatial planning, originate from mathematic-statistical, technical, legal and -- last, but not least -- financial problems. Nevertheless, it seems that the times of traditional map production are over and the computer-based systems will be the rule in the immediate future (GABERT, this volume).

Suggestions for work to be undertaken

There is a great need to make the public aware of mutual relationships and dependence of human mankind upon geological conditions. This is only possible if geologists are willing to meet the challenge and to accept the responsibility. Exaggerated self-demands ("ideal map") have to be eliminated. Instead, geoscientists should learn to live with the "real world" and to make assessments and judgements with the information available, not always satisfactory. Also, they should make a clear presentation of their position and recommendations, so that their know-how is understood by planners, decision-makers and the public. This implies, among other tasks:

* Increased cooperation in ecology, regional, economic and environmental planning. Geologists could not only help, but take an active part in decision making (LÜTTIG 1987c).

* The foundation has to be laid by a practically oriented, interdisciplinary education. The geologist must know more about environmental planning, but the planner should also be informed about the position of modern geology and its possibilities for contributing to the problems of spatial planning. This can be achieved by a well-founded geological basic training. In general, the curricula have to be adjusted, because and intensified cooperation between geologists and planners is only possible on the basis of improved information (LÜTTIG 1987c).

* Participation of the geoscientist as an equal partner in the dialogue with the planner and incorporation of geoscientific maps from the beginning of the planning process. This requires mutual willingness to compromise. In order to suceed in cooperation, the geologist has to enter planning teams. This is only possible on the basis of sound geoscientific data and good professional advice, but a convincing image and presentation to the public are also important (LÜTTIG 1987c).

* On the basis of well-founded geological knowledge, data and existing information has to be combined in order to provide valid statements. A permanent maintenance and supplementing of missing data are the basis of successful planning.

* The cooperation on a national and international level, concerning questions of training and research, harmonization and standardization of types of maps and development of general legends should be improved. Different models could

be tested, discussed and their applicability for the production of geoscientific maps for planning could be optimized by cooperation and exchange of ideas (LÜTTIG 1987c).

Supranational projects and studies, especially the development of small-scale maps for international planning rank high in the list of tasks to be undertaken.

Synthesis

Planning on the basis of an optimal utilization of the natural potential is only possible by gathering and evaluating the right geoscientific information. This can be obtained by:

*   Developing an appropriate cartographic method on the basis of a set of criteria, which enables a comparison different parts of a country or a greater region, using common methods and standards.
*   Drafting simple, distinct, readable and understandable maps, adaptable and capable of development and quantification with a general legend that ensures continuity and comparability of the presentation, even at different scales.
*   Using a form of presentation referring to natural resources and demonstrating various ways of data utilization or combinations thereof, which are appropriate for a given area. One or more maps, but as few as possible, should be produced for particular topics. The scale has to correspond to the given base for planning. Too large-scaled presentations lead to too specific interpretation on the basis of insufficient data; too small-scaled maps are only suitable for a rough, general idea in regional planning.
*   Testing critically existing material for planning purposes concerning its meaningfulness to geoscientific topics. It is necessary to obtain an overview about information gaps and to close them.

* Translating geoscientific information for planning purposes. The legend as a storage medium for this information should not be too extensive, but as complete as possible, in order to allow an estimation of the exposure to danger or the degree of assessment. Statements and phrasing should be intelligible to the planner.

In the development of geoscientific maps for land-use planning certain priorities must be set:

* Areas, sensitive to conflicts, and non-renewable natural potentials, which are particularly easily exposed to irreversible processes, should be given high priority (e.g. water potential).
* The investigation of the Natural Environment's Potential should also have high priority for the spatial planning of areas where characteristic land-use conflicts arise, or where serious local strains are present, e.g. coastal areas (CHARLIER 1987).

The methodology and procedure concerning the registration, maintenance and evaluation of natural environment's potentials has been summarized by ÖROK (1988):

* Collection of relevant data (literature, archives, interpretation of satellite and aerial photography, specific field work).
* Presentation of different contents and evaluation of partial potentials, depending on the priority (e.g. potential for nature conservation, raw materials, water, waste disposal).
* Preparation of maps at different scales, containing the fundamental information for planning. The maps must be easily reproducible.
* Construction of systematic map systems.
* Data-processing based on uniform or integrable systems. Digitalisation of data and maps and transfer to central systems of documentation and information, which form the basis for a GIS. The systems should be compatible in order to

enable the development of a data network.

* Continuation, extension and stepwise integration of the information systems into one comprehensive system.

* Decentralized up-dating by experts in specialized institutions.

* Organisation of facilities for the utilization of documentation and information-systems by internal and external users. Appropriate organisation and construction of ADP systems.

* Selection of appropriate types of models for the evaluation of natural environment's potentials with instructions for precise application.

* Appropriate evaluation of the natural potentials by assessment of qualities, such as:

-- capability,

-- sensitivity,

-- strain and exposure to danger.

* Provision of funds by the government or financing by public budgets, in some cases by special funds, in order to maintain the systems. This requires the understanding and acceptance by the public.

In the author's opinion, the definition of the term "**Environmental Geological Mapping**" given by MONRO & HULL (1987: 111) explains very clearly the importance and the purpose of maps of the natural environment's potential and their key-position for future environmental protection:

"Environmental Geological Mapping is the representation in map form of geological factors which are relevant to human activity, health and safety."

Planning the utilization of the earth's surface in a satisfactory way, to get maximum advantage of its potential, with a minimum damage, can only be done by taking the right geoscientific information into account.

References

ABAD, J.F., FRESNO, F.L. & PENA, J.L.P. (1982): Geoscientific map of the natural environment of the Province of Almeria. -- Proc. 4th Congr. internat. Assoc. engng. Geol. **1**: I 47 - I 56; New Dehli.

AKADEMIE FÜR RAUMFORSCHUNG UND LANDESPLANUNG (1969): Untersuchungen zur thematischen Kartographie. -- ARL Forsch. Sitz.ber. **51**: 161 pp.; Hannover.

ALBERTZ, J. (1986): Remote sensing of the Earth's surface: The production of satellite image maps. -- Universities **28**, 3: 163 - 168; Tübingen.

ALGER, C.S. & BRABB, E.E. (1985): Bibliography of United States landslide maps and reports. -- US geol. Surv. Open-File Rep., **85**-585: 1 - 119; Reston.

ALLAGLO, L. K., AREGBA, A., D'ALMEIDA, N.C., GU-KONU, E.Y., KOUNETSRON, K. & SEDDOH, K.F. (1987): Togo, its geopotential and attempts for land-use planning -- A case study. -- In: [ARNDT, P. & LÜTTIG, G.W.]: Mineral resources extraction, environmental protection and land-use planning in industrialized and developing countries: 243 - 270; Stuttgart (Schweizerbart).

ALLEMANN, F. (1977): Geologische Risikokarte 1:5 000, Gemeinde Triesenberg. -- Triesenberg/Liechtenstein.

AMIR, S. (1984): Israel's coastal program: resource protection through management of land-use. -- Coast. Zone Managem. J. **12**, 2/3: 189 - 223; Washington.

ARBEITER-CZERNY, I. (1983): Naturraumpotentialkarten der Steiermark, Bezirk Radkersburg. -- 24 maps.; Graz/Styria.

[ARCHER, A.A., LÜTTIG, G.W. & SNEZHKO, I.I.] (1987): Man's dependence on the Earth. The Role of the Geosciences in the Environment. -- 216 pp, 83 fig., 15 pl.; Stuttgart (Schweizerbart), Nairobi & Paris.

[ARNDT, P. & LÜTTIG, G.W.] (1987): Mineral resources extraction, environmental protection and land-use planning in industrialized and developing countries. -- 337 pp, 108 fig., 24 tab., 2 folders; Stuttgart (Schweizerbart).

ARNOULD, M., BROQUET, J.F., DEVEUGHELE, M. & USSEGLIO POLATERRA, J.M. (1979): Cartographie geotechnique de la ville de Paris -- premieres realisations (13eme, 19eme et 20eme Arrondissements). -- Bull. internat. Assoc. engng. Geol. **19**: 109 - 115; Krefeld.

AUST, H. & BECKER-PLATEN, J.D. (1984): Umweltschutz. -- Geol. Jb. A, **73**: 401 - 406, 1 tab.; Hannover.

AUST, H. & BECKER-PLATEN, J.D. (1985): Angewandte Geowissenschaften in Raumplanung und Umweltschutz. -- In: [BENDER, F.]: Angewandte Geowissenschaften **3**: 1 - 136; Stuttgart (Enke).

ÅKERBLOM, G. (1987): Investigations and mapping of radon risk areas. -- Norg. geol. Unders. spec. Publ. **2**: 96 - 106; Trondheim.

BARDINET, C., GABERT, G. MONGET, J.-M. & YU, Z. (1988): Application of multisatellite data to thematic mapping. -- Geol. Jb. B **67**: 3 - 47; Hannover.

BECKER-PLATEN, J.D. (1983): Geowissenschaftliche Karte des Naturraumpotentials. -- Forsch. dt. Landeskde. **220**: 119 - 164; Trier.

BECKER-PLATEN, J.D. & LÜTTIG, G.W. (1980): Naturraumpotentialkarten als Unterlage für Raumordnung und Landesplanung. -- Arb.-Mater. Akad. Raumforsch. Landespl. **27**: 1- 60; Hannover.

BECKER-PLATEN, J.D., DORN, M. & LOOK, E.-R. (1987): An introduction to the legend of the Geoscientific map of the Natural Environment's Potential (GMNEP) of Lower Saxony and Bremen. -- In: [ARNDT, P. & LÜTTIG, G.W.]: Mineral resources extraction, environmental protection and land-use planning in industrialized and developing countries: 119 - 126, Stuttgart (Schweizerbart).

BECKER-PLATEN, J.D., HOFMEISTER, E., KLEMZ, B. & STEIN, V. (1986): Landnutzungskarten -- Ein Versuch zur Darstellung der Flächenbeanspruchung. -- Raumforsch. Raumordn. **44**, 6: 217 - 234; Hannover, Bonn.

BERGSTROM, R.E., PISKIN, K. & FOLLMER, L.R. (1976): Geology for planning in the Springfield Decatur Region, Illinois. -- Ill. St. geol. Surv. Circ. **497**: 1 - 76 pp,; Urbana/Ill.

BØLVIKEN, B. (1986): Geokjemisk kartlegging nå og i fremtiden. -- Geologie **38**, 10: 260 - 263; Helsinki.

BOSSE, H.-R., BRINKMANN, K., LORENZ, W. & ROTH, W. (1982): Karte der Bundesrepublik Deutschland 1:1 000 000. Gebiete mit oberflächennahen mineralischen Rohstoffen. -- 19 pp.,2 tab., 1 map; Hannover.

BRABB, E.E. (1984): Innovative Approaches to Landslide Hazard and Risk Mapping. -- Proc. 4th internat. Symp. Landslides Toronto 1984, **1**: 307 - 324, Toronto.

BROWN, R.D. JR. & KOCKELMANN, W.J. (1983): Geologic principles for prudent land-use. -- US geol. Surv. prof. Pap. **946**: 97 pp.; Washington.

BROWN, L.F., FISHER, W.L., ERXLEBEN, A.W. & MCGOWEN, J.H. (1971): Resource capability units, their utility in land and water-use management, with examples from the Texas Coastal Zone. -- Bur. econ. Geol. Circ. **71**, 1: 22 pp., Univ. of Texas/Austin.

BUENAVENTURA, A. (1987): Mineral exploration, geological mapping and land-use planning in Colombia. -- In: [ARNDT, P. & LÜTTIG, G.W.]: Mineral resources extraction, environmental protection and land-use planning in industrialized and developing countries: 313 - 315; Stuttgart (Schweizerbart).

(BGR) BUNDESANSTALT FÜR GEOWISSENSCHAFTEN UND ROHSTOFFE UND GEOLOGISCHE LANDESÄMTER (1983): Die Geowissenschaften im Dienste der Daseinsvorsorge, des Umweltschutzes und der Sicherung des Naturraumes. -- 44 pp.; Hannover.

BUSTAMANTE, M. & HERMELIN, M. (1988): Aplicacion de la geologia en el plan de ordenamiento del Valle de Aburra. -- Mem. V. Congr. colomb. Geol. **1**: 496 - 515; Bogota.

CENDEREO, A. (1975): Environmental Geology of the Santander Bay Area, Northern Spain. -- Environm. Geol. **1**: 97 - 114; New York (Springer).

CENDEREO, A. & DIAZ DE TERAN, J.R. (1987): The environmental map system of the University of Cantabria, Spain. -- In: [ARNDT, P. & LÜTTIG, G.W.]: Mineral resources extraction, environmental protection and land-use planning in industrialized and developing countries: 149 - 182; Stuttgart (Schweizerbart).

CENDEREO, A., DIAZ DE TERAN, J.R. & SAIZ DE OMENACA, J. (1976): A technique for the definition of environmental units and for evaluating their environmental value. -- Landscape Planning **3**: 35 - 66; Amsterdam.

CENDEREO, A., DIAZ DE TERAN, J.R., FERNANDEZ, O., GARROTE, R., GONZALES LASTRA, J.R., INORIZA, J., LÜTTIG, G., OTAMENDI, J., PEREZ, M., SERRANO, A. & GRUPO IKERLANA (1987): Detailed geological hazards mapping for urban and rural planning in Vizcaya (Northern Spain). -- Norg. geol Unders. spec. Publ. **2**: 25 - 41; Trondheim.

CENDRERO, A. NICTO, M., ROBLES, F., SANCHEZ, J., BIGUES, J., HERRERO, M.I., NAVARRETE, J.J., DIAZ DE TERAN, J.R., FRANCES, E., GONZALES-LASTRA, J.R., BOLUDA, R., GARAY, P., GUTIENEZ, G., JIMENEZ, J., MARTINEZ, J., MOLINA, M.J., OBARTI, J., PEREZ, A., PONS, V., SANTOYO, A. & STUBING, G. (1976): Mapa Geocientifico de la Provincia de Valencia. - 71 + 350 pp, 7 maps, Valencia (Diputacion Prov. de Valencia).

CHARLIER, R.H.L. (1987): Planning for coastal areas. -- Norg. geol. Unders. spec. Publ. **2**: 2 - 24; Trondheim.

CHERNEGOV, Y.A. (1987): Land-use planning accounting for geological factors. -- In: [ARNDT, P. & LÜTTIG, G.W.]: Mineral resources extraction, environmentalprotection and land-use planning in industrialized and developing countries: 225 - 232; Stuttgart (Schweizerbart).

CHRISTIAN, C.S. (1982): The Australian approach to environmental mapping. -- US geol. Surv. prof. Pap. **1193**: 298 - 316; Washington.

DANIELS, C.H. VON & LÜTTIG, G.W. (1982): Geowissenschaftliche Karten des Naturraumpotentials als Unterlage für Raumordnung und Landesplanung. -- In: Energierohstoffe im Alpen-Adria-Raum, Symp. Montanuniv. Leoben 13./14.10.1980: 151 - 168; Graz.

DOTTIN, O. (1987): The CGMW and the International Small-Scale Maps in the Geoscience. -- Norg. geol. Unders. spec. Publ. **2**: 112; Trondheim.

EGE, J.R., GRIFFITTS, W.R. & OVERSTREET, W.C. (1985): Engineering geology site appraisal for the federal capital of Nigeria. -- Bull. internat. Assoc. engng. Geol. **31**: 71 - 79; Krefeld.

FISHER, W.L., BROWN, L.F., MCGOWEN, J.H. & GROAT, C.G. (1973): Environmental Geologic Atlas of the Texas Coastal Zone: Beaumont-Port Arthur Area. -- Bur. econ. Geol.; 93 pp., 9 maps; Univ. of Texas/Austin.

FLOYD, J.D., AITKEN A., BALL, D.F., LAXTON, J.L. & LONG, D. (1983): Environmental Geology Maps of South-East Edinburgh. -- Brit. geol. Surv. Open File Rep., SI 83/5; Edinburgh (BGS).

FORSTER, A. & CULSHAW, G. (1987): Engineering geological maps as an aid to planning. -- Norg. geol. Unders. spec. Publ. **2**: 52 -57; Trondheim.

FROELICH, A.J., HACK, J.T. & OTTON, E.G. (1980): Geologic and Hydrologic Map Reports for Land-Use Planning in the Baltimore-Washington Urban Area. -- US geol. Surv. Circ. **806**: 1 - 26; Washington.

GABERT, G. (this volume): Computer assisted methods for the construction, compilation and display of geoscientific maps.

GARNÅSJORDET, P.A., LARSEN, J., LONE, O. & WOLFF, F.CHR. (1981): Tynset, arealressurskart 1619 I, M 1:50 000. -- Trondheim (Norg. geol. Unders.).

GATTINGER, T.E. (1980): Geowissenschaftliche Naturraumpotential-karten: Ein Instrument der Raumordnung und Raumplanung. -- Verh. geol. B.A. Wien **1980**, 3: 229 - 240; Wien.

[GEOLOGICAL SURVEY OF INDIA] (1980): A decade of environmental geoscientific studies. -- Spec. Publ. **9**: 180 pp; Calcutta.

(GSJ) GEOGRAPHICAL SURVEY OF JAPAN (1979): Thematic mapping in Japan. -- In: World Cartography (XV) **15**: 28 - 44; New York (UN).

[GEOMORPHOLOGICAL SERVICES LTD] (1985): The GSL Report on Environmental Geology Mapping. -- Marlow, Bucks.

GODEFROY, P. & HUMBERT, M. (1983): La cartographie des risques naturels lies aux mouvements de terrain et aux seismes. -- Hydrogeol. Geol. Ing. **2**: 69 - 90; Paris.

GONZALES-LASTRA, J.R., DIAZ DE TERAN, J.R. & GONZALES-LASTRA, J. (1980): Ensayo de un metodo de prediccion y cartografia de riesgos geologicos. Aplicacion a los deslizamientos superficiales. -- Geol. ambient. y Ordenac. Terit., Reun. nat., Commun. Santander: 1 - 11, 7 fig; Santander.

GRÄF, W. (1986): Naturraumpotentialkarten im Dienste einer umweltbewußten Rohstoffsicherung, dargestellt am Beispiel der Steiermark. -- Mitt. österr. geol. Ges. **79**: 15 - 29, Wien.

GRIGGS, G.B. & GILCHRIST, J.A. (1977): The Earth and Land Use Planning. -- 492 pp., Belmont/Calif. (Duxbury).

HAAG, K., SCHACKNIES, G. & WAGNER, B. (1987): Digitale Kartographie. -- COM **22**, 4: 46 - 49; Berlin, München.

HAASE, G. & SCHLÜTER, H. (1980): Zur inhaltlichen Konzeption einer Naturraumtypenkarte der DDR im mittleren Maßstab. -- Petermanns geogr. Mitt. **124**, 2: 139 - 151, 2 fig., 2 tab.; Gotha & Leipzig.

HADITSCH, J.G. (1980): Gedanken zur Erarbeitung von Naturraumpotentialkarten für das Land Steiermark. -- Natur Land **66**, 4: 106 - 108; Graz.

HAFDI, A. (1987): Approach of a methodology for drawing up a habilitability map. -- In: [ARNDT, P. & LÜTTIG, G.W.]: Minerals resources extraction, environmental protection and land-use planning in industrialized and developing countries: 271 - 278; Stuttgart (Schweizerbart).

HAGEMAN, B.P. (1984): Geological information, a vital element in environmental planning with emphasis on coastal plains. -- Geol. Jb. A, **75**: 93 - 123, 20 fig; Hannover.

HENDERSON, G.A.M. (1980): Commentary on the Coppins Crossing 1:10 000. Engineering Geology Sheet, Canberra, Australian Capital Territory. -- 34 pp., 1 map; Canberra (BMR).

HERMELIN, M. (1984): Riesgos geolgicos en el Valle de Aburra. -- Mem. Conf. sobre Riesgos geol. en el Valle de Aburra: 22 pp; Medellin.

HESJEDAL, O. (1983): Norsk kartplan 2, tematiske kart og geodata. -- Norg. offentl. Utredn. **46**: 1 - 117; Oslo.

HOFMANN, G.W. (1976): Mapping for urban land-use planning in Southeast Queensland - - a first approach. -- Bull. internat. Assoc. engng. Geol. **14**: 113 - 117; Krefeld.

HOFMANN, G.W. (1977): Geology related to regional planning -- Australian experience. - - Queensland Governm. min. J. **78**, 913: 541 - 544; Brisbane.

HÖNIG, H.G. (1984): Sichtung und Bewertung der wichtigsten vorliegenden Naturraumpotential-Modellentwicklungen im In- und Ausland. -- 204 pp.; Graz (Inst. Umweltforsch.).

HUBER, E. (1979): Thematic mapping in Switzerland. -- In: World Cartography (XV) **15**: 52 - 62; New York (UN).

HUGDAHL, H. & WOLFF, F.CHR. (1987): Development of Geotope Maps in Norway. - - Norg. geol. Unders. spec. Publ. **2**: 109; Trondheim.

HUMBERT, M. & VOGT, J. (1983): Le fichier d'informations sur les mouvements de terrain en France et ses applications. -- Hydrogeol. Geol. Ing. **2**: 91 - 101, 3 fig; Paris.

INGEMISA (1985): Estudio de los riesgos naturales en Vizcaya. Fase piloto. Elantxobe. Memoria -- 106 pp; Bilbao.

[INSTITUTO GEOLOGICO Y MINERO DE ESPANA/IBERGESA](1982a): Mapa geoscientifico del medio natural (1:100 000). -- Tomo 1: materiales, recursos. -- 52 pp; Madrid.

[INSTITUTO GEOLOGICO Y MINERO DE ESPANA/IBERGESA](1982b): Mapa geoscientifico del medio natural (1:100 000). -- Tomo 2: procesos, riesgos. -- 32 pp.; Madrid.

JACOBSEN, G. (1977): Environmental and urban geology in Australia: Workshop meeting of Government Engineering Geologists, November 1976. -- BMR J. austr. Geol. Geophys. **2**: 73 - 76; Canberra (BMR).

JAPAKASTR, T. (1987): A study on environmental and Quaternary geology of Thailand. - - In: [ARNDT, P. & LÜTTIG, G.W.]: Mineral resources extraction, environmental protection and land-use planning in industrialized and developing countries: 299 - 302; Stuttgart (Schweizerbart).

JOURNAUX, A. (1975): Legende pour une carte de l'environnement et de sa dynamique. - - 15 pp., Caen (Fac. Lettr. Univ.).

KAUTSKY, G. (1986): Nordkalotte Geology and Mineral Deposits -- A cartographic Synthesis. -- Terra cognita **6**, 3; Oxford (Blackwell).

KAVOUN, K.P. (1983): The significance and importance of geological factors in planning. -- Ms., 31 pp.; Moscow.

KEMPTON, J.P. (1981): Three-dimensional geologic mapping for environmental studies in Illinois. -- Envir. Geol. Not. **100**: 1 -43; Campaign/Ill.

KOCKELMANN, W.J. (1976): Use of USGS Earth Science Products by County Planning Agencies in the San Francisco Bay Region, California. -- Open File Rep. US geol. Surv.: 76 - 547; Menlo Park.

KÖNIGSSON, L.-K. (1987): Geology, biology and Man -- A discussion of ecology, paleoecology and the planning of the geoenvironment. -- In: [ARNDT, P. & LÜTTIG, G.W.]: Mineral resources extraction, environmental protection and land-use planning in industrialized and developing countries: 15 - 28; Stuttgart (Schweizerbart).

KOZLOWSKI, S. (1973): [The plan for safeguarding Poland's landscape and its initial accomplishment] (in Polish). -- Zakl. Odr. Przyrody polsk. Akad. Nauk, Ochroina Przerody **38**: 61 - 83; Warszawa.

KRAUSE, K.-H. (1988): Konzeption einer komplexen Flächennutzungskartierung unter ökologischen Aspekten im Maßstab 1:50 000. -- Hall. Jb. Geowiss. **13**: 67 - 89; Gotha.

LABORATORIUM VOOR GRONDMECHANICA DELFT & RIJKS GEOLOGISCHE DIENST HAARLEM (1984): Ingenieursgeologische Kaarten van Nederland. -- 21 pp.; Delft, Haarlem.

LARSEN, J.I. (1973): Geology for Planning in Lake County, Illinois. -- Circ. Ill. St. Geol. Surv. **481**: 43 pp.; Urbana/Ill.

LESER, H. (1984): Methodische und inhaltliche Probleme von Karten des Naturraumpotentials. -- Ber. dt. Landeskde. **59**, 1: 267 - 283; Trier.

LOOK, E.-R. (1984): Raumordnung und Landesplanung im Niedersächsischen Landesamt für Bodenforschung. -- Geol. Jb. A, **73**: 395 - 400; Hannover.

LOZINSKA-STEPIEN, H. (1979): Engineering geology maps at a scale of 1:25 000 for regional planning purposes. -- Bull. internat. Assoc. engng. Geol. **19**: 69 - 72; Krefeld.

LUNDGREN, L. (1986): Environmental Geology. -- 16 + 576 pp.; Englewood Cliffs /N.J. (Prentice Hall).

LÜTTIG, G.W. (1972): Naturräumliches Potential I, II und III. -- In: Niedersachsen, Industrieland mit Zukunft. -- 9 -10; Hannover (Niedersächs. Minist. Wirtsch.).

LÜTTIG, G.W. (1975): Geoscience and the potential of the natural environment. Geoscientific Studies and the Potential of the Natural Environment: 29 - 40; Köln (Dt. UNESCO Komm.).

LÜTTIG, G.W. (1977): Die Rolle der geowissenschaftlichen Kartographie in der vorausschauenden Umweltforschung. -- Kartogr. Nachr. **27**, 3: 81 - 89; Bonn-Bad Godesberg.

LÜTTIG, G.W. (1978a): Geoscientific Maps for Land-Use Planning. A Certain Approach how to communicate by New Types of Maps. -- Internat. Yearb. Cartogr. **18**: 95 - 101; Bonn-Bad Godesberg.

LÜTTIG, G.W. (1978b): Geoscientific maps of the environment as an essential tool in planning. -- Geol. en Mijnb. **57**, 4: 527 - 532; s'Gravenhage.

LÜTTIG, G.W. (1979): Geoscientific maps as an basis for land-use planning. -- Geol. Fören. Stockh. Förh. **101**, 1: 65 - 69; Stockholm.

LÜTTIG, G.W. (1983): Zur Theorie des Naturraumpotentials und seiner theoretischen Anwendung. -- Verh. dt. Geogr.tag **43**: 141 - 142; Wiesbaden.

LÜTTIG, G.W. (1987a): Large scale maps for detailed environmental planning. -- Norg. geol. Unders. spec. Publ. **2**: 71 - 76; Trondheim.

LÜTTIG, G.W. (1987b): The Subcommission of Maps on Environmental Geology (SC-NMEG) of the IUGS Commission for the International Map of the World and its work in industrialized and developing countries. -- In: [ARNDT, P. & LÜTTIG, G.W.]: Mineral resources extraction, environmental protection and land-use planning in industrialized and developing countries: 319 - 331; Stuttgart (Schweizerbart).

LÜTTIG, G.W. (1987c): Geology versus mineral, groundwater and soil resources' management -- Approach to the public -- Education and training questions -- types and acceptance of geopotential maps. -- In: [ARNDT, P. & LÜTTIG, G.W.]: Mineral resources extraction, environmental protection and land-use planning in industrialized and developing countries: 319 - 331; Stuttgart (Schweizerbart).

[LÜTTIG, G.W.] (in print 1990): International guidelines on geology and land-use planning, Volume III. -- In: [KOZLOVSKY, E.A.]: UNEP/UNESCO Project "Geology and Environment", Nairobi & Paris (UNEP/UNESCO).

LÜTTIG, G.W. & PFEIFFER, D. (1974): Die Karten des Naturaumpotentials. Ein neues Ausdrucksmittel geowissenschaftlicher Forschung für Landesplanung und Raumordnung. -- N. Arch. Niedersachsen **23**, 1: 3 - 13; Göttingen.

MARKER ,B. & MC CALL, j. (1989): Environmental Geology Mapping. -- In: [MC CALL, J. & MARKER, B.]: Earth Science Mapping: 201 - 236, 12 fig., 11 tab.; London (Graham & Trotman).

MATSUDA, H. (1980): Digital National Land Information. -- GeoJournal **4**, 4: 313 - 318; Wiesbaden.

Mc CALL, J. & MARKER, B. (1989): Earth Science Mapping for Planning, Development and Conservation. -- X + 268 pp.; London (Graham & Trotman).

MC CARTNY, W.D. & MATHESON, A.H. (1974): Mineral capability maps and land-use planning in B.C. -- Canad. min. J. **95**, 8: 47 - 53; Montreal.

MEINE, K.-H. (1971): Gedanken über Standpunkte und Wesensverschiedenheiten der Thematischen Geographie in Europa. -- IGU europ. Reg. Conf. Budapest 1971: 1 - 25; Budapest.

MEINE, K.-H. (1979): Thematic mapping: present and future capabilities.-- In: World Cartography (XV) **15**: 1 - 16; New York (UN Publ.).

MELNIKOV, E.S. (1979): The main principles of procedures for the National Engineering Geological Survey of the USSR. -- Bull. internat. Assoc. engng. Geol. **19**: 93 - 95; Krefeld.

MERLA, A., MERLO, G. & OLIVIERI, F. (1976): Detailed engineering geology mapping in selected Italian mountain areas: methodolgy and examples. -- Bull. internat. Assoc. engng. Geol. **14**: 129 - 135; Krefeld.

MINELL, H. (1987): Thematic Geological Mapping in Northern Sweden. - Norg. geol. Unders. spec. Publ. **2**: 120; Trondheim.

MINGDE, W. (1987): Elements of geological mapping for urban planning. -- Expert Working Group Meeting, CUM Workshop: 1 -11; Bangkok.

MOLDAN, B. (1987): Environmental geologic mapping in Czechoslovakia. -- Norg. geol. Unders. spec. Publ. **2**: 82 - 83; Trondheim.

MOLDAN, B., CICHA, I., BEDNAR, J., CADEK, J., JETEL, J., JOVANDA, J., OPLETAL, M.,POKORNY, J., TOMSEK, M., VESELY,J., VOLSAN, V., ZEMAN, M. & ZIKMUND, J. (1987): 1:50 000 Maps of the Natural Environmental Potential of the CSSR. -- In: [ARNDT, P. & LÜTTIG, G.W.]: Mineral resources extraction, environmental protection and land-use planning in industrialized and developing countries: 183 - 196; Stuttgart (Schweizerbart).

MONRO, S.K. & HULL, J.H. (1986): Environmental Geology in Great Britain. -- In: [BENDER, F.]: Geo-Resources and Environment: 107 - 124; Stuttgart (Schweizerbart).

MONRO, S.K. & HULL, J.H. (1987): Environmental Geology in Great Britain. -- Norg. geol. Unders. spec. Publ. 2: 11; Trondheim.

MONTEIRO, J.H. (1987): Coastal and Continental Shelf Natural Resource Maps and Planning. -- Norg. geol. Unders. spec. Publ. 2: 108; Trondheim.

DE MULDER, E.F.J. (1987): Recent development in environmental geology in The Netherlands. -- In: [ARNDT, P. & LÜTTIG, G.W.]: Mineral resources extraction, environmental protection and land-use planning in industrialized and developing countries: 127 - 136; Stuttgart (Schweizerbart).

MURTHY, M. (1982): Environmental mapping in India. -- US geol. Surv. prof. Pap., N 1193: 292 - 297; Washington.

NICHOLS, D.R. (1982): Application of Earth Sciences to Land-Use problems in the United States with Emphasis on the role of the U.S. Geological Survey. -- In: [WHITMORE, F.C.Jr. & WILLIAMS, M.E.]: Resources for the Twenty-First-Century, US geol. Surv. prof. Pap. 1193: 283 - 291; Washington.

NICKLESS, E.F.P. (1982): Environmental geology of the Glenrothes district, Fife Region. Description of 1:25 000 sheet No. 20. -- Rep. Inst. geol. Sci. 82, 15.

[ÖROK (ÖSTERREICHISCHE RAUMORDNUNGSKONFERENZ)] (1988): Empfehlung Nr. 21 -- Empfehlungen zur Erstellung von Naturraumpotentialkarten. -- 10 pp.; Graz.

OTTESEN, R.T. & BØLVIKEN, B. (1987): Use of geochemical maps in regional planning. -- Norg. geol. Unders., spec. Publ. 2: 46 - 51; Trondheim.

PATTY, E.J. & WONGSOSENTONO, S. (1987): Land-use, environment and construction materials. -- In: [ARNDT, P. & LÜTTIG, G.W.]: Mineral resources extraction, environmental protection and land-use planning in industrialized and developing countries: 303 - 311; Stuttgart (Schweizerbart).

RADO, S. & PAPP-VARY, A. (1976): Maps for planning. -- In: [COMPTON, P.A. & PECSI, M.]: Regional development and planning. -- Stud. geogr. hung. 12: 47 - 53; Budapest.

RAJU, K.C.C. (1975): Operation Anantapur. -- Oper. Anantapur Bull. 1: 1 -14, 8 maps; Hyderabad (Geol. Surv. India).

RAJU, K.C.C. (1987): Geoscientific maps in land-use planning in India. -- In: [ARNDT, P. & LÜTTIG, G.W.]: Mineral resources extraction, environmental protection and land-use planning in industrialized and developing countries: 281 - 298; Stuttgart (Schweizerbart).

RAJU, K.C.C., KAREEMUDDIN, M.D. & PRABHAKARA, R.P. (1979): Operation Anantapur. -- Misc. Publ. **47**: 67 pp; Madras (Geol. Surv. India).

[RIJKS GEOLOGISCHE DIENST, STICHTING VOOR BODENKARTERING & DIENST GRONDWATERVERKENNING TNO] (1986): Subsoil uncovered. -- 36 pp., Haarlem.

[ROMARIZ, C., PRATES, S. & ANDRADE, C.] (1989): Program, abstracts and field guide. -- Intern. Symp. applied envir. Geol: 138 pp.; Lisbon (Univ. of Lisbon).

SAYADO, J.M. (1982): Las unidades geomorfologicas como base para la evaluacion integrada del paisaje natural. -- Acta geol. illioana **16**, 1: 169 - 180.

SILIWONCZUK, Z. (1969): Objasniemia dio przegladowej mapy surowcow skalnych polski. -- Arkusz Zamoss 1:300 000.-- 37 pp.; Warszawa.

SPANGLE, W. & ASSOCIATES, LEIGHTON, F.B. & ASSOCIATES & BAXTER, MC DONALD & COMPANY (1976): Earth Science Information in Land-Use Planning. Guidelines for Earth Scientists and Planners. -- US geol. Surv. Circ. **721**: 33 pp., Arlington.

STENESTAD, E. (this volume): Standards in environmental geology.

TURNER, A.K. & COFFMANN, D.M. (1973): Geology for planning: a review of environmental geology. -- Quart. J. Colo. School Mines **68**, 3; Golden/Colo.

[U.S. DEPARTMENT OF HOUSING AND URBAN DEVELOPMENT & U.S. DEPARTMENT OF THE INTERIOR] (1971): Environmental planning and geology. -- 204 pp., Washington.

VINKEN, R. & VOSS, H.-H. (1983): Die automatische Datenverarbeitung -- ein neues Werkzeug bei Aufnahme und Entwurf geowissenschaftlicher Karten. -- 43. dt. Geogr.-tag Mannheim 5.-10.10.1981, Tag.-Ber. wiss. Abh.: 133 - 135; Wiesbaden.

VORACEK, V. (1984): Map of the environment of the Czech Socialist Republic. -- Sborn. pr. CSAV geogr. Ustav **1984**, 5: 71 - 72; Praha.

WADGE, A. (1987): Geological Mapping and Environmental Planning in Urban Areas. -- Norg. geol. Unders. spec. Publ. **2**: 121; Trondheim.

WALTERS, M. & WIGAND, H.J. (1987): Naturraumpotentialauswertungsmodelle. Auswahl von Auswertungsmodellen und Überprüfung der Verfügbarkeit der dafür erforderlichen Naturraumpotentialdaten. Gutachten im Auftrag der ÖROK. -- 126 pp.; Graz.

WILLE, V. (1987): Examples of maps for the planning process. -- In: [ARNDT, P. & LÜTTIG, G.W.: Mineral resources extraction, environmental protection and land-use planning in industrialized and developing countries: 83 - 97; Stuttgart (Schweizerbart).

WILSON, D. & SMITH, M. (1985): Planning for Development; Thematic geological map, Bridgend area -- Open File Rep. for DoE, Aberystwyth (BGS).

WOLFF, F.CHR. (1977): Trondheim. Geoscientific Map of the Natural Environmental Potential Utilization. Scale 1:250 000. -- Trondheim (Norg. geol. Unders.).

WOLFF, F.CHR. (1986): Mapping in the Service of Environmental planning. -- Episodes 9, 2: 113 - 114; Ottawa.

WOLFF, F.CHR. (1987a): Development of geological maps for land-use planning in Norway -- In: [ARNDT, P. & LÜTTIG, G.W.: Mineral resources extraction, environmental protection and land-use planning in industrialized and developing countries: 137 - 147; Stuttgart (Schweizerbart).

[WOLFF, F.CHR.] (1987b): Geology for environmental planning. -- Norg. geol. Unders. spec. Publ. 2: 121 pp; Trondheim.

ZHAMOIDA, A.L., BASKOV, E.A., GANESHIN, G.S., IL'IN, K.B., MEZHELOVSKY, N.V., NIKIFOROVA, G.YA., SOKOLOV, R.J. & UNKSOV, V.A. (1984): Geologic Maps Exhibition -- 84 (Geokarta 84). -- 27th Sess., internat. geol. Congr. gen. Proc.: 147 - 160; Moscow.

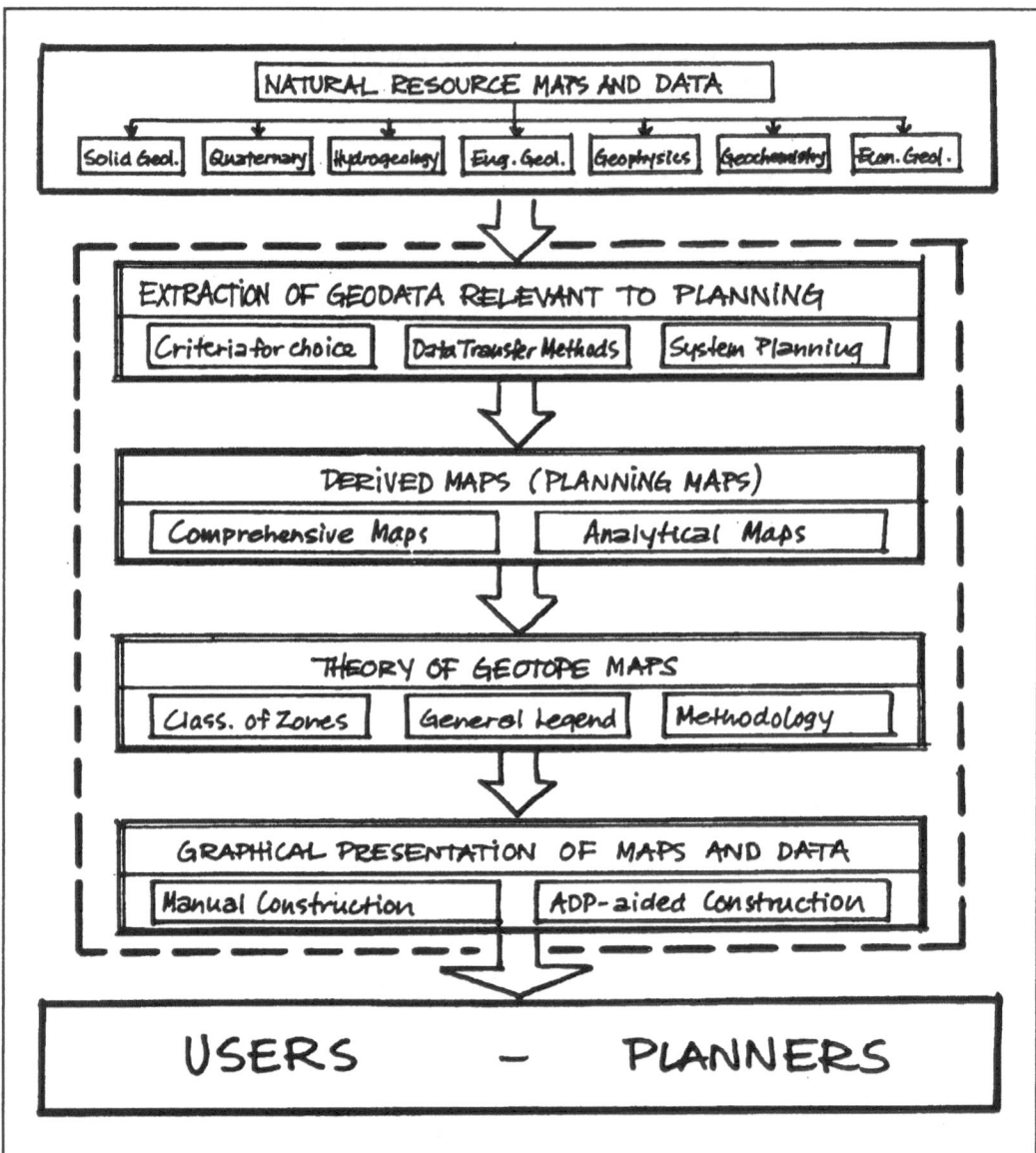

Figure 1: Flow chart for transferring geoscientific maps into useful maps for planners (HUGDAHL & WOLFF 1987).

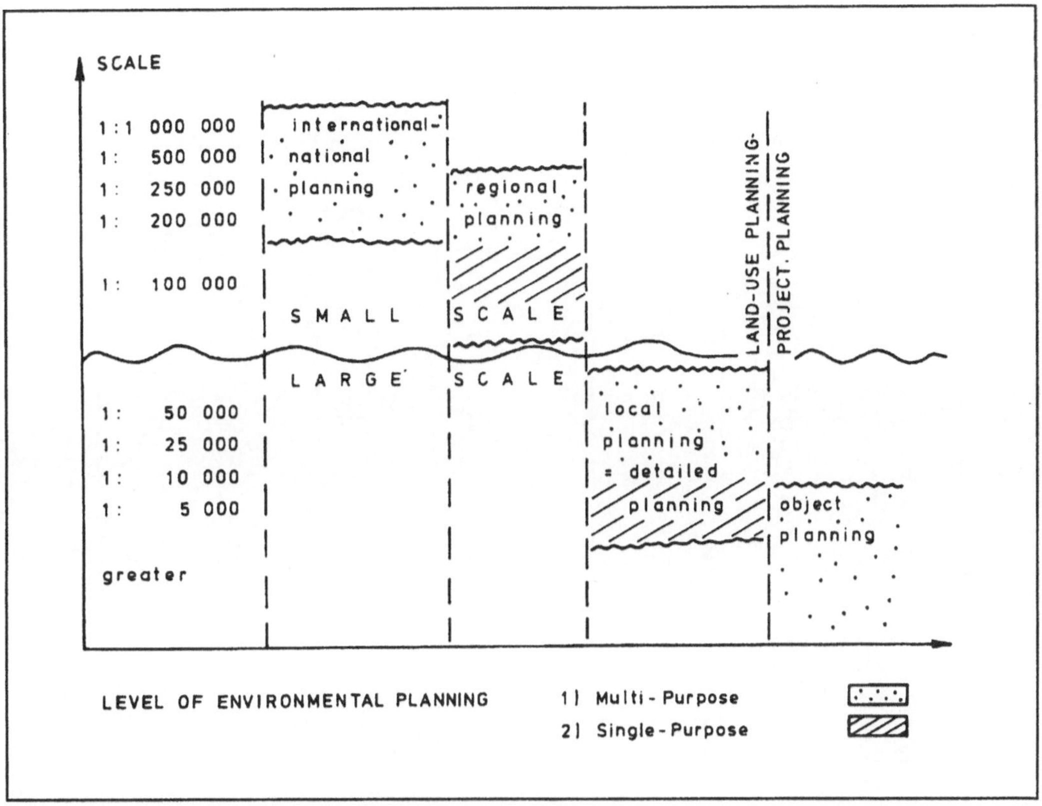

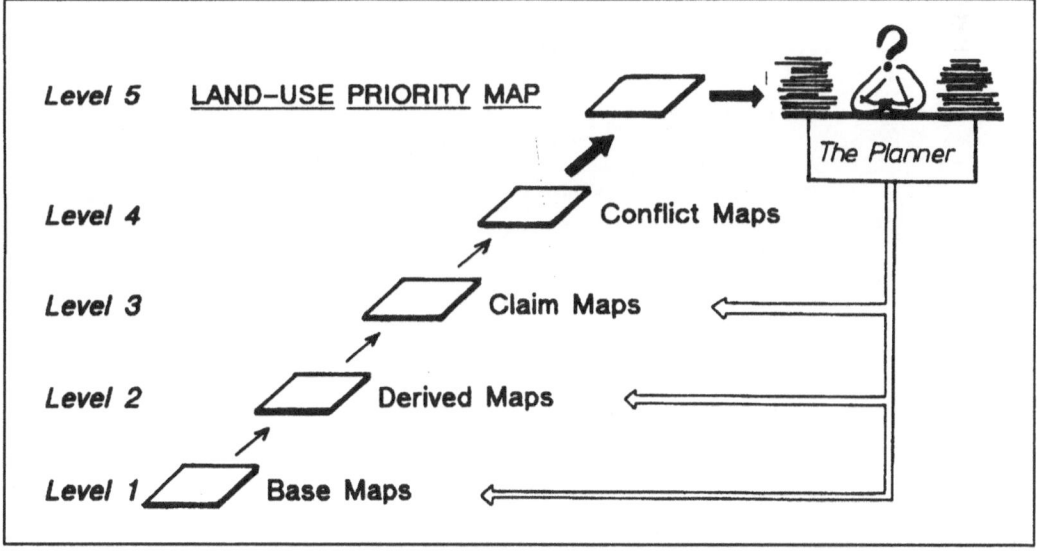

SCALE

1:1 000 000
1: 500 000
1: 250 000
1: 200 000

1: 100 000

international-
national
planning

regional
planning

SMALL SCALE

LAND-USE PLANNING-
PROJECT. PLANNING

LARGE SCALE

1: 50 000
1: 25 000
1: 10 000
1: 5 000

greater

local
planning
= detailed
planning

object
planning

LEVEL OF ENVIRONMENTAL PLANNING

1) Multi-Purpose
2) Single-Purpose

Figure 2: Graphic representation of relation between level of planning and scale of maps (LÜTTIG 1987a).

Level 5   LAND–USE PRIORITY MAP

The Planner

Level 4   Conflict Maps

Level 3   Claim Maps

Level 2   Derived Maps

Level 1   Base Maps

Figure 3: The organization of the Geoscientific Map of the Natural Environment's Potential (GMNEP).

Computer-assisted Methods for the Construction, Compilation and Display of Geoscientific Maps

GOTTFRIED GABERT *

Abstract:

The paper reviews modern methods for map construction, compilation and display on the basis of current applications at the Geological Surveys of the Federal Republic of Germany and Lower Saxony.

The graphical representation of geoscientific data, for example mapping and exploration results, is generally done in the traditional way of analog maps. Different possibilities to produce digital maps exist:

-- map construction directly from geological field data,
-- digitization of existing maps, especially manuscript maps,
-- conversion of remotely sensed data into raster or vector maps.

For map construction directly from geological field data, the establishment of databases (borehole and outcrop data) is a prerequisite. In addition, appropriate systems for

* Author's address: Dr. G. GABERT, Federal Institute for Geosciences and Natural Resources, P.O.B. 510153, D-3000 Hannover 51, FRG., Secretary-Treasurer of COGEO-DATA.

data collection, storage, retrieval and further processing must be available. For map digitization, autocartographic systems, optical and video scanners are used, and for the evaluation of remotely sensed data, image processing systems are required.

The information contained in a conventional, analog map is of course restricted as it represents only part of the knowledge available. Digital maps substantially increase the scope of information if they contain basic data from a variety of different sources. From such a digital database specific data can be selected and integrated.

The solution for interdisciplinary problems requires the integration of multi-disciplinary data. For large amounts of data simple overlay techniques are not sufficient for carrying out sophisticated geoscientific interpretations. Here, complete data integration in raster mode is very useful and applies intersection techniques for line and area data. These techniques are particularly applicable for maps on land-use planning and environmental protection where large amounts of data form the basis for the recognition of possible conflict areas. The integration of raster and vector data provides excellent examples of modern techniques of map compilation and display.

Introduction

In view of ever increasing environmental problems, geosciences are confronted with demands to provide meaningful data for the solution of a large variety of different problems. Examples are data on the availability and use of mineral and energy resources, of industrial minerals and groundwater, on soil and bedrock geology as well as data on geogenetic and anthropogenetic geochemical factors. Most important now, and in the future, are answers regarding questions of environmental protection: Where are suitable locations for waste repositories, by which means can old, forgotten repositories be relocated? What is the degree of contamination of soils with toxic metals, which part of it is geogenetic, and which is anthropogenetic? How can we define areas of competing demands for both mineral resource potential and environmental protection needs? If answers to these questions can be provided, and if solutions to such problems can be found, they are always based on data displayed on geoscientific maps, either in analog form, or stored in digital graphical databases.

The graphical representation of geoscientific data, for example mapping and exploration results, is generally done in the traditional way of analog maps. Different possibilities to produce digital maps exist:

-- map construction directly from geological field data,
-- digitization of existing maps, especially manuscript maps,
-- conversion of remotely sensed data into raster or vector maps.

For map construction directly from geological field data, the establishment of databases of borehole and outcrop data is a prerequisite. In addition, appropriate systems for data collection, storage, retrieval and further processing must be available. For manual map digitization, autocartographic systems are used, and for automatic digitization, optical and video scanners are in use; for remotely sensed data, image processing systems are required.

The information contained in a conventional, analog map is of course restricted as it represents only part of the knowledge available. Digital maps substantially increase the scope of information if they contain basic data from a variety of different sources. From such a digital database specific data can be selected and integrated.

The solution for interdisciplinary problems requires the integration of multi-disciplinary data. For large amounts of data simple overlay techniques are not sufficient for carrying out sophisticated geoscientific interpretations. Here, complete data integration in raster mode is very useful and applies intersection techniques for line and area data. These techniques are particulary applicable for maps on land-use planning and environmental protection where large amounts of data form the basis for the recognition of possible conflict areas. The integration of raster and vector data provides excellent examples of modern techniques of map compilation and display.

Map construction, compilation and display

Map construction is based on the spatial distribution of geometric elements according to the required scale; it is based on geoscientific raw data, either alphanumerical data obtained in the field, or digital data from satellites. Map compilation, on the other hand, is based on the digitization of existing analog map data like manuscript maps, published maps, cross sections, etc.

If required, the products of both constructed maps and compiled maps can be merged. It results in

--   the display of borehole or outcrop locations and cross-sections (for the geological map),
--   the construction of profile types and the map legend (for sequence maps),
--   the construction of the isopaches of the Quaternary base and the subcrops of the pre-Quaternary (for the Quaternary base map), and
--   the construction and display of columns and their point locations (for columnar maps).

Constructed maps can be stored in a graphical database in the same way as digitized maps, and the map data can be further processed in an interactive mode.

For geoscientific applications, the ammount and variety of descriptive alphanumerical data are typically large. The description of the content of point, line and area data can be of such a complexity that simple file structures are insufficient for data storage and handling. Here, sophisticated databank and database management systems have to be applied which meet the requirements of the geosciences, as do, for example, relational systems. They are flexible and dynamic with regard to data structures and selective retrieval. Most of the geoscientific alphanumerical data stored in database are point-related data, like borehole, outcrop or well data. Line-related data may represent faults or geomorphological features. Area data are mostly derived from point-related data, and the area data can be reinterpreted and compiled on demand.

The construction of geoenvironmental maps in particular involves the evaluation of large amounts of geoscientific data. Here, computer-based systems are used with advantage for data collection, storage, retrieval and processing for graphical display. This requires the establishment of data files, the use of efficient database management system and sophisticated graphics software. Basic information is collected from existing archival records and through field mapping activities.

The minimum database requirements for the construction or compilation of geo-environmental maps are

-- mineral resource data like sand, gravel, clay, peat, salt and coal deposits,
-- hydrogeological data of groundwater aquifers and cover sediments,
-- geological mapping data with regard to existence and extent, thickness and depth of lithostratigraphic units,
-- soil data,
-- engineering geology data.

The data can be organized in three hierarchical levels: borehole-related data, strata-related data and sample-related data.

Sources of information are:

-- borehole and geological strata descriptions from the archival records,
-- new field data in form of outcrop and borehole descriptions,
-- laboratory results (from the Geological Survey and outside sources),
-- special files from different branches within the Geological Survey (like palae-
   ontological and radiometric age determinations of samples).

As an example, the map series 1 : 200 000 of Lower Saxony which displays the natural environment's potential is based on the geological maps 1 : 25 000 and comprises

-- soil maps displaying the agricultural potential,
-- geotechnical maps with engineering geology data,
-- hydrogeological maps showing groundwater use and groundwater potential,
-- deep-lying mineral and energy resource maps displaying metals, coal, industrial
   minerals, salt and hydrocarbons,
-- near-surface mineral resource maps displaying mineral and energy resource deposits
   and reserves,
-- maps of protected areas to control and preserve near-surface resources,
-- land-use planning maps displaying geoscientific priorities in cases of competitive
   user requirements,
-- a general legend for the whole map series.

The database management system originally developed for the project "maps of the natural environment's potential of Lower Saxony" is a system which includes a component for the handling of geoscientific data (from boreholes, outcrops, stratigraphic records, laboratory samples), and a component for graphical data processing and plotting of various maps, profiles and lists from basic data (DASP: KÜHNE 1983).

This relational database management system also permits the linking of fied data with sample and analytical data, and it provides interactive capabilities and the facility to link files of different structures.

Data from the borehole database can be further processed for graphical display. There exist a number of graphics programs to construct

-- point symbol maps, for instance borehole location maps with number, depth of base and top of a marker horizon,

-- borehole sections on columnar form with standardized signatures for the lithology. These columns can be constructed as single columns, or as a number of columns aligned along a predefined line, as a columnar section, or in the form of a columnar map.

-- Retrieval from the database can be further processed to construct profile type or sequence maps which display the location, symbol and the number of the respective sequence of geological strata. The profile types are computed from the sequence of type-forming lithological or genetical units. These units, in turn, have to be predefined by means of adequate search queries which provide a classification of a given data set.

-- A legend is then constructed that informs about the sequence, frequency and maximum number of sequence types.

With this method, identical sequences of strata derived from outcrop and borehole records characterize identical profile types (defined sections) which are delineated on the map. The application of this method permits display in a readable form, sequences of an unlimited number of strata down to a predefined base horizon.

The numbering and color system of the profile types follow rules that conform with the age relations of the strata. By means of the numbers and colors of the profile types the user is able to read from the legend and see the schematic profile of which rock units are combined in a specific sequence.

One can thus clearly identify the relevant profile types at any point on the map. While the information depth of the profile type map is much greater than that of the conventional geological map, i.e. 50 m vs. 2 m on the average, the observation density of profile type maps in Lower Saxony is only 1 borehole per square kilometer.

In areas with flat-lying strata, the display of geological data in the form of profile type or sequence maps is useful for the demonstration of the three-dimensional, spatial position of the sedimentary rock sequences.

Not only defined lithological, biostratigraphical and facies sequences can be represented on the profile type map, but also defined geophysical or hydrogeological units and sequences. As manual evaluation work for the construction of profile type maps is extremely time-consuming, computer-aided methods are an advantage to use.

Complementary to the profile type map is the columnar section map which displays in columnar sections, true to scale, the thickness and petrographic composition of the different layers as well as the thickness of the complete sequence of the strata at each borehole location. The columns are plotted automatically from geological data of a borehole database.

The basic information on each borehole or outcrop is contained in the borehole location map which displays the number of the borehole, its depth and the top or base of a marker horizon.

From all these data, sub-surface and structural maps can be constructed which reveal, for example, in form of contour lines the base of the Holocene or Quaternary, or the top of the Mesozoic hard rock units.

An example for the transformation of geoscientific point data into graphical area data is a program developed by PREUSS and VOSS (1983). An interactive mode leaves possibilities for the geoscientist to influence the definition and construction of the boundary lines and to manipulate them.

The program works by means of area and line generation on a raster basis. Hereby, overlays of several geological units can be created in a rather simple way, using not only vector, but also raster plotters. The program thus offers access to raster-based ink jet plotters or electrostatic color plotters to produce colored computer maps, for example Quaternary base maps.

This program can process point information, for instance from boreholes, for areal display of geological units. Areas of presence, absence and incomplete information can be differentiated. Existing line information, for instance morphological or tectonic features, can be taken into account and integrated in the map construction. The constructed areas can thus be stored in a graphical database as the basis for a digital geological map.

The processing of area data for the construction of geoscientific maps is thus advantageously carried out by means of raster-oriented systems. Especially for the graphical intersection of areas from different maps, raster techniques are far superior to vector techniques. Both raster and vector techniques have advantages and drawbacks. The application of raster techniques, for example, requires extremly large storage capacities; raster graphics are not as precise as vector graphics; however, modern image processing allows very quick manipulation of large raster data sets.

Therefore, geoscientific cartography would prefer to use integrated, or hybrid, raster-vector systems; difficulties arise from differences in the internal structures: vector graphics are based on line-data as chains of sequentially linked pairs of coordinates. In raster graphics the information content is represented by isolated pixels, and point, line and area data are not directly retrievable, in contrast to vector graphics.

Within the last 10-15 years, raster graphics have become more and more important in the geosciences. For satellite data, raster-based image processing systems have been developed. Raster data are also obtained by optical scanners and by video-scanning of vector graphics.

Hybrid systems using vector and raster techniques require storage capacity for both types of data and fast conversion facilities of line data from raster to vector format. These requirements are met by the recently developed Graphic Interactive Raster-Oriented System GIROS (PREUSS 1988, 1989). It is designed for reading vector data and raster images, for raster-vector conversion and area construction using intersection techniques. It used advanced raster techniques for the storage of the vector data in internal raster structures without loss of information and precision. Alphanumerical geoscientific data can be linked to a point within an area by means of a "point-in-polygon routine". This linking of alphanumerical attributes and graphical data results in the generation of coded areas and a map legend of an almost unlimited number of units. A total of 50 search queries allows both data classification and its display on screen in the form of an area mosaic. It is obvious that inumerable possibilities exist for the construction of derived thematic maps.

GIROS is used for tasks of geological mapping as well as land-use and environmental planning; current applications include the construction of soil maps which form the basis for a soil information system, and the construction of sections and maps for the geotectonic atlas of NW-Germany.

Autocartographic systems are applied advantageously if the same digital data can be made available for a variety of applications, like map compilation and production, as well as the generation of derived maps. Especially the preparation of thematic maps for production puposes is a major cost-effective factor, and that is also true with regard to maps derived from a once established digital database.

In the Geological Surveys of the Federal Republic of Germany and of Lower Saxony, autocartographic systems have been in operation for more than 10 years. They are applied mainly for map digitization and the construction of digital databases, for map compilation, evaluation and the construction of derived maps, and for the preparation of map production. The subjects covered by thematic maps are geology, mineral resources, soil mapping and land-use planning, geochemistry and geophysics. For the geoscientific map series of the Federal Republic of Germany of the scale 1 : 1,000,000, both digital topographic and gescientific data in vector format are merged in the process of map preparation; the digital topographic data are derived from the Euro-Database 1 : 1,000,000, the digital geoscientific data from maps existing in analog or digital form.

An example for map construction on vector format from alphanumerical geological data is the series of the Geological Map of Lower Saxony 1 : 25,000. The geological base map of the scale 1:25,000 is the classical geological map that displays the geological situation of the uppermost two meters with fair precision based on 3000 - 5000 drill-holes per map sheet.

During geological field mapping of the scale 1 : 25,000, carried out in a computer oriented way, geological data are collected from boreholes, outcrops and samples. The data form the basis for the computer-assisted construction of sections and profiles, maps and legends.

The profile type map reveals the geological succession of the unconsolidated sediments of the Quternary to a depth of 50 meters. Different colors and index numbers characterize the specific succession of strata at any point on the map. Schematic sections and legends in matrix form provide instructive information on the composition of the Quaternary. The construction of the map requires a great number of boreholes up to 50 m depth, one borehole per quare kilometer on the average. To indicate the reliability of the map construction, the relevant borehole locations are shown on the map.

In the columnar map the columns display to scale the petrographic composition of the layers at each single borehole location. The map thus provides point information on the thickness of the different layers and the total succession of Quaternary sediments. The columns are plotted automatically from borehole records according to standarized petrographic symbols (DIN 4220).

The sub-surface map reveals the pre-Quaternary rocks, mainly the hard rocks of the Mesozoic. It is based on the same boreholes that lead to construction of the profile type and columnar maps. In addition, contour lines show the morphology of the base of the Quaternary.

This set of maps provides maximum information, but retains its clearness and readibility and is thus well suited also for non-geoscientists. It is supplemented by a detailed explanation dealing with geological and applied aspects of mineral exploration, groundwater use, building foundations and soil science.

An example for map construction in raster format from geochemical data is the Geochemical Atlas of the Federal Republic of Germany 1 : 2,000,000 (FAUTH et al. 1985). For the preparation of the geochemical atlas of the FRG, the sample density is 1-2 samples per km . It was essential to maintain homogenous procedures for sampling, sample preparation and analyses to provide maximum compatibility of results. A total of 80 000 stream sediment and water samples were analyzed: the water samples for Pb, Zn, Cd, Cu, Ni, Co, U and F, and, in addition, the stream sediment samples for W, Ba, Cr, Sr, V, Li and Sn.

The sample data are stored in digital form in the geochemical database GEOMUL-DAT from which the data are retrieved and displayed in form of sample location points and analytical results. For the display of the results an Applicon color plotter was used by which two- and three-dimensional element maps were produced by means of UNIRAS software packages.

The basic analytical information is displayed in the form of sample location points: maps of the scale 1 : 50,000 to 1 : 200,000 show every sample point, in contrast to the small-scale maps of the atlas of the scale 1 : 2 000 000 where the arithmetic mean and the maximum value are displayed in form of squares 3 x 3 km.

Contamination of the environment comprises two different components, the geogenic and the anthropogenic contamination. In order to recognize the anthropogenic factors and to estimate their influence, the geogenic component must be known first. The appropriate medium for display of the geogenic factors is the geochemical atlas which, in turn, is a tool for planning new exploration activities for still undiscovered mineral deposits.

Regional geochemical background values can be specified and areas with anomalous element distribution can be isolated. These anomalies are caused by differences either in lithology or mineralization. Here again a distinction can be made between known and unknown mineralized areas. Important information with regard to environmental contamination provides those maps which display the distribution of the pH-values and conductivity values in surface waters. Low pH-values enhance the solubility and mobility of base metals which then contaminate the groundwater and influence crops and forests.

The existing database of the geochemical atlas serves the needs of many users, and is an appropriate instrument for planning purposes like

-- exploration for base metals and uranium,
-- groundwater exploration,
-- recognition of soil contamination in agricultural and groundwater use,
-- inventory of the geochemical status in the FRG,
-- measures for the protection of the environment,
-- connections between geology and medicine.

A comparison of Pb in stream sediments and surface waters, for example, shows that high Pb-values in waters are more diffusely distributed than those of stream sediments which are concentrated in mining areas. Indications of environmental contamination is followed up by special research programs.

An example for map compilation on vector format from digitized map data and alphanumeric mineral resource data is the Map of Near-Surface Raw Materials of the Federal Republic of Germany 1 : 1 000 000.

The areas displayed comprise reserves and resources, actual mining areas and potential areas of building materials, ceramic and volcanic materials, gypsum, peat, quartz sands, kaoline, bentonite, feldspar, pozuolane, diatomite, talcum, bauxite, calcite, lignite, oil shales and oil sands.

These commodities are displayed in different colors, and, in addition, identified by alphanumeric symbols. If several commodities occur within one area, then the symbols are arranged in priority order representing the color of the commodity with the highest priority. Based on geological map data, areas with an important resource potential are delineated by horizontal hatching. All the geological and resource data, as well as the graphic information, are stored in digital form.

This database is the first of a series of geoscientific and mineral resource maps of the Federal Republic of Germany of the scale of 1 : 1 000,000. Other databases of this map series 1 : 1 000,000 have been completed like the geological map and the soil map, or are in preparation, like the mineral resources map and the tectonic map of the Federal Republic of Germany.

A new map series in digital form is the Map of the Near-Surface Raw Materials in the FRG 1 : 200 000 which is compiled from alphanumeric and graphical data.

An example of the integration of remotely sensed data in raster format and geological map data in vector format is provided by the international multidisciplinary CODATA project which used a combination of multisensor data gathered on different scales by various satellites in order to map the geomorphology, geology and land-use in Tanzania, East Africa, on a regional scale (BARDINET et al. 1987, 1988).

The main objectives of this project were to

-- differentiate between the soils and rocks on the basis of their reflective and thermal properties,
-- record the morphology and structural features,
-- characterize the relations between rocks, soils and vegetation cover.

Under favourable conditions, the combination of thermal data at low resolution from METEOSAT, NOAA 7 and TIROS N with reflective data at high resolution from LANDSAT MSS led to the identification of soils and geomorphology, lithology and major structures. The thermal and multi-spectral signatures of vegetation, soils and rocks were correlated with geological and geomorphological field data as an example of multi-disciplinary application.

Ground control was necessary in order to prepare an interpretation grid for the multi-spectral signatures of the remotely sensed objects taking into account the spatial and spectral resolutions of METEOSAT, NOAA 7, TIROS N and LANDSAT. The classification of the multi-satellite data was guided by the interactive use of lineaments, faults and geological boundaries.

The computer maps were prepared on two scales:

-- 1 : 2 500 000 geometrically rectified on the World Map and covering the land mass of Tanzania,
-- 1 : 500 000 geometrically rectified on a geological map covering the Iringa-Njombe area in south-central Tanzania.

For the map of the scale 1 : 2 500 000, data from NOAA 7, TIROS N and METEOSAT were used, for the map of 1 : 500 000 data from METEOSAT, NOAA 7 and LANDSAT MSS.

Multi-satellite thematic mapping provides a means for creating a multi-temporal inventory of the natural environment at scales depending on the satellites used. Vegetation, soil type and rock lithology may be recognized through indirect measurements based on the temperature cycles of the albedo in the visible light and the near-infrared, or through a combination of these data.

In the interdisciplinary international GARS Kibaran Project in Africa, remote sensing data of LANDSAT MAA, LANDSAT TM and SPOT and sophisticated image processing techniques were applied to geological problems of the proterozoic Kibaran and Ubendian fold belts in Central and East Africa. The main image analyses applied were channel combinations and supervised classifications based on both digital geological and spectral databases. The results of this project serve as an aid to geological mapping of these fold belts, and to an improved interpretation of their lithology, structure and mineralization (GABERT 1987, LAVREAU & BARDINET 1988).

Conclusions

Of increasing importance in the future is the

-- construction of digital maps derived from alphanumeric and graphical databases,
-- integration of interdisciplinary data in raster and vector format for compilation of thematic (derivative) maps,
-- design and application of geoinformation systems.

Modern types of geoinformation systems are characterized by interactive facilities, multi-disciplinary databases, efficient database management systems and sophisticated graphical display techniques. Geoinformation systems are effective tools for geological research. Skilled users are able to collect, manipulate, analyze, update and display digital spatial data of different sources, resolutions, scales and projections.

Knowledge-based techniques will play an increasing role for the comprehensive evaluation of large databases stored in information systems of the future.

The integration of digital data from topographical terrain models, geoscientific maps and satellite imagery will replace 2-dimensional maps by 3-dimensional representations of geological models in their spatial setting.

References

BARDINET, C., GABERT, G. & MONGET, J.-M. (1987): Multisatellite Thematic Mapping in Tanzania - a CODATA Project. In: GLAESER, P.S. (Ed.), Computer Handling and Dissemination of Data: 300-303, Amsterdam, North-Holland, 1987.

BARDINET, C., GABERT, G., MONGET, J.-M. & YU, Z. (1988): Application of Multisatellite Data to Thematic Mapping. - Geol. Jb. (B) **67**: 1-74, 12 Fig., 6 Tabl., 5 Maps, Hannover 1988.

FAUTH, H., HINDEL, R., SIEWERS, U. & ZINNER, J. (1985): Geochemischer Atlas Bundesrepublik Deutschland: 1-79. - Hannover, BGR 1985.

COGEODATA-DFG (1989): International Colloquium on Digital Maps in Geosciences: Abstracts; Würzburg 1989.

GABERT, G. (1982): Handling of Geological Field and Map Data. - Nat. Res. Dev., **15**: 21-40, Tübingen 1982.

GABERT, G. (1987): The GARS Program: Geological Applications of Remote Sensing to Proterozoic Fold Belts in Central Africa. - In: MATHEIS, G. & SCHINDEL-MEIER, H. (Eds.), Current Research in African Earth Sciences: 337-338, Rotterdam 1987.

GABERT, G., HINZE, C. & KOCKEL, F. (1984): Geologische Karten und Kartierungen. - Geol.Jb.(A) **73**: 123-134, Hannover 1984.

KÜHNE, K. (1983): DASP - Ein System zur Verwaltung und Auswertung geowissenschaftlicher Daten. - Geol.Jb.(A) **70**: 41-59, 9 Abb., Hannover 1983.

LAVREAU, J. & BARDINET, C. (1988): Image Analysis, Geological Control and Radiometric Survey of Landsat TM Data in Tanzania. - Mus. roy. Afr. centr. Tervuren (Belg), Ann. Sér. in-8 , Sc. Geol. **96**: 1-190, Tervuren 1988.

PREUSS, H. (1988): Map Construction using Advancved Raster Techniques. - Geol.Jb.(A) **104**: 187-195, 7 Fig., Hannover 1988.

PREUSS, H. (1989): Neue Methoden der graphischen Datenverarbeitung im Vorfeld der geowissenschaftlichen Kartographie. - In: J. ENCARNACAO & H. KUHLMANN (eds.), Graphik in Industrie und Technik: 174-192; Berlin-Heidelberg, Springer 1989.

PREUSS, H. & VOSS, H.- H. (1983): Das "Grenzlinienprogramm Geologie" - ein Programm zur Konstruktion, Definition und Manipulation von geowissenschaftlichen Grenzlinien. - Geol.Jb.(A) **70**: 211-231, 12 Abb., Hannover 1983.

VINKEN, R. (1983): Die Automatische Datenverarbeitung als Hilfsmittel bei der Aufnahme und Konstruktion geowissenschaftlicher Karten. - Geol.Jb.(A) **70**: 1-306, Hannover 1983.

VINKEN, R. (1988): Construction and Display of Geoscientific Maps Derived from Databases. - Geol.Jb.(A) **104**: 1-475, Hannover 1988.

Geomorphological Hazards and Environmental Impact: Assessment and Mapping

MARIO PANIZZA *

Abstract

In five sections the author develops the methods for the integration of geomorphological concepts into **Environmental Impact and Mapping**. The first section introduces the concepts of *Impact* and *Risk* through the relationships between "Geomorphological Environment" and "Anthropical Element". The second section proposes a methodology for the determination of *Geomorphological Hazard* and the identification of *"Geomorphological" Risk*. The third section synthesizes the procedure for the compilation of a *Geomorphological Hazards Map*. The fourth section outlines the concepts of *Geomorphological Resource Assessment* for the analysis of the *Environmental Impact*. The fifth section considers the contribution of *geomorphological studies and mapping* in the procedure for *Environmental Impact Assessment*.

* **Author's address:** Prof. M. PANIZZA, Instituto Geologia, Università de Modena, Corso Vittorio Emanuele, 59, I-41100 Modena, Italy.

1. **Introduction: Relationships between "Geomorphological Environment" and "Anthropical Element"**

The relationships between "**Geomorphological Environment**" and "**Anthropical Element**" can be outlined as shown in Figure 1.

The "**Geomorphological Environment**" may be subdivided into *"Geomorphological Resources"* (for example, a "geomorphological asset", such as a mountain peak (Fig. 2) or a sea cliff) and *"Geomorphological Hazards"* (or "geomorphologically instable situations", such as a landslide (Fig. 3) or a river bank subject erosion).

The *"Anthropical Element"* is defined as *"human activity"*, i.e. as something that is the specific action of man (such as the construction of a road or a breakwater), or as a situation of *"area vulnerability"*, i.e. the complex entirety of the population, buildings and structures, infrastructures, economic activity, social organization, and any expansion and development programs planned in an area.

In the relationship between "**Geomorphological Environment**" and "**Anthropical Element**", the position of *"Geomorphological Resources"* is regarded to be mainly a *passive* component in relation to the *"Human activity"*. In other words, a resource may be altered or destroyed by "man" (e.g., a mountain landscape that has been modified by a bulldozer).

Considering again the relationship between "**Anthropical Element**" and "**Geomorphological Environment**", in the subdivision of the former the position of *"Geomorphological Hazards"* is considered to be *active* with respect to the *"Area vulnerability"*. A hazard may alter or destroy "human activity of work" (e.g., the Vajont dam landslide in Italy or river erosion that caused a road to collapse).

Two more concepts can be defined within this context: RISK as the consequence of

"Geomorphological Hazards" on a situation of *"area vulnerability"*; IMPACT as the consequences of *"Human activities"* on a *"Geomorphological Resource"*.

2. Geomorphological Hazard Assessment and the Analysis of Geomorphological Risk

A logical scheme of a methodology for the determination of "**Geomorphological hazard**" and the identification of "**Geomorphological Risk**" is summarized in Figure 4 (PANIZZA, 1987).

The term "**Geomorphological hazard**" is defined as the "probability that a certain phenomenon reflecting geomorphological instability will occur in a certain territory in a given period of time". For example, in any one area, the possibility of a landslide occurring over a 50-year time span, can be evaluated. And the concept of "**geomorphological instability**", which has also been introduced, considers that an unstable landform is a landform which "shows a lack of equilibrium with the natural environment and which is evolving towards equilibrium through modification or by means of particularly dynamic processes". Instability is therefore concerned with forms that evolve in a manner which is disturbing for man's environment. Examples include a slope subject to landsliding or an evolving river meander (PANIZZA, 1978).

To determine geomorphological hazards, it is therefore necessary to identify the degree of instability of each landform. From a conceptual point of view, instability should be identified by following two procedures:

a)   an analysis of the *causes* of instability: i.e. whether natural, man-induced or both;

b)   a study of the *effects* of instability: in other words, what is required is a study of both past geomorphological history and present geomorphological dynamics.

Moreover, instability should be considered in relation to the phenomena that determine it. The term, *"instability"*, should not be used in an absolute sense because an area could prove to be unstable in relation to one specific process (e.g. a landslide), but stable with respect to an other (e.g. rill wash). Indication of the type of instability offers, most importantly, elements that are essential for the selection of the most appropriate and effective remedies (Panizza, 1978). Degrees of instability vary with different types of utilizations. The stability required for a nuclear plant is quite different in degree and type from that which may be sufficient for a road.

The **"Geomorphological Risk"** must be from the "geomorphological point of view", like a natural risk connected with a *geomorphological* hazard: it refers to the probability that the economic and social consequences of a particular phenomenon reflecting geomorphological instability will exceed a certain threshold. Therefore, "geomorphological" risk equals the "product" of geomorphological hazard and a territory's social and economic *vulnerability*. This vulnerability refers in particular to an area's buildings and its economic and social infrastructures. Assessment of risk must therefore pass through an analysis of all man-made structures existing in a certain area and take into consideration to what extend the existing reality is more or less vulnerable with regards to a certain geomorphological event (exposure and vulnerability). For example, erosion is a hazard that may not involve any risk in a desert area, whereas in a densely populated or highly industrialized area, it could represent a high risk. However, the assessment of risk must also be made with the prospect of new buildings and infrastructures in mind (program plan and vulnerability). Thus, before proceeding to an engineering solution, an accurate investigation of a territory's geomorphological stability is indispensable.

Vulnerability should also be considered in relation to the existing degree of social organization, i.e. environmental education, civil defense, and prediction and monitoring techniques. Where people know and safeguard their territory, where the bodies governing intervention are efficient, and where advanced warning systems of major instabilities exist (e.g. river-spates, sea storms, etc.) and finally, where such a level of social organization is high, there will certainly be lower levels of vulnerability.

As for "Geomorphological" risk *mitigation*, it is possible to intervene against the hazard as well as against vulnerability. For example, in the case of a building that has been built or that is about to be built on a degraded slope affected by solifluction, it is possible to intervene with land-upgrading works or with deep building foundations. In the case of lateral river erosion, it is possible to implement hazard mitigation through diversions or rectifications of the flow (chanellizations) or through a reduction of its energy (dikes). One could also procede by reinforcing banks or, in other words, by means of a reduction of vulnerability. Nevertheless, in most of the more hazardous cases, such as large landslides, mitigation of the hazard is the only intervention possible.

Therefore, mitigation of geomorphological risk must come to terms with exposure, program planning and organizational problems, i.e., with problems of a social and economic nature as well as those involving political administration. In fact, risk mitigation operations, which are of a preventive nature, imply an evaluation of costs and benefits. Some guarantees and levels of protection must be determined for both man and the environment. In some cases a risk threshold will have to be established and the higher it is, the more onerous and expensive it will be. In other cases a priority scale will have to be established between several environmental risks or between risk mitigation and other programs.

3. Mapping of Geomorphological Hazards

In order to carry out a **map** of **Geomorphological Hazards**, it's necessary to operate according to the scheme depicted on the left side of Figure 4; that is, keeping separate, on the one hand, the *causes* of instability (natural context and anthropic activities) and, on the other hand, the *effects* (geomorphological history and geomorphological dynamics) (Panizza, 1988). The procedure is synthesized in the scheme of Figure 5.

Among the natural *causes* which favor or prevent instability, the following could be cited: geological, hydrogeological, topographic, climatic, and forest-related causes. Included among anthropic causes are all human activities, such as stock-rearing, agriculture, engineering works, and so on, as they all have an impact on the environment. The analysis of all these parameters that predispose to instability could be shown on an **Integrated Analysis Map**. Both the natural and anthropic contexts will also indicate the limits within which geomorphological instability may develop, taking into account that some parameters remain constant (e.g. lithology, the force of gravity, etc.) and others vary (e.g. human activities, neotectonics, etc.). In other words, it will be possible to define the range of variability, e.g., for forseeable landslides in a certain area, over a specified time period, both in terms of volume and in terms of quantity. It will thus be possible to estimate that in the course of 10 years, 1 to 5 landslides could occur, with a maximum volume of 100 m^3 each. As another possibility, one could predict that soil loss in a year due to rill wash on a particular slope will reach 100 tons per km .

The analysis of *effects* is based on a geomorphological investigation that offers an overall picture of the forms and processes of instability, both past and present. This should result in a **Geomorphological Dynamics Map**. Such an analysis should also provide a reconstruction of the geomorphological history of the most disastrous events that have occurred in the area. A study of gemorphological dynamics is concerned both with forms of erosion and forms of accumulation. These two types of landforms may be considered separately or in their interaction. The study of past events should lead to a better understanding of the magnitude, frequency and impact of phenomena which reflect geomorphological instability. A comparison of present-day processes with past geomor-phological events will lead to an estimate of an evolutionary trend for instability. For example, a history of cliff retreat could be compared with present-day rates of coastal erosion or the recurrent flooding of a plain could be compared with present-day rates of fluvial sedimentation.

The comparison and critical discussion of documents deriving from the analysis of both causes and effects should lead to a synthesis concerning instability and therefore to geomorphological hazard assessment for a territory: the **Geomorphological Hazard Map**.

4.   Geomorphological Resource Assessment in the Analysis of Environmental Impacts

The growing interest in **cultural resources,** witnessed at the level of public opinion and legislative initiatives regarding the identification, protection, salvaging, accessibility, and utilization of this heritage, urges a reflection on the topic of the so called *"natural" assets* and in particular *geomorphological assets.*

It must be pointed out that natural and environmental assets, and therefore geomorphological assets as well, make up part of cultural assets. Nature and Environment are an integral part of what can be defined as *"Culture".*

Amongst natural non-biological assets, the most widespread and spectacular are precisely **geomorphological resources**: a river gorge, a mountain peak, a natural bridge, a sea cliff and the like. They have always been of interest because of their scenic appeal. However, these scenic qualities are not the only attributes that should confer value to such landscape elements, but it is also and sometimes mainly their cultural significance as testimonies of geomorphological events in the past.

Legislative initiatives to safeguard and utilize natural resources in general, and geomorphological assets in particular, imply, in certain cases, an appraisal of their *"value"* in quantitative terms. This is especially the case in *"Environmental Impact Assessment".* Such problems are being posed with increasing urgency, which is due to the rapid increase in the number of projects for urbanization, civil works, etc. in many areas. The growing human intervention on the earth's surface is progressively increasing environmental deterioration and the social demand for recreation areas, uncontaminated localities, parks, protected zones. Nevertheless, the appraisal of the above-mentioned "value" and impact is a difficult problem to solve, linked as it is to the difficulties of providing an evaluation of the various geomorphological aspects of the landscape that is both objective and quantifiable.

The studies that must be conducted involve the concept of *measure*, that is, the necessity of assigning to each "environmental marker" (and therefore, also to geomorphological assets) a "weight", which in turn, permits an assessment in quantitative terms of their "environmental value" and a comparison with other markers, to permit the definition of a hierarchy of priorities as well. The criteria that can be used for an approach to the assignment of this "weight", that is, for a "measurement" of landscape features, may be of two kinds: one of an *esthetic* type, related to their spectacularity, but also one of a *scientific* type.

The first approach is to a great extend of an *intuitive* nature. In this case, the approach to nature depends upon the individual contemplating it and his/her state of mind at that time. It is derived from feelings, and personal perceptions are highly subjective and, therefore difficult to classify and to compare with feelings and perceptions of other persons.

Several procedures for such evaluations exist in literature (i.e.: Linton, 1968; Fines, 1968; Leopold, 1969). Some of these present morphometric methods of measuring several landscape components that have been judged as representative of the scenic quality of the landscape. Other procedures appear to be more subjective and they concern the perception of the landscape as a whole in qualitative terms. Numerical rating scales are proposed. The limitations appear to be considerable in all cases as well, due to the subjectivity inherento judgements conditioned by both the sensivity and the experience of the observer, or because several important components are omitted and the vision of the landscape as a whole is neglected, or because, by breaking up and dividing the resource itself, it is distorted and definitively stripped of its wholeness and essence.

The scientific type of approach is, on the other hand, based precisely on the *scientific* knowledge of the natural resource, the perception of the laws regulating its evolution and the awareness of its significance for humans. Therefore, this is a task that can only be performed by well trained geomorphologists, who can accurately identify and evaluate those attributes.

It is of fundamental importance instead trying to present and approach the resource in its totality, without pre-established tables and matrices, because there are other intersecting or combining variables which will inevitably intervene and augment or decrease its value. We are referring here to human forces and particularly to socio-economic, cultural, and political groups, which may select a given resource as the most significant as a tourist attraction, for example, or as the best scientific expression of certain environmental processes, and so on.

A landform becomes a *geomorphological resource* only if it has a social implication, that is, only if other parameters, external parameters, intervene to confer value to it (see Figure 6).

As long as a given river, or a given landscape is studied by and known only to scientists and researchers, it remains "private" knowledge and its potential as a resource is not materialized. However, if the scientist and the researcher popularize it, making its cultural and environmental significance known to the public, and thereby showing its social implication, then the landform becomes a geomorphological resource in the perception of the society (Panizza & Piacente, 1989).

But where and when a *geomorphological recource* has a *social implication* (see human activity: e.g. tourism) an *environmental impact* can occur (see Figure 1).

Environmental impact studies have appeared at a time in which the ideological evolution of the post-industrial society is bringing about an increase in critical and negative attitudes towards technology and its effects on nature and the environment (opposition to nuclear energy, genetic engineering, polluting industries, etc.). But the pragmatic application of these ideological trends to the day-to-day social and political organization is difficult without a scientific basis.

One of the needs of this post-industrial age is, perhaps, the return to "neohumanism", a holistic concept of culture. Environmental impact studies provide the means to combine scientific, economic, social and cultural aspects in a harmonious and integrated concept of the landscape, in order to achieve a balanced relationship between human beings and their environment. this kind of holistic approach to the environment should provide a compehensive view of environmental problems, including the basic elements needed by policy- and devision-makers for correct choises.

5. Geomorphological Mapping and Environmental Impact Assessment

The contribution of geomorphological studies to **Environmental Impact Assessment** must contemplate two main topics: on the one hand, geomorphological research on the *environment* within the area a given project involves: its description, history, diagnosis of present state and trends; on the other hand, the identification of the likely geomorphological *effects of the project* on the environment, either positive or negative; i.e., the expected evolution of the geomorphological systems. There are two geomorphological elements to be considered in the environment: geomorphological resources and geomorphological stability (Panizza, 1988).

Figure 7 represents schematically a methodological approach for the incorporation of geomorphology into these kinds of studies. Figures 8 - 13 show an example of an application of this geomorphological procedure for the Environmental Impact Assessment: it concerns a project of a road in an alpine landscape in Italy.

The *Map of Geomorphological Resourse-Assets* (Fig. 9) is derived directly from a Geomorphological Map (fig. 8), through a selection of its contents. The *Map of Geomorphological Hazards* (Fig. 10) is obtained according to the methodology outlined in paragraph 3.

The comparison of the map of *geomorphological assets* (Fig. 9) with the work foreseen for the construction and the operation of the project will supply the elements required for an evaluation of whether such assets could be lost or altered. The results can be shown on a map of the negative effects on geomorphological assets: a geomorphological **Impact Map** (Fig. 11).

The comparison between the map of *geomorphological hazards* and the *vulnerability* of the area (existing buildings and infrastructures, economic and social activities, and plans for expansion) will indicate the situations of present and potential geomorphological risk. These data will be transferred to a *map of "geomorphological" risk* (Fig. 12). The comparison between the map of geomorphological stability and the consequences expected to accompany the construction work and the operation of the project will lead to the *map of induced "geomorphological" risk* (Fig. 13). From these two maps we have a "geomorphological" **Risk Map**.

The results of these studies should then be compared with the other components of the environment: physical, biological, social, cultural, historical, esthetic, etc. This means evaluating and inserting in each case the most significant and determing geomorphological components within the integral system of **Environmental Impact Assessment** (Leopold et al., 1971). Subsequent steps include weighing the importance of each geomorphological component, evaluating the time spans of their evolution, inserting them into precise and integral hierarchical organization, comparing the various environmental components and trying to establish the interactions among them.

References

BOLT, B.A., HORN, W.L., McDONALD, G.A. & SCOTT, R.F. (1975): Geological Hazards. - 300 p. Springer Verlag, Berlin.

COATES, Donald R. (1976): Geomorphology and Engineering. - 360 p. Allen & Unwin, London.

COOKE, Ronald U. & DOORNKAMP. John C. (1974): Geomorphology in Environmental Management. - 413 p. Clarendon Press, Oxford.

CRAIG, Richard G. & CRAFT, Jesse L. (1982): Applied Geomorphology.- 253 p. Allen & Unwin, London.

DREW, David (1983): Man-Environmental Processes. - 135 p. Allen & Unwin, London.

FINES, K.D. (1968): Landscape Evaluation: a Research Project in East Sussex. - Reg. Studies, 2(1), 41-45.

FLAGEOLETT, Jean C. (1989): Les mouvement de terrain et leur prévention. - 114 p. Masson, Paris.

FOOKES, P.G. & VAUGHAN, P.R. (1986): A handbook of Engineering Geomorphology.- 343 p. Surrey Univ. Press, New York.

HAILS, John R. (1977): Applied Geomorphology. - 418 p. Elsevier, Amsterdam.

LEOPOLD, Luna B. (1969): Landscape Aesthetics. - Nat History, Oct. '69, 36-45.

LEOPOLD, Luna B., CLARKE, Frank E., HANSHAW, Bruce B. & BASLEY, James R. (1971): A procedure for Evaluating Environmental Impact. - U.S. Geol. Surv., circ. 645, 13 p., Washington.

LINTON, David L. (1968): The Assessment of Scenary as a Natural Resource. - Scott. Geogr. Mag., 84, 219-238.

PANIZZA, Mario (1978): Analysis and Mapping of Geomorphological Processes in Environmental Management. - Geoforum, 9(1), 1-15, Pergamon Press, Oxford.

PANIZZA, Mario (1987): Geomorphologocal hazard Assessment and the Analysis of Geomorphological Risk. - Intern. Geomorph., Manchester 1985, Part I, 225-229 J.Wiley & Sons, London.

PANIZZA, Mario (1988): Geomorfologia Applicata.- 342 p. La nuova Italia Scientifica, Roma.

PANIZZA, Mario & PIAZANTE, Sandra (1989): Geomorphological Assets Evaluation. - Proc. Intern. Geomorph., Frankfurt 1989, in press.

PESCI, Marton (1985): Environmental and Dynamic geomorphology.- 220 p. Akad. Kiadò, Budapest.

SCHEIDEGGER, Adrian E. (1975): Physical Aspects of Natural Catastrophes. - 289 p. Elsevier, Amsterdam.

TRICART, Jean (1978): Géomorphologie Applicable. - 204 p. Masson, Paris.

VERSTAPPEN, Hermann Th. (1983): Applied Geomorphology. - 437 p., Elsevier, Amsterdam.

WATHERN, Peter (1988): Environmental Impact Assessment. - 332 p. Unwin Hymand, London.

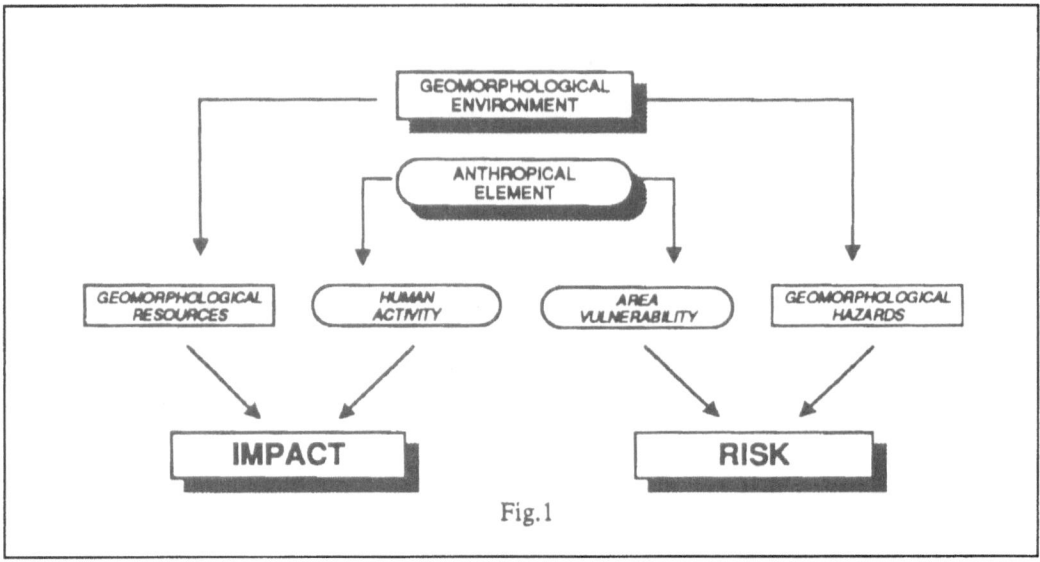

Figure 1:   Impact and Risk from the relationship between Geomorphological Environment and Anthropical Element.

Figure 2:   Geomorphological resource: the Tofana di Rozes, an asset in the Dolomites (Cortina d'Ampezzo, Italy).

Figure 3:   Geomorphological hazard: the Valtellina landslide
that fell in the Adda Valley (Sondrio, Italy) the 28th
of July, 1987.

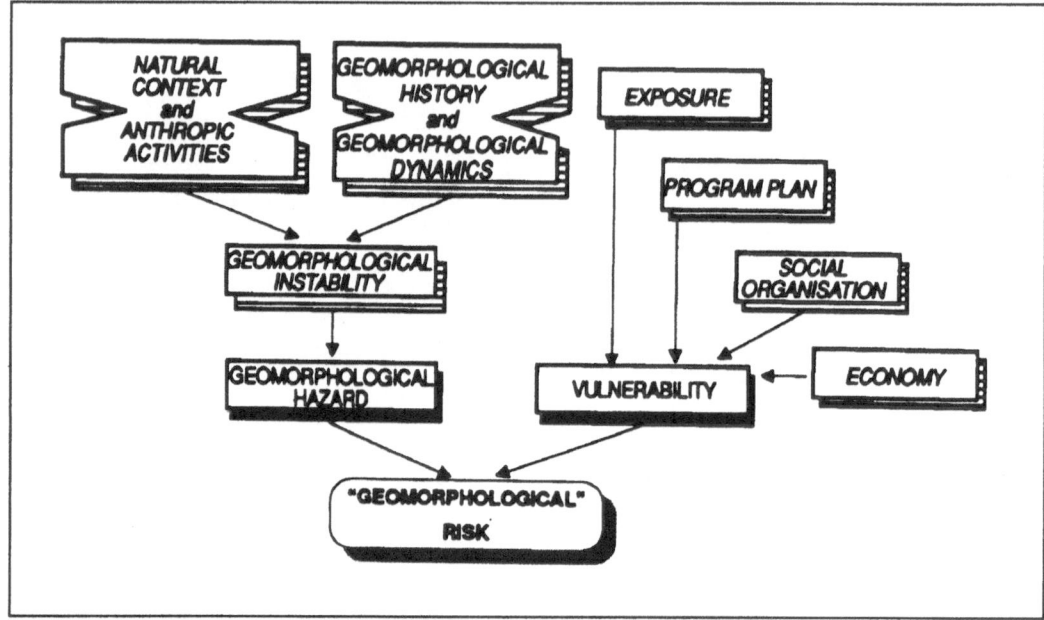

Figure 4:   Logical scheme of the methodology for the determination of Geomorphological Hazard and the identification of "Geomorphological" Risk.

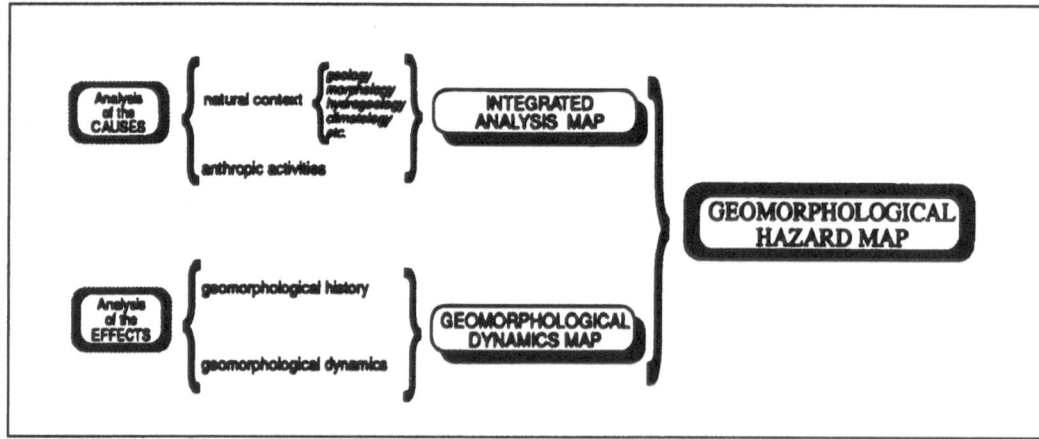

Figure 5:   Conceptual scheme of the methodology for the Geomorphological Hazard
Map.

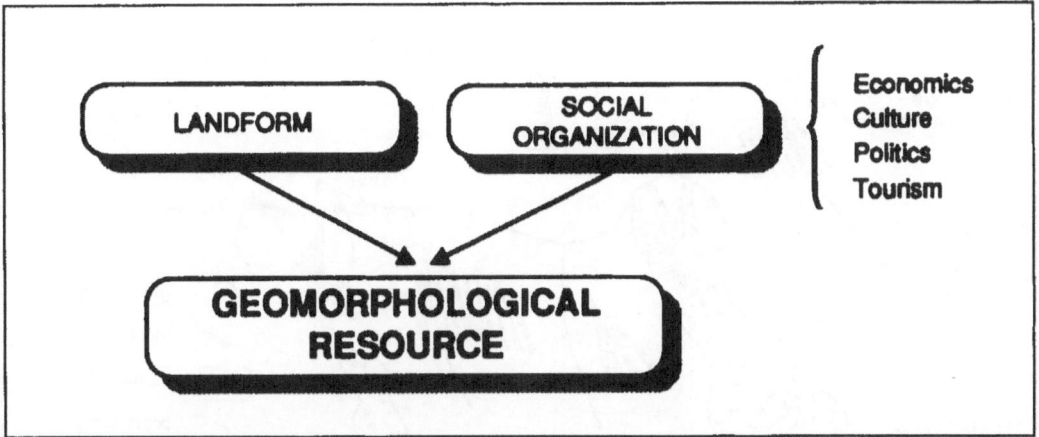

Figure 6: Conceptual relationships between Landforms and Geomorphological Resources/Assets.

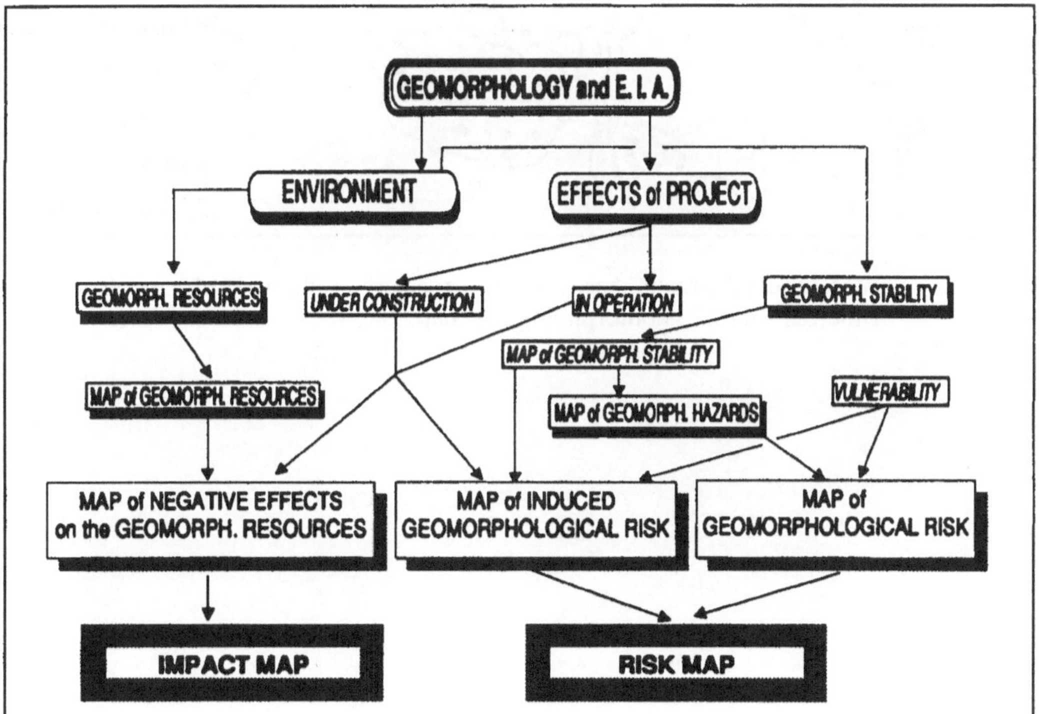

Figure 7: Conceptual and methodological scheme on the geomorphological contributions to Environmental Impact Assessment.

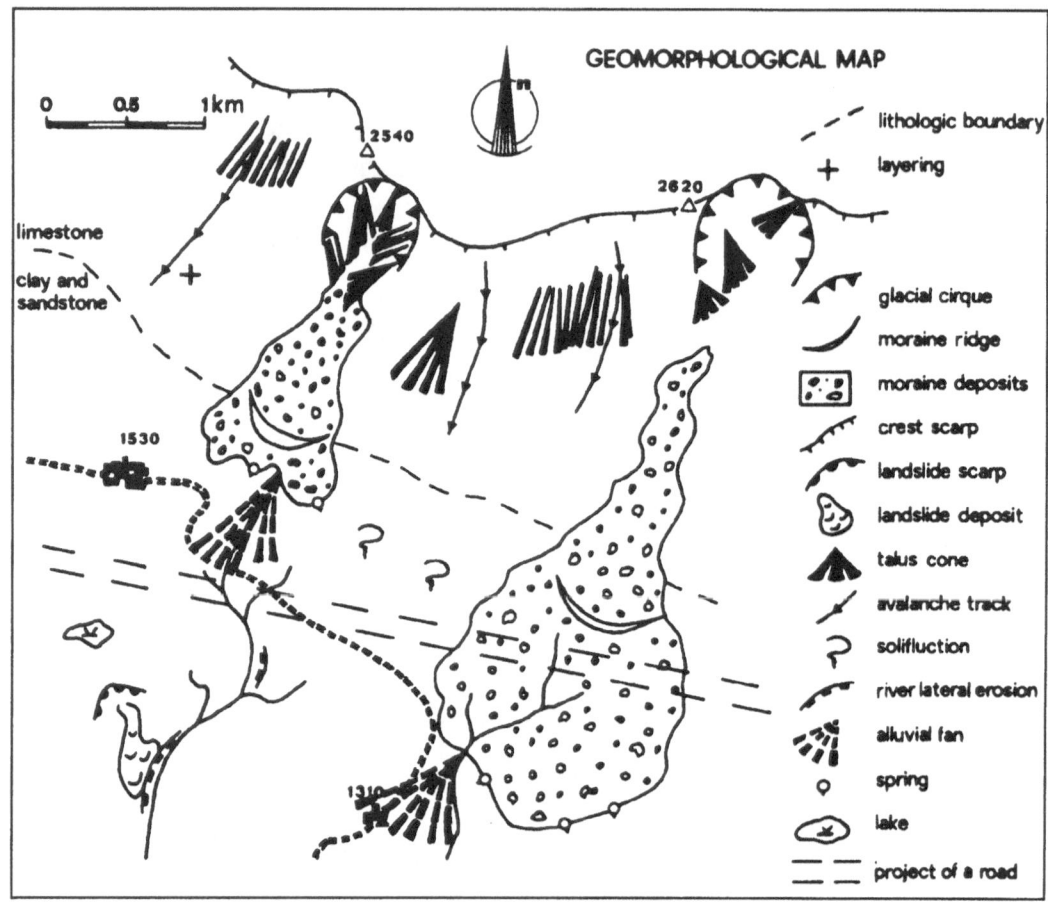

GEOMORPHOLOGICAL MAP

- - - - lithologic boundary

+ layering

glacial cirque

moraine ridge

moraine deposits

crest scarp

landslide scarp

landslide deposit

talus cone

avalanche track

solifluction

river lateral erosion

alluvial fan

o spring

lake

project of a road

limestone

clay and sandstone

2540

2620

1530

1310

Figure 8: Environmental Impact Assessment: a road in an alpine landscape. The basis of the research: a geomorphological map.

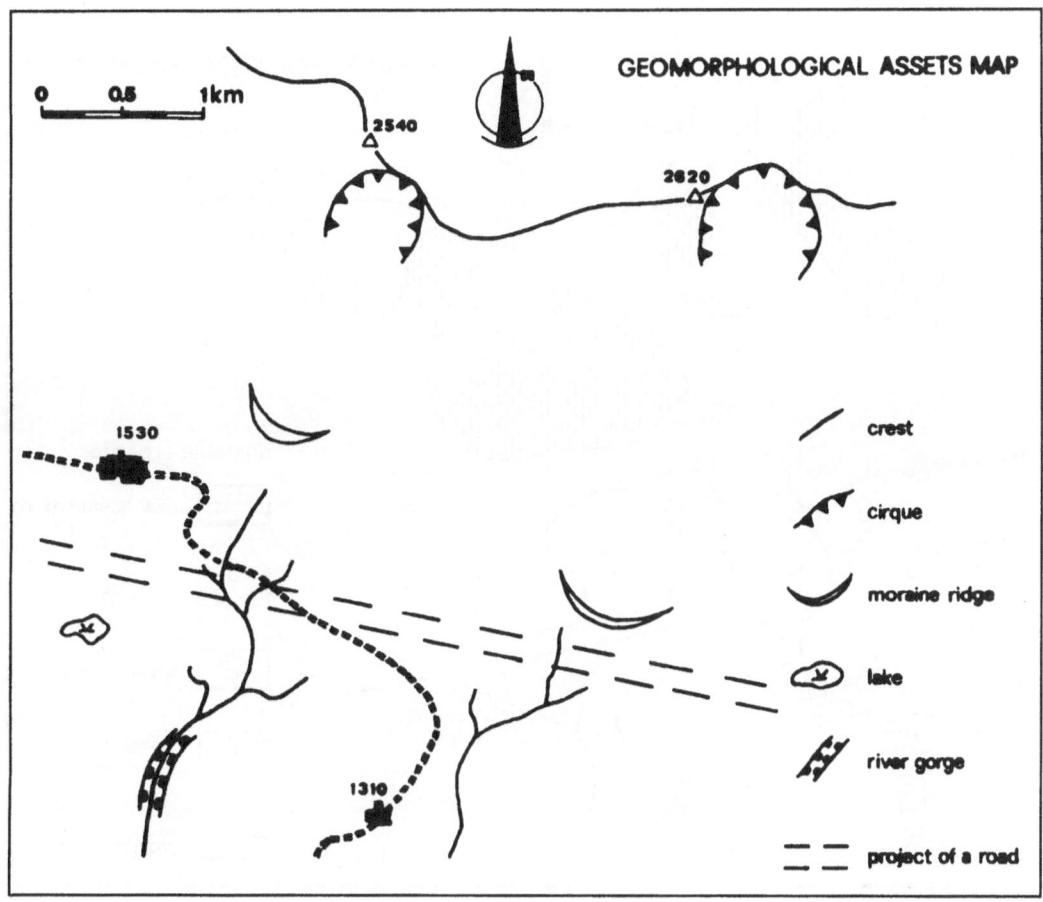

Figure 9:   Environmental Impact Assessment: a road in an alpine landscape. The geomorphological assets map, derived from the geomorphological map of Figure 8.

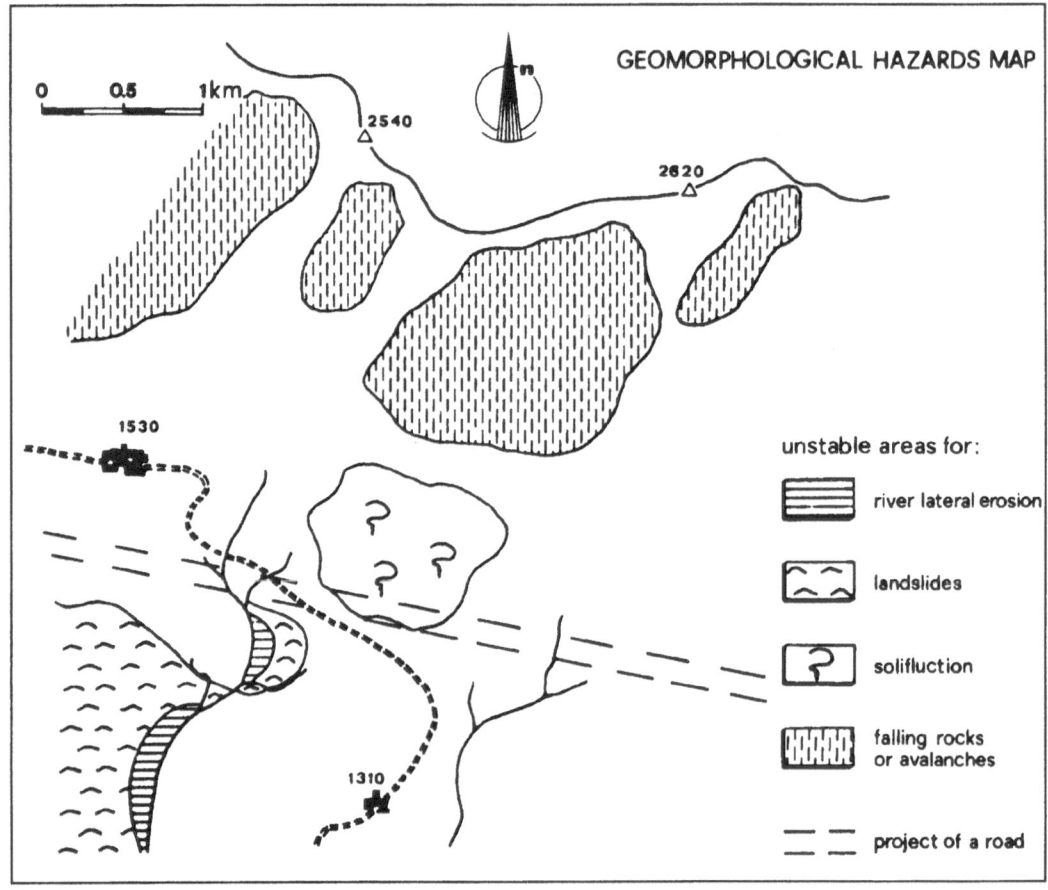

Figure 10: Environmental Impact Assessment: a road in an alpine landscape. The geomorphological hazards map, derived from the geomorphological map of Figure 8.

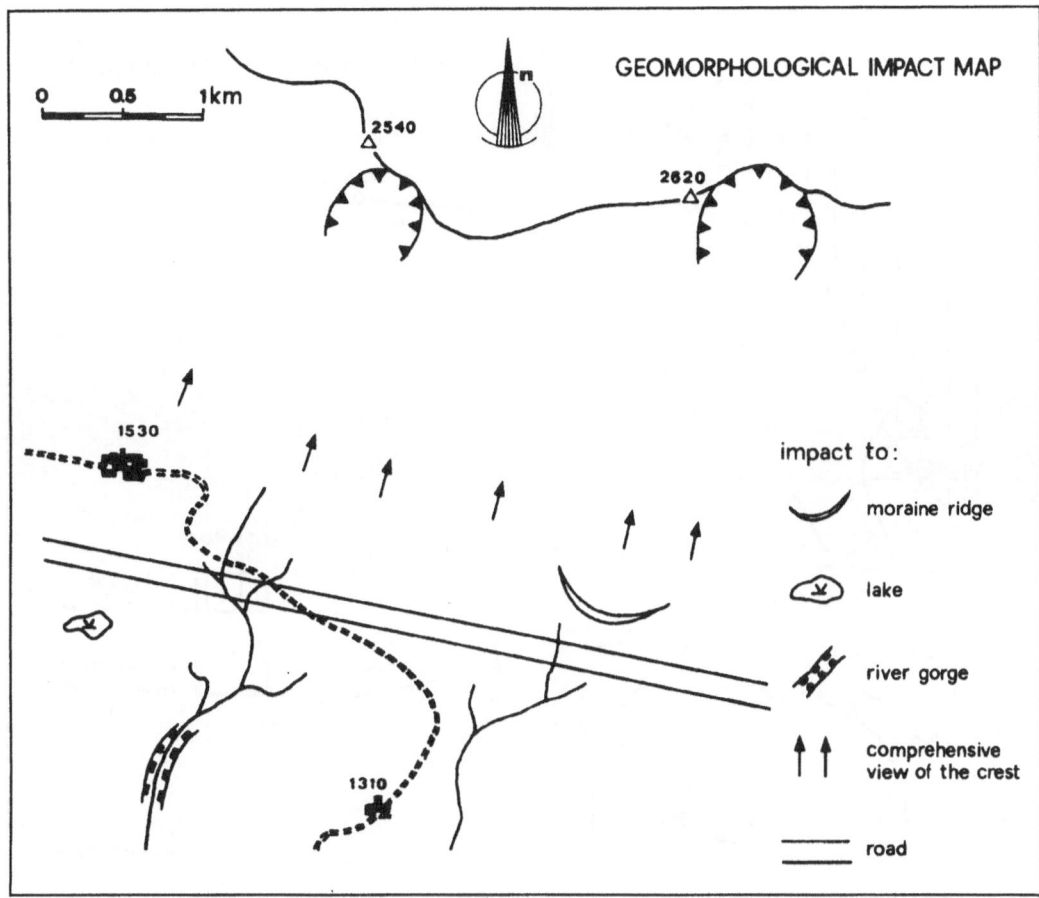

Figure 11: Environmental Impact Assessment: a road in an alpine landscape. The geomorphological impact map, derived from the comparison of the geomorphological assets map (Figure 9) with the negative effects of the road.

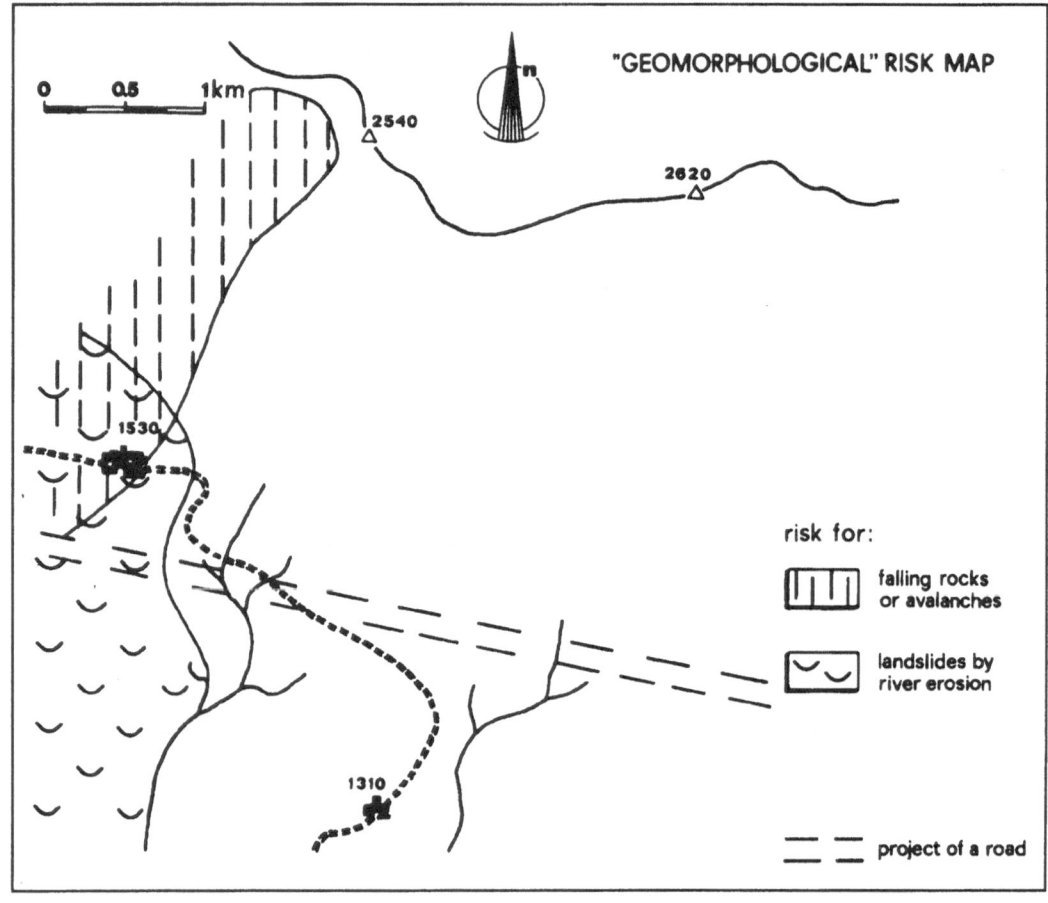

Figure 12: Environmental Impact Assessment: a road in an alpine landscape. The "geomorphological" risk map, derived from the comparison of the geomorphological hazards map (Figure 10) and the vulnerability of the area.

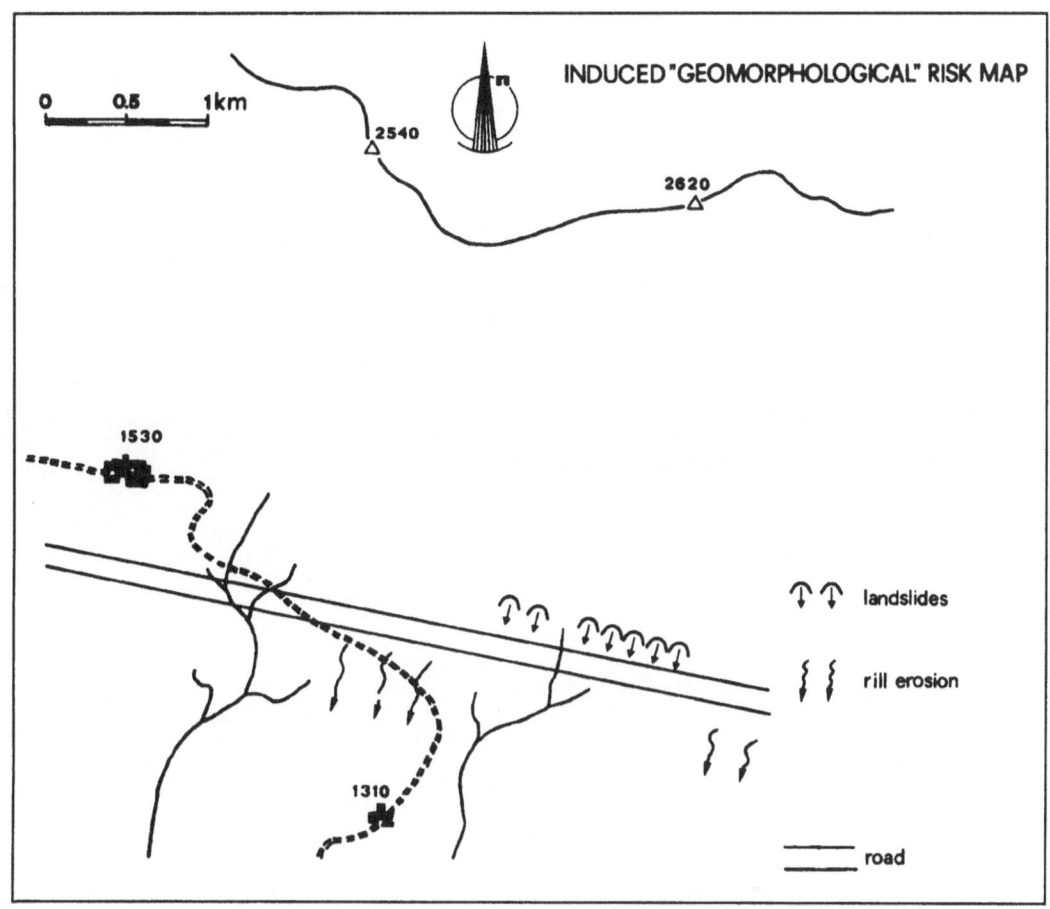

Figure 13: Environmental Impact Assessment: a road in an alpine landscape. The induced "geomorphological" risk map derived from the comparison of the geomorphological stability with the negative effects of the road.

Urban Geology: Present Trends and Problems

ED F.J. DE MULDER [*]

Abstract

Urban geology is a relatively new branch of the earth sciences that came into being in the past fifty years. This paper deals with the availability of data in urban areas, site investigations, and the correlation of engineering geological data in cities. Recent developments in the field of urban geology are: the growing awareness of the discipline among geoscientists, the developments in project acquisition and the overall automation in munincipal administration. Some recommendations are given at the end of the paper.

1. Introduction

Most cities owe their origin and present existence to geological and geographical factors that control the position of the coastline, the river courses, the situation of corridors through mountain ranges, the foundation conditions, the vulnerability to natural hazards, and the distribution of natural resources. In the course of time, man has modified the natural conditions in these places tremendously. The rapid increase in human population

[*] Author's address: Dr. E.F.J. de MULDER, Geological Survey of the Netherlands, P.O.B. 157, 2000 AD Haarlem, The Netherlands.

in the past decades took place primarily in urban centers. This population explosion caused a growing demand for natural resources such as fresh groundwater and construction materials (SCHNEIDER & SPIEKER 1973, LÜTTIG 1987). Furthermore, the urban population became exposed to a number of serious threats such as flooding, slope failure, seismic catastrophes, construction failures because of foundation instability, groundwater pollution and soil pollution. City planners and decision makers now begin to realize that earth sciences play a key role in providing the desired resources and in mitigating natural and man-induced hazards (PRINYA NUTALAYA 1988). Gradually, a new branch of geological science came into existence: Urban Geology.

Urban geology can be considered as the field of Applied Geology that deals with major population centres. This discipline combines those branches of the earth sciences that can assist in urban management and development. Urban geology is one of the most interdisciplinary fields of the earth sciences, it covers parts of engineering geology, environmental geology, and land management. In addition to conventional geological disciplines such as stratigraphy and tectonics, geotechnics (rock-/soil mechanics) and geohydrology are of major importance in urban geology.

The first urban geological activities that can be traced in literature date from the end of the nineteentwenties. In Germany special soil maps were made to support urban planning (HOYNINGEN-HUENE 1931, STREMME 1932). Towards the end of the thirties the Preußische Geologische Landesanstalt in Germany (BRÜNING 1940) combined detailed maps on a scale of 1:10,000 and 1:5,000 with maps indicating the suitability for various kinds of land-use and compiled these in a "Bodenatlas" for urban expansion. As a result of the population explosion and economical revival immediately after World War II the amount of urban geological activities increased dramatically in many countries especially in Europe and in North America. New systematic geological mapping programmes were set up in a.o. Germany, Czechoslovakia, and in The Netherlands. These were intended to support physical planning and pay attention to physical properties and to the succession of strata in the subsurface (HAGEMAN 1963). An excellent example of such urban geological activities was the detailed (scale 1:5,000) mapping of foundation conditions in the city of Prague. The numerous data and maps of this city are constantly updated (LEGGET 1973). For more than a dozen German cities atlasses with a number of thematic land-use maps were prepared (MÜCKENHAUSEN & MÜLLER 1951). Although those maps were obviously intended for urban planning, their readibility was generally poor. To economize on the high (colour) printing costs, too much information was crammed into too few maps. Information about soil properties and suitability for various kinds of land-use was presented on the maps and in the explanatory notes in qualitative terms only. This was undoubtly due to the very limited availability of geotechnical and geohydrological in-situ and laboratory test results.

As a result of the explosive economic growth in the United States of America after World War II and the subsequent urban expansion the number of geologists concerned with urban geology grew tremendously. For example, at the end of the sixties about 150 geologists were employed in this field in the city of Los Angeles alone (MCGILL 1973). In the same period a breakthrough in the use of geological data for urban planning and management was achieved in Canada mainly because of the publications of LEGGET (Refs. 1973, WHITE 1989). In the industrial countries concern about our natural environment grew and the dangers of pollution caused by large-scale waste disposal predominantly in and around urban centres, became apparent in the seventies. Detection, immobilization and restoration of polluted areas, and selection of appropriate sites for waste disposal opened up a new field of interest and created a new challenge for urban geologists. Geochemistry was added to their expertise and Environmental Geology became the subject of a fast growing number of new studies. This resulted in new ways of informing planners about the potentials and the limitations of the soil. "Geopotential maps", in which preferential land-use based on earth-scientific and related disciplines is indicated, were first introduced in the Federal Republic of Germany (LÜTTIG 1978). This mapping system was later applied in several other countries. Many urban geological maps were made and printed by the U.S. Geological Survey during this decade (e.g., MCGILL 1973, BASKERVILLE 1981, MERGUERIAN & BASKERVILLE 1987). In various European countries special studies were carried out in order to investigate the most appropriate way to present (processed) geological data for urbanized areas on maps (e.g., MONROE & HULL 1987, FORSTER et al. 1987, DE MULDER 1986).

In Spain, geotechnical maps for urban planning on a scale of 1:25,000 were made by the Instituto Geologico y Minero de España for the cities of Huelva, Granada, Palma de Mallorca, Almeria, Malaga, Cordoba, Alcoy, Valladolid (INSTITUTO GEOLOGICO Y MINERO DE ESPAÑA 1984, CENDRERO et al. 1987) and for the Madrid region on scales of 1:400,000 to 1:100,000 by the Instituto Technológico GEOMINERO de España (AYALA CARCEDO 1988).

The increased use of geohydrological and geotechnical models at the end of the seventies and at the onset of the next decade made it possible to predict and to quantify the effects of human interference in the geosphere. A striking example was the prediction of the harmful effects of land subsidence resulting from land reclamation in The Netherlands (CLAESSEN et al. 1988). This study showed not only that the construction of a new polder would cause great damage to buildings in the surrounding land area but indicated also the amounts of forthcoming damage claims. The use of electronic data files facilitated the presentation of data on maps substantially. Thematic maps became more accessible to planners, decision makers, and engineers by delecting all kinds of marginal information which would not be understood directly by the target group and by presenting these data in a more quantified manner (GEOLOGICAL SURVEY OF THE NETHER-LANDS et al. 1986, DE MULDER 1986). Recently, in The Netherlands some more fundamental studies were carried out concerning the relationships between the geological, geotechnical and gehydrological conditions of the subsurface and the costs of construction and maintenance of infrastructural work, such as roads and sewerage systems. These studies show that a certain relationship exists between the costs and the vulnerability to settlement of the compressible Holocene beds under the cities in the west of The Netherlands. Apart from the vulnerability to settlement the change in groundwater level necessitated by these construction activities proves to be related to the above-mentioned costs (DE MULDER & HILLEN, in press [b]). Considerable attention is now given to the development of a criterium by which these relationships may be expressed with respect to the costs to the cities' munincipal councils. After development of such a criterium, the subsurface of many Dutch cities will have to be mapped in detail to characterize these in terms of vulnerability to settlement and change in groundwater level.

Outside Europe and North America urban geology began to develop recently in Southeast Asia and the Pacific mainly because of the stimulating activites of the Economic and Social Commission for Asia and the Pacific (ESCAP, United Nations) since the mid eighties. Three volumes of the Atlas of Urban Geology compiled by J. RAU and published by ESCAP in Bangkok comprise the results of special urban geological studies and state-of-the-art reports from the peoples Republic of China, Bangladesh, Fiji, Indonesia, Malaysia, Nepal, Pakistan, the Republic of Korea, Sri Lanka, Thailand and Viet Nam. The Association of Geoscientists for International Development (AGID) has organized several LANDPLAN seminars about the role of geology in the planning and development of urban centres in Southeast Asia (e.g. TAN & RAU 1986). A few decades earlier, urban geological studies for the city of Calcutta (India) were carried out and published (DASTIDAR & GOSH 1967) followed by a number of urban geotechnical studies scattered all over this subcontinent (RAJU 1987). In Africa no major urban geological studies or geological studies for land management apart from Togo and Morocco (ALLAGLO et al. 1987, HAFDI 1987) are known to the author.

It is obvious that urban geological studies bring about very specific problems. These are discussed in section 2, while some of the present trends in this field are presented in section 3. With reference to the specific problems and learning from the present trends, some recommendations for future research and other activities are given in section 4.

2. Problems in Urban Geology

In conducting geological studies in urbanized areas, certain technical legal and political problems are often encountered. Some of these problems are summarised below.

Availability of observation points

The concentration of construction activities in urbanized areas generally requires a greater number of data points concerning subsurface conditions than in rural areas. However, the extent that the results of these site investigation are available to (engineering) geologists depends largely on the way that site investigation, data collection and urban management are organized on that city. In countries like The Netherlands, most data for urban geological studies can be obtained relatively easily form the munincipal departments of Public Works or Housing. Municipal administrations of the larger cities generally have organized, computerized and maintained their data files very well, whereas in some smaller cities the data files consist of a loose brunch of bore-hole descriptions and Cone Penetration Test (CPT) graphs in a drawer. The Geological Survey of The Netherlands also has access to data files of the principal private or semi-private geotechnical organizations and companies.

Although many site-investigation studies have been performed in cities in many countries, it may take a considerable effort to collect these data because the munincipalities may have no responsibility for data collection or certain city departments might even fear adverse technical or political reactions to their willingness to deliver subsurface data to (engineering) geologists.

Without municipal cooperation, the urban geologist has to rely on data from geotechnical advisory companies, drilling companies, insurance companies, tunnel, railway and road construction companies, real estate bureaus, architects, etc. for subsurface data.

The number of actual data for urban geological studies depends upon the scale at which maps should be produced. Because land-use in the urban areas often changes within geographically short distances, maps based upon interpreted subsurface data should, preferably, be made on detailed scales such as 1:10,000 to 1:5,000. The number of observation points necessary in geologically less complicated areas is (on empirical grounds) assumed to be between some 25 - 50 for maps at a scale of 1:5,000 and 10 - 25 for maps at a scale of 1:10,000. Besides these relatively detailed maps it might be wise to produce some small scaled maps in order to draw attention to i.e. certain hazards which may threaten the city.

Site investigation in urban areas

When the number of data points available is not sufficient, additional data should be collected by site investigation. If exposures are lacking, which is normally the case in urbanized areas, site investigation will generally consist of drilling bore holes, or in-situ tests, such as CPTs or SPTs. Most surface geophysical measurement methods cannot be applied in cities because of the presence of electrically conductive materials or contaminations in the topmost metres of the ground or because of the inconvenience to the public, as will often be the case when using seismic methods. The only geophysical methods, which might provide good results under these circumstances, are georadar and geoelectrical measurements with a floating cable in canals or in other fresh-water courses. Older air photographs of the area made prior to urban expansion often provide a very helpful tool in mapping the subsurface conditions here.

For the execution of borings or in-situ tests, permission has to be granted by the munincipal authorities. Furthermore, execution will often be hampered by limitations to the choice of suitable locations. Finally, in urbanized areas the topmost metres of the subsurface often contain fill material, such as rock fills, wood, demolition debris etc., which may hamper penetration.

Engineering geological correlation problems in urbanized areas

Specific correlation and mapping problems occur in urbanized areas. The topmost bed is often man-made and frequently consists of fill material. The distribution of this "bed" may be very irregular both in horizontal and vertical senses. Sometimes, the engineering geologist may rely on data about the bed's distribution and thickness from municipal sources (e.g. the Public Works Department), but often, especially in the older parts of the city, no data are available at all. Geological mapping in urbanized areas might be done in a very indirect way as well. This might be the case when a geological boundary line has to be traced in a built-up area. Then, the presence or absence of construction failures visible in buildings could be a good indicator for the situation of such a boundary line.

Another correlation and mapping problem often occurs in cities because of the variation in surface load exerted on the underlaying beds and because of the results of local groundwater extraction. Both may lead to sudden lateral changes in thickness of compressible beds such as peat and clay, and, consequently, also to dramatic lateral changes to the geotechnical and geohydrological properties of these beds. This creates an extra uncertainty factor to engineering-geological mapping of urban areas.

The reliability of boundary lines on (engineering) geological maps is a matter of great concern for geoscientists, the city municipality and the citizens. From such lines conclusions are drawn with respect to areas of potential risks, future maintenance works or property values which might in turn, cause speculation activites. At present no real satisfactory solutions have been found to avoid justifiable or injustifiable concern of the public and still indicate engineering-geological boundaries on large scaled maps of urbanized areas.

3. Trends in Urban Geology

Growing awareness among geoscientists

During the past decade many geoscientists climbed down from their ivory towers and became involved in geological aspects of environmental development and management. The rapid growth of Engineering Geology as a new discipline in universities in many countries, and the large number of members and publications on this subject in for example the Journals of the IAEG (International Association of Engineering Geologists) and AAEG (American Association of Engineering Geologists), and the recent creation of a Steering Committee on Environmental Geology in IUGS (International Union of Geological Sciences) demonstrate the growing awareness of the present problems in society, especially in densely populated areas.

Development in project acquisition

Initiatives for urban-geological activities may be taken either by the municipal authorities or geoscientists. In some cases, as in The Netherlands, provincial or even national government agencies or ministries might generate such activities. Unless natural disasters or catastrophic shortages of natural resources have occurred, geoscientists generally face problems in convincing the (municipal) authorities to sanction urban-geological activites. Occasionally, municipal authorities may be convinced by pointing out the financial benefits. An attempt to make a cost/benefit analysis of urban geological activities has been made for the city of Amsterdam (DE MULDER 1988b).

The rapid development in urban-geological activities in South-east Asia can be considered as a major trend in project acquisition. In this region, both the Committee for Co-ordination of joint Prospecting for Mineral Resources in Asian Offshore Areas (CCOP) and the Economic and Social Commission for Asia and the Pacific (ESCAP) of the United Nations, and the Association of Geoscientists for International Development (AGID) have organized various workshops and seminars with geoscientists, planners and decision makers on geological boundary conditions for urban development. At present, eleven countries in South-east Asia participate in the ESCAP programme "Geology for Urban Development". In addition to these activities, an explosive growth in geological studies of SE Asian urban centres took place in the eighties. In a number of publications (a.o. ESCAP 1985, ESCAP 1987, ESCAP 1988a, ESCAP 1988b) urban geological information is presented from a.o. Bangkok, Kuala Lumpur, Kuching, Cebu, Manila, Sydney (Australia), Bombay, Hong Kong, Jakarta, Surabaya, Karachi, Singapore, Rabaul (PNG), Katmandu, Shanghai, Tianjin and many other cities in the Peoples Republic of China. In some countries in this area special sections in the Geological Surveys have been installed in order to pay continuous attention to the field of Urban Geology.

Automation in municipal administration

In the past few years, all kinds of data-files from municipal administrations have been stored on automated databases, especially in developed countries. Taking advantage of the lessons learned in developed countries automation has now started in various developing countries. After storage of population data and other geologically irrelevant data, many cities have now reached the stage that hydrological, hydrogeological and geological data have been or are about to be stored in databases. Advantagous effects of this process, and of vital importance to the development of urban geological activities are

1.) the need for the municipality to collect as many data as possible and to expose files that were previously not accessible;

2.) the necessity to store data in a standarized format;

3.) the ability to process such data rapidly and to produce all kinds of single-value maps (for example distribution maps or isopach maps) or to combine groundwater, geotechnical and geological data in thematic and special maps (for example on the vulnerability to subsidence); and

4.) the simplification of new data entry and data correction, so that file maintenance is guaranteed much better.

The rapid availability of various kinds of fancy coloured but not printed maps might facilitate the process of convincing city authorities of the value of this type of information to their physical planning and decision-making. This, in turn, might elicit more political and financial support for an increased effort in the expansion or improvement of the data files. The introduction of Geographical Information Systems (GIS) in Geological Surveys, Provincial and Planning Organizations and even in municipal bodies of some major cities have stimulated and facilitated this process considerably.

Due to the development of geohydrological and geotechnical modelling prediction/-prevision studies on the impact of man's activities in and above the ground can be conducted quite well nowadays. The results of such studies prove to be of major importance for planners and decision makers.

4. Recommendations

Since one of the objectives of the Santander meeting is to anticipate on future developments in the various fields of applied geology and to stimulate planners, decision-makers and our collegue earth scientists in these fields of interest, some recommendations for further activities are given below.

1.) Before starting urban geological mapping of (parts of) cities it is strongly recommended that a board of representatives of the municipality and of various disciplines be established. Such a board will have to control progress of mapping activities, monitor the quality of the products and integrate the results in the city's policy (DE MULDER, in press).

2.) Urban geological studies are interdisciplinary by definition. Therefore, it is strongly advised that these studies be conducted by a team of specialists in geology, geotechnics (soil/rock mechanics), and geohydrology (groundwater quantity, groundwater quality, groundwater modelling). Preferably, knowledge of geochemistry, pedology, and climatology should also be represented.

3.) All geological, geotechnical and groundwater data should preferably be stored in a relational computer data-base. This approach will result in uniform data entry, fast data processing, map production and map up-dating and will substancially facilitate data-base management.

4.) Geological mapping of urban areas should be done in a purely lithostratigraphic sense; in the distinction between various units time boundaries should not dominate lithologic characteristics. The units discerned should be characterized by distinct geohydrological and geotechnical properties that coincide with lithologic differences.

5.) The major portion of the results of urban-geological studies will be produced as maps. Maps are valuable tools in communication with non-geologists. Therefore, information on the maps should be presented as clearly as possible.

6.) Maps printed in the classical way are generally less appropriate for the presentation of results of urban geological studies than maps produced by computer plotters. The main reason for this is that urban geological studies by definition never result in final products since new data become availble daily and maps should be updated periodically. This is done most easily by using a Geographical Information System (GIS) installed as software on a Personal Computer, a Mini Computer or (preferably) on a Mainframe. Nowadays, high-quality graphical products - including multicolor maps - can be obtained from computer plots via GIS. "Real" printed maps are meant only for production and distribution of relatively large quantities. Since the availability of large-scale (= detailed) maps to the public might easily lead to confusion and misinterpretation, the production of "real" printed maps is only recommended for large numbers of small-scale (= less detailed) maps of urban areas.

7.) Since the reliability of map images is of great relevance in urban areas (see paragraph 2) major effort should be expended on developing methods for expressing the reliability of lines on maps.

8.) Since the problems depicted in paragraph 2 and the trends outlined in paragraph 3 are not restricted to specific countries or areas with particular geological conditions and because the number of workers in the field of urban geology, or geology for urban development, should exchange their views on these matters, preferably at an international level. Therefore, it is recommended that an International Working Group on Urban Geology be created.

9.) One of the activities of such a Working Group might be to make an inventory of actual and potential hazards in megalopoli, preferably in developing countries. Based on such an inventory of hazards in combination with an inventory of geoscientific data of the city recommendations for a programme of a hazard mitigation for the respective cities could be formulated. Such an activity might fit well in the programme of the coming Decade on Natural Disaster Reduction.

References

ALLAGLO, L.K. et al. (1987): Togo, its geopotential and attempts for land-use planning. A case study.- In: [ARNDT, P. & LÜTTIG, G.W.]: Mineral resources' extraction, environmental protection and land-use planning in the industrial and developing countries: 243-270, Stuttgart (Schweizerbart).

AYALA CARCEDO, F.J. et al. (1988): Atlas Geocientifico del Medio Natural de la Comunidad de Madrid.- Madrid (Instituto Technológico GEOMINERO de España).

BASKERVILLE, C.A. (1981): The foundation geology of New York City.- Geol. Soc. Amer. Rev. eng. Geol. **5**: 95-117, Chicago.

BRÜNING, K. (1940): Bodenatlas von Niedersachsen.- Göttingen (Wirtschaftswiss. Ges. Stud. Nieders.).

CENDRERO, A. et al. (1987): Detailed geological hazards mapping for urban and rural planning in Viscaya (Northern Spain).- Norg. geol. Unders., spec. Publ. **2**: 25-41, Trondheim.

CLAESSEN, F.A.M. et al. (1987): Secondary effect of the reclamation of the Markerwaard Polder.- Geol. & Mijnb. **67**: 238-291, Dordrecht.

DASTIDAR, A.G. & GOSH, P.K. (1967): Subsoil conditions of Calcutta.- Journ. Inst. Eng. (India) **48**, 3, CI 2: 692, Calcutta.

ESCAP (1985): Geology for Urban Planning; selected papers on the Asian and Pacific Region.- ST/ESCAP/394, 41 pp., Bangkok.

ESCAP (1987): Geology and Urban Development; Atlas of Urban Geology, 1.- Bangkok.

ESCAP (1988a): Urban Geology in Asia and the Pacific; Atlas of Urban Geology, 2.- ST/ESCAP/586, 228 pp., Bangkok.

ESCAP (1988b): Urban Geology of coastal lowlands in China; Atlas of Urban Geology, 3.- ST/ESCAP/624, 168 pp., Bangkok.

FORSTER, A. et al. (1987): Environmental geology maps of Bath and the surrounding area for engineers and planners.- Geol. Soc. eng. Geol. spec. Publ. **4**: 221-235, Chicago.

GEOLOGICAL SURVEY OF THE NETHERLANDS et al. (1986): Subsoil Uncovered.- 36 pp., Haarlem, Delft, Wageningen.

HAFDI, A. (1987): Approach of a methodology for drawing up a habilitability map.- In: [ARNDT, P. & LÜTTIG, G.W.]: Mineral resources' extraction, environmental protection and land-use planning in the industrial and developing countries: 271-278, Stuttgart (Schweizerbart).

HAGEMAN, B.P. (1963): A new method of representation in mapping alluvial areas.- Verh. kon. ned. geol. mijnb. Gen., Geol. Serie 21, 2, Jub. Conv. 2: 211-219, Amsterdam.

HOYNINGEN-HUENE, P.F. VON (1931): Übersichtskartierung im Gebiet der Meßtischblätter Kempen, Krefeld, Viersen, Willich nebst Randgebieten.- "Briefe" Landesplanungsverb. Düsseldorf, 21, Berlin.

INSTITUTO GEOLOGICO Y MINERO DE ESPAÑA (1984): Mapa Geotechnico par ordenacion territorial y urbana de Valladolid.- Madrid.

LEGGET, R.F. (1973): Cities and Geology.- 624 pp., New York (McGraw Hill).

LÜTTIG, G.W. (1972): Naturräumliches Potential I, II und III.- In: Niedersachsen, Industrieland mit Zukunft: 9-10, Hannover (Nds. Min. Wirtsch. öff. Arb.).

LÜTTIG, G.W. (1978): Geoscientific maps of the environment as an essential tool in planning.- Geol. & Mijnb. 57, 4: 527-532, Amsterdam.

LÜTTIG, G.W. (1987): Approach to the problems of mineral resources' extraction.- In: [ARNDT, P. & LÜTTIG, G.W.]: Mineral resources' extraction, environmental protection and land-use planning in the industrial and developing countries: 7-13, Stuttgart (Schweizerbart).

MCGILL, J.T. (1973): Growing Importance of Urban Geology.- In: [TANK, R.W.]: Focus on Environmental Geology: 378-385, New York (Oxford University Press).

MERGUERIAN, C. & BASKERVILLE, C.A. (1987): Geology of Manhattan Island and the Bronx, New York City.- Geol. Soc. Amer. Cent. Field Guide - NE Section, New York.

MONROE, S.K. & HULL, J.H. (1987): Environmental Geology in Great Britain.- Norg. geol. unders. spec. Publ. 2: 111, Trondheim.

MÜCKENHAUSEN, E. & MÜLLER, E.H. (1951): Geologisch-bodenkundliche Kartierung des Stadtkreises Bottrop i.W. für Zwecke der Stadtplanung.- Geol. Jahrb. 66: 179-202, Hannover.

MULDER, E.F.J. DE (1986): Applied and Engineering Geological Mapping in The Netherlands.- Proc. V Int. Congr. I.A.E.G. 6: 1755-1759, Buenos Aires.

MULDER, E.F.J. DE (1988a): Thematic geological maps for urban management and planning.- Proc. int. Symp. urb. Geol.: 46-59, Shanghai + Bangkok (ESCAP).

MULDER, E.F.J. DE (1988b): Engineering geological maps: a cost-benefit analysis.- In: [MARINOS, P.G. & KOUKIS, G.C.]: The Engineering Geology of Ancient Works, Monuments and Historical Sites: preservation and protection: 1347-1357, Rotterdam (Balkema).

MULDER, E.F.J. DE (1989): Thematic Applied Quaternary maps: a profitable Investment of expensive wallpaper? - In: [MULDER, E.F.J. DE & HAGEMAN, B.P.]: Applied Quaternary Research: 105-117, Rotterdam (Balkema).

MULDER, E.F.J. DE (in press): Engineering Geology for Urban Planning.- Proc. Sem. on Geology & Land-use Planning, Kuching + Bangkok (ESCAP).

MULDER, E.F.J. DE & HILLEN, R. (in press [a]): Preparation and Application of Thematic Engineering and Environmental geological Maps in The Netherlands.

MULDER, E.F.J. & HILLEN, R. (in press [b]): Recent developments in Engineering Quaternary Geology in the Geological Survey of The Netherlands.

PRINYA NUTALAYA (1988): Geologic and hydrologic hazards and thier effects on urban development in Southeast Asia.- Proc. int. Symp. urban Geol.: 60-71, Shanghai + Bangkok (ESCAP).

PRINYA NUTALA & RAU J.L. (1981): The sinking Metropolis.- Episodes 4: 3-8, Ottawa.

RAJU, K.C.C. (1987): Geoscientific maps in land-use planning in India.- In: [ARNDT, P. & LÜTTIG, G.W.]: Mineral resources' extraction, environmental protection and land-use planning in the industrial and developing countries: 281-298, Stuttgart (Schweizerbart).

SCHNEIDER, W.J. & SPIEKER, A.M. (1973): Water for the Cities - the Outlook.- In: [TANK, R.W.]: Focus on Environmental Geology: 385-392, New York (Oxford University Press).

STREMME, H. (1932): Die Bodenkartierung als wichtigste Vorarbeit der Generalplanung.- In: [MAUESMANN, A.]: Die Umstände im Siedlungswesen, Berlin.

TAN, B.K. & RAU, J.L. (1986): Landplan II; Role of Geology in Planning and Development of Urban Centres in Southeast Asia.- Rep. Ser. 12: 1-92, Bangkok (AGID).

WHITE, O.L. (1989): Quaternary Geology and urban planning in Canada.- In: [MULDER, E.F.J. DE & HAGEMAN, B.P.]: Applied Quaternary Research: 165-175, Rotterdam (Balkema).

Figure 1: ... too much information was crammed into too few maps.

Figure 2:   The introduction of Geographical Information Systems has facilitated temporary map production for urban planning purposes.

Landslide Risk Assessment and Zoning

JORDI COROMINAS [*]

Abstract

Landslide risk maps are increasingly used as documents. Mapping at large scales requires a good accuracy in determining unstable areas and their extent. A brief review of methods and techniques used in landslide risk assessment is presented.

Identification of unstable areas is previous to any risk assessment because landslides mainly occur in areas showing previous instability. The analysis of the landscape, landslide deposits and activity indicators by means of remote sensing, sedimentological techniques and monitoring, have revealed to be very useful. The hazard assessment has to consider the landslide source and the expected runout. In a simple approach, threshold slopes and morphological indexes are used in landslide source localization although multidata treatment is a more accurate empirical technique. Some plots relative to the travelled distances covered by landslides, show a clear relationship between the volume of both large and small landslides and their runout distances. 2-D and 3-D models have been recently developed for predicting landslide behaviour. The potentially of the landslide occurence may be analyzed by means of statistical and indirect empirical methods (most of them through relationships with rainfall intensity-duration records). Monitoring techniques provide specific data to undertake short term failure prediction.

[*] Author's address: Prof. J. COROMINAS, Geotechnical Engineering Department, Technical University of Catalunya, Jorge Girona Salgedo, 31, E - 08034 Barcelona, Spain.

Zoning methods are considered through the basis of simple landform interpretation, combining simple landslide related parameters, data treatment, modelling unstable areas and monitoring.

Introduction

Landslide risk mapping has experienced a great advance over the last decades. These maps divide the country into areas or dominas where damage caused by hazardous mass movements is expected to be of the same order of magnitude and based on slope stability assessment and subsequent zoning. Hazard assessment tries to estimate the localization, type, magnitude and occurence of landslides. A substancial progress has been made in the development of stability analysis methods of a definitive slope as can be seen in the field of three-dimensional analysis (GENS et al. 1988), of the behaviour of unsaturated soils and of the effect of soil suction in increasing slope stability (FREDLUND 1987), among others. Nevertheless, the assessment of large areas using slope stability analysis methods is not possible at present because of the huge costs that acquiring data for such analysis would represent. Thus, specific methods for slope reconnaisance of large areas have recently been developed. The reader interested in methods of stability analysis is referred to some excellent reports as those from LLORET et al. (1984) or MOSTYN & SMALL (1987).

The aim of this paper is to present some of the relevant aspects of landslide hazard assessment and mapping developed recently from both geomorphological and the engineering geological points of view. The terms used in the text for landslide description are based on VARNES' classification (1978). Although snow avalanches may imply a certain quantity of debris mobilization from the slope, they have not been considered here.

Identification of Slope Instability

A basic aspect previous to landslide hazard assessment is the identification of unstable areas. This aspect is often underestimated although it may cause serious problems. Many roads and tunnels in mountainous regions cross undetected old landslides which may be reactivated by the works carried out at their foot (GONZALEZ DE VALLEJO et al. 1988). Landslide misinterpretation has also been observed in geomorphological maps and even in hazard maps (ANTOINE 1977), This misunderstandings are sometimes due to the scale of some large movements, the magnitude of which is undetected by local recognition.

These failures in identifying slope instabilities cause concern because of the fundamental principle that characterizes most hazard assessment studies, which is the principle of actualism (VARNES 1984). This means that natural slope failures in the future will most likely be in geological, geomorphological and hydrological situations that have led to past and present failures. Several research works confirm this statement, i.e. KOJAN et al. (1972), who found that 81 % debris flows triggered by the 1969 rains in Sta Inez - San Rafael mountains, California, occurred in locations with previous failures, as identified from air photographs. This does not mean that failures will not occur in unaffected slopes as in the case of bank undercutting, weathering of slope material or mountain uplifting. Slope instability can be detected by study of landform analysis, landslide deposits and activity indicators (COROMINAS 1986).

Landform analysis

RIB & LIANG (1978) and CROZIER (1984) have provided excellent synthesis on criteria used in reconnaisance of morphology related to landslide activity and vulnerable locations. The use pf panchromatic black-and-white and infrared aerial photographs for identifying arquate head and lateral scarps, hollows and related water impoundments, seepage areas and gullying, has proved as a very useful technique. Satellite imaginery has furnished a general sight of the earth's surface although at a scale too small for intensive practice in landslide assessment. Nevertheless, recognition of large movements, identification of main scarp at crown edge and hummocky ground associated to slide deposits may be possible with the Landsat MSS resolution (SOUTHWORTH & SHULTZ 1986). At present, difficulties in identifying the spectral characteristics specific to the unstable areas remain because of the masking effect of the great variations in topographic, lithologic, structural and vegetational conditions (WEBER 1984). SPOT satellite images will provide an improvement of the observation scale, but a lesser number of bands hinders quick progress in spectral characterization of landslides.

Morphometric analysis of topographic maps and surveying at an adequate scale in relation to the mass movements, allows us to obtain characteristic profiles of landslides (COROMINAS 1986). It is possible to derive, either by hand or automatically, slope maps on which arquate forms and reverse slopes indicating tilting in rotational slides can be easily identified (PITTS 1979).

Sedimentological features

Spatial disposition, textural characteristics, colour and tone features are the main factors in identifying landslide deposits. Important differences exist among these deposits depending on the type of movement they are generated from. Continuous rockfall in periglacial environments originates coarse grained clast supported scree deposits, whereas rock avalanches, debris and mudflows produce poorly sorted matrix supported, elongate deposits. In spite of such differences, slide deposits may be recognized from other colluvial, glacial or related deposits. COSTA & JARRET (1981) proposed several criteria for differentiating water floods and debris flows based on sedimentological and morphological features of deposited material like, the sorting coefficient which was found to be comprised from 3.9 to 11.5 in debris and mudflows and from 1.8 to 2.7 in water floods. Additional characteristics of debris flows are missing or inverse grading, lateral levees and terminal lobes. Preferred clast orientation has sometimes been found in solifluction lobes and debris flows (MILLS 1984, GATES 1987). Using statistical analysis it is possible to generate a rough discrimination of different sediments (see figure 1, NELSON 1985).

Activity indicators

Field reconnaisance yields a great quantity of information on mass instability. Signs of recent movements may be identified on an indirect way by observing structural damage in walls, buildings, channels, roadways or tunnels. The rigidity of the structure on front of ground displacement may lead to openings of cracks, tilting, settlements or to other observable structural deformations.

Vegetation may also be an useful indicator. At permanent water ponds caused by rotational movements, junks and black poplars (*Populus nigra*) grow easily. In scree slopes, vegetation avoids the activity areas and main tracks for rockfalls. In periglacial environment alpine grassland (*Festuca skia, Festuca supina*) are adapted to very shallow clast displacement triggered by frost activity and solifluction processes in the soil. Long garlands are developed when clasts override the grass and forces plant to grow downslope (SOUTADÉ 1980). When a soil layer is moving fast, scars may develop in the grassland. In forested slopes, tilted trees are characteristic of creep processes.

Monitoring of a potentially unstable slope is both a precise and expensive way for present activity detection. Because of its costs, monitoring has strong constraints, and adequate selection of the site to be surveyed is needed. This means that not only the selected points to be monitored have to be representative of the whole slope movement but instrumentation must be adequate to the foreseen movement type and rate of displacement. Conventional surveying methods are based on displacement measurements at both surface and subsurface of slope. The wide range methods available are described in WILSON & MIKKELSEN (1978), GILI (1989) and KOVÁRI (1990).

Hazard Assessment

Hazard estimation is the main aspect in risk mapping, sometimes not well solved. ANTOINE (1977) suggests for the use of an "ignorance coefficient" in order to reduce the optimistic conclusions even if they are well founded. Indeed it is doubtful that catastrophic events such as the Mont Granier rock avalance that took place in 1248 in the French Alps, where more than 200 million cubic meters overrode a surface of 15-20 km could have been foreseen and correctly mapped. Anyhow, this possible ignorance only may be counteracted by providing more detailed analysis and implementation of monitoring and control devices. Hazard assessment have to give answer to some basic matters as type of mass movement, location, runout distance, velocity and time of failure (HARTLÉN & VIBERG 1988). We will analyse some of these aspects.

Localization of potentially unstable areas

Morphological characteristics of the slope: The estimation of slope stability by analogy with other slopes of the same lithology and geographic region, has been used in the past. RENEAU & DIETRICH (1987) show that concave topography forces colluvial debris to accumulate in hollows which concentrate subsurface runoff favouring debris flow generation. SKEMPTON (1952) analyzing London Clay natural slopes found a critical slope angle. Where ground water level reached the surface in winter, slips can develop in 10 degree slopes, whereas flatter slopes were stable although subject to creep. This critical angle was in agreement with stability analysis based upon laboratory analysis strength of soil samples. COLIN-ROUSE & FARHAN (1976) also found limiting angles for natural slopes in South Wales. CROZIER (1984) showed an envelope enclosing stable and unstable angles for a particular slope heights, indicating the decrease of stability with increase in height. Nevertheless generalization based on this specific relationship is difficult due to the changing conditions on the slope. Deforestation by timber harvesting or burning or slope undercutting by present fluviatile erosion, may cause a latter instability. SIDLE et al. (1985) claimed that a lower limit of slope gradient for various small movements is difficult to specify because slope steepness perhaps underestimates the importance of soil thickness which may also be considered as a threshold.

From an accurate analysis of 177 stable and failed slopes in Hong Kong, BRAND & HUDSON (1982) considered a great variety of factors such as geometry, lithology, laboratory tests, vegetation and preventive works and concluded that stability of slopes cannot be predicted on the basis of a few easily measurable parameters. Plotting height versus slope angle (figure 2), previous rules established for differenciating stable and unstable cutslopes were revealed inadequate. Dotted line corresponding to 1982 is the locus of slope geometries for which safety factor is 1.4 as calculated by conventional equilibrium analysis. This dotted line has a good acceptance as a lower bound envelope of the final geometry of failed slopes, although some slopes that have stood stable for more than 20 years fall above the line. This line then, should be considered as a long term stability slope behaviour.

Several attempts have been made in order to relate catchment area with failure susceptibility. Topographical characteristics provide some critical index which allow to separate failed and unfailed slopes. Indexes proposed by HATANO & OKIMURA have been cited by OYAGI (1984) and have the following form:

HATANO'S index $\qquad$ $F = a^{0.33} \tan \Theta,$

OKIMURA'S index $\qquad$ $F = A^{0.22} \tan \Theta,$

where A is the catchment area for a selected point, a is the average length of the catchment area (it is obtained dividing the area by the width of the foot line or the toe of a slide failure surface) and Θ, the slope angle. In the Takedaira area of the Mizunami region, HATANO'S critical value index for a slide zone generally exceeds 3.4, and OKIMURA'S index at Tennodani in Rokko Mountain was found to be 2.96. More recently, OGAWA & KAWAMURA (1989) have proposed a landslide prediction method in which a global topographical index is introduced to quantitatively evaluate groundwater runoff influence.

JACKSON et al. (1987) using morphometric data relative to fan areas, fan slope, drainage basin area and drainage basin height managed to discriminate debris flow fans from alluvial fans (Figure 3). The former have a small basin (first or second-order streams) and steep fans. Fluvial fans have large basins (third-order or greater streams) and less steep fans. Basin area (A_b) and basin height (H_b) were combined as a measure of basin ruggedness in the following manner:

$$R = H_b A_b^{-0.5}$$

Most of the fluviatile fans have $R < 0.3$ and slopes < 2.5 , and debris flow fans have $R > 0.25\text{-}0.3$ and slopes > 4 .

Multidata techniques: Because of slope stability depends on several factors acting at the same time, some efforts have been directed towards the acquisition of simply and quickly determined parameters. STEVENSON (1977) using scored factors proposed a method for evaluate relative landslide risk in clay slopes. The final risk is obtained as a product of three terms:

$$R = (P + 2w)(S + 2C)\ U,$$

where P is a clay factor obtained from the range of available values of plasticity index; W is a water factor, considered as a maximum water table relative to typical failure plane; S is slope angle; C is slope complexity (if it has been previously failed) and U, land use. The range of values of each factor is divided in three parts which vary from 1 to 3, except land use that is from 1 to 1.5. Risk values higher than 60 suggest failure, and values between 50 and 60 mean possible instability.

Discriminant analysis provide a more accurate stability assessment. A classical work using statistical techniques is that from JONES et al. (1961) on landslides in Pleistocene terrace deposits of Columbia River. 160 slump-earthflow movement and additional 160 stable slopes were considered. Qualitative and quantitative factors influencing sliding were searched. A final analysis using the discriminant-function method was performed considering as influencing factors: original slope (x_1), submergence percentage (x_2), terrace height (x_3) and groundwater (x_4). The reduced discriminant function was given as follows:

$$y = 0.00216247 \log x_1 + 0.00334811 \log x_2 + 0.00944030 \log x_3 + 0.00673126\ x_4.$$

Calculations of the discriminant functions for the 320 landslide and stable slopes gave values ranging from -0.0019 to +0.0404. Lower values represented stable slopes, and higher values represented landslides. Less than one percent of the landslides had a discriminant value below 0.0106 and 0.142 were considered as a relatively stable because of less than 5 % of landslide fell between these values. A similar approach was used by NEULAND (1976) who used the principle component - analysis to select the influencing variables from an initial data set of 31 parameters. Final discriminant function was obtained after rejecting some variables in order to avoid complexity. The discriminant function was:

$$y = 0.9222 * 10^{-5}\ S^2 + 0.7926 \log (A + 10) - 0.6098 \log (d + 10),$$

where S is the slope gradient; A is the catchment area and d, a soil density related parameter. Values of 0.1934 mean unstable slopes and values of 0.1781 stable slopes. The model predicted correctly 94 % of analyzed slopes.

Expected Volume and Runout Distance

Destructive effects of mass movements depend on their magnitude and mobility. There exist failure mechanisms that must not be interpreted as catastrophic phenomena. Superficial, predominantly seasonal soil creep (see HUTCHINSON 1988) for example, has a very low velocity of mm or cm per year that reduce damage potentiallity even though it can concern the whole slope. Conversely, soil and rock masses moving at great speed may reach sites far from the landslide source causing important morphological changes and injuries. Rock avalance from Nevado Huascaran in Peruvian Andes, which demolished the villages of Yungai and Rancahirca in 1970, travelled 16 km at a mean speed of 280 km/h (PLAFKER & ERIKSEN 1979).

Determination of the mobility of landslides is difficult. Nevertheless, a relationship has been found between volume of fallen mass and reach. HSÜ (1975) observed a decrease of the equivalent coefficient of friction (maximum drop height divided by the maximum horizontal distance travelled) with the rockfall volume (Figure 4a). This friction decrease is more evident in biggest rockfalls. HUTCHINSON (1988) found a new envelope for chalk failures and slow slides of coal mine waste different from those proposed by HSÜ. He observed a different reach for talus formation derived from chalk rockfall from that attained by chalk flows (Figure 4b). Specific lithology and type of movement seems to have more important influence in volume reduction. COROMINAS et al. (1988) observed a different behaviour depending on the type of failure mechanism for less voluminous movements (Figure 4c). This fact may be explained by considering that rock falls disipate a lot of energy by hitting the substratum and breaking into small rock fragments, whereas shallow slab slides should only dissipate the energy as friction in the shear surface. Debris flow presenting a thick basal shear zone or dispersivity of their clastic components (grain flow), will occupy an intermediate place. The use of this empirical relationship is still limited due to the lack of more representative data.

Computer modelling of the path of fallen mass have been developed recently. LOPEZ (1982) proposed a bidimensional model for rockfall which by analyzing the path followed by the blocks (free fall, bouncing, rolling), calculated the acquired speed and both path and reach. Three-dimensional models allow path analysis when it is subjected to remarkable lateral diversion: thalweg mouth, defence works, hollows (ROCHET 1987).

Most of proposed methods and models about landslide reach consider an initial volume of moved mass. The predictability of the volume to be mobilized is very difficult unless the failure has started and visible scars bound the landslide. ZHANG et al. (1985) have established a relationship between the maximum peak discharge of debris flow and rainfall intensity in a debris flow prone area in Yuannan Province, China. In this region debris flows occur when rainfalls reaches a threshold intensity of 1.8 mm in 10 minutes, but larger debris flows (more than 100 m^3/s) usually occur when rainfall intensity exceeds 5 mm in 10 minutes. However antecedent wetness and debris flow material stored in the channel have also significant affect. In the Pyrenees, COROMINAS et al. (1988) found some relationship between catchment area and volume of shallow movements (Figure 5). In small catchment areas movements also tend to be of small dimensions. Conversely larger basins yield bigger movements which, considering the volume-runout distance relationship, tend to go farther.

Failure prediction

Forecasting the time of occurence of a slope failure is more difficult than assessing the degree of stability. Uncertainity of failure forecasting reduces the accuracy of hazard assessment because this parameter is directly considered in hazard rating. Several approaches have been postulated in last few years (see ALONSO 1987) even though only those based on movement monitoring provide an adequate time-forecasting.

Probabilistic methods: Series of historical landslides are scarce because these movements usually take place in inhabited mountainous areas. In Kärkevagge area, Sweden, rock falls were found to be concentrated during thawing periods (RAPP 1960). A more detailed frequency-analysis of rock fall was performed in Canadian Rocky Mountains covering a seven-year period (GARDNER 1983) where a grid of 500 m cells was overlapped on a 1:50,000 scale topographic map. If a rockfall was observed, its cell location, time of occurrence and size were recorded. A total of 1076 rockfalls were inventored with an average frequency of 0.49 rockfalls/hr.

Longterm data series may be completed using dating techniques as radiocarbon, lichenometry or dendrochronology. By dating debris flow activity in the last 500 years it has been possible by means of determining growth rates of *Rhizocarpon* spices in Scotland (INNES 1983). Damage on trees caused by rockfall or debris flow may be also used (HUPP 1983, BRAAM et al. 1987). Evidence of movement is obtained from study of annual growth ring of trees. Corrosion scars in trunks due to impact of blocks or anomalies in ring width due to tilting trees, are easily identified (see Figure 6). In estimating return periods of movements much care has to be taken in using these geo-botanical dating techniques. Not all the events have to leave signs or marks in the trees, nor do debris flow processes cover the whole extension of a debris fan area. These facts may lead to an underestimating of the activity, and more additional research is needed when defining a regional debris flow frequency.

Indirect empirical methods: In areas affected by generalized shallow movements like debris flows, debris slides, mudflows or rockfalls, correlations with landslide triggering factors may be useful. In this case the exactly location is unknown. In several countries like Japan or Hong Kong, large data series of mass movements are available, and correlations with rainfall have been made (ONODERA et al. 1974, LUMB 1975). Landslide density and severity depend on both intensity and antecedent rainfall. The main factor for landslide occurence is maximum rainfall intensity though the effect of antecedent rain is to reduce the threshold intensity for triggering the movements.

The seasonal antecedent rainfall has a great influence in failure forecasting. Based on empirical observations of debris slides and debris avalanches, CAINE (1980) provides a basis for predicting shallow failures using the critical rainfall intensity which has the following form:

$$I = 14.82 \, D^{-0.39},$$

where   I = rainfall intensity (mm/h) and
         D = duration (h).

SIDLE et al. (1985) used CAINE'S threshold relation in order to calculate the recurrence period of landslides. Because of the data of rainfall intensity - duration - frequency relationship were well known at Reefton, New Zealand, the authors overlaid these graphs on CAINE'S threshold relation. They would identify return periods for several combinations of intensity-duration of precipitation that would probably induce debris slides. The fit is not perfect because the effect of seasonal antecedent rain is not considered on CAINE'S relation. Thus, as shown in Figure 7, rainfall intensity-duration combination of frequent occurrence can easily exceed the critical line in spite of its low intensity (i.e. with 1 hour storm of 14.8 mm rainfall, slopes ought to fail), and this does not agree with the observed behaviour. Adding another threshold relationship like whether soil saturation could physically occur for different intensity-duration combinations, new critical lines are obtained. The analysis of the graphs of moisture needed for soil saturation (see Figure 7) shows that under relatively dry antecedent conditions like estimated 125 mm to achieve soil saturation, CAINE'S relationship should be compared only with observed rainfall intensity-duration frequency graphs for durations exceeding 30 hrs (SIDLE et al. 1985). These kind of relationships may be used as a criteria for establishing warning systems (YANO & SENOO 1985).

Long term predictions using series of climatic data variability related to solar cycles has been established in the Black Sea Coast (SHEKO 1977). The graphs of the mean annual landslide activity by parts of the cycles are drawn and the probability of the movements related to a definite cycle part is calculated. Positive anomalies of rainfall were predicted for 1986-87 and 1997-2000 periods.

Slope displacement analysis: Failure of soils and rocks is often preceeded by some quantity of strain. Systematic monitoring of displacements in a wall at Kensal Green, showed that movements tended to accelerate when they were close to the failure (SKEMPTON 1964). Likewise the bigger the unstable mass is, the greater the preceeding movements are. The analysis of such deformations allowed to develop several failure models which yield a time-forecasting of Chuquicamata mine failure in Chile. Among the variety of tested functions, exponentials seem to give a better fit. Failure took place in the earliest predicted data (VOIGHT & KENNEDY 1979).

Extrapolations of such curves offer some difficulties. Heavy rain or earthquake shaking may accelerate movements, and simple forecasting methods may no longer be applicable. At the failure in progress at La Clapière, the French Alps, a relationship between the rate of deformation and time has been established (RAT 1988). The rate of deformation was found to change with seasonal periods of snow thawing, reflected by the increase of the Tinée river discharge (Figure 8a). Nevertheless, a residual trend towards acceleration is showed when considering a period of several years (Figure 8b).

Creep curves have been analyzed to determine the mathematical form that they adopt, and particularly when accelerating rate of deformation leads to failure. From laboratory experimental data, SAITO (1965, 1969) proposed empirical relationships between the rate of secondary and tertiary creep and time of failure. Since this early work, several expressions have been proposed. FUKUZONO & TERASHIMA (1985) analyzing an experimental slope failed by means of artificial rain found during the late deformation stage before failure that the logarithm of acceleration of surface displacements is proportional to the logarithm of velocity. They proposed the following expression:

$$\frac{d^2x}{dt^2} = a \left(\frac{dx}{dt} \right)^\alpha$$

where x = surface displacement,

          t = time,

          a and α = constants.

VOIGHT (1989) proposes a general law as follows:

$$\Omega_1^{-\alpha} \, \Omega_2 - A = 0,$$

where Ω is an approximate measurable quantity such as a longitudinal or shear strain, geodetic length or angle change. A and α are constants. He proposes a basic equation for landslide time prediction, for α not 1, such as:

$$t_f - t = (\Omega^{1-\alpha} - \Omega_f^{1-\alpha}) / A \, (\alpha - 1),$$

where t_f = time of failure,

          t = observed time,

          Ω, Ω_f = observed and failure rates of deformation,

          A, α = constants.

A and α are constants deduced from graphs like those of Figure 9 concerning Mount St. Helens movements. Data from EDM measurements and electronic tilt, give independent α values and inverse rate-time curves, but these curves converge towards a common event occurence time. The steady decrease of inverse rate allows the extrapolation of event time to be acceptable.

Mass Movement Zoning

The accuracy of zoning depends on the adequacy of map scale with respect to considered phenomena, means availability and time requirements. Location of unstable sites, spatial distribution and relative hazard evaluation are the basic parameters for zoning, even though the main difficulty is to find objective criteria for extrapolating. BRABB (1984), CARRARA (1984), HANSEN (1984) and VARNES (1984) provide a very comprehensive review of methodologies for hazard zoning and mapping. We are going to briefly indicate different zoning methodologies.

Zoning based on landform interpretation: Simple landslide inventories are included in this group. Zoning may be performed taking into account the presence or absence of landslide phenomena. The main advantage is that this type of zoning may be carried out by interpreting aerial photographs with a minimum of field checking. Therefore, they are qick and inexpensive. Geomorphological maps correspond to more elaborate documents in which zoning may be inferred from several homogenous landforms and associate processes. Additional work has to be made in transforming geomorphological maps into hazard maps.

Zoning based on landslide susceptibility: Relative stability is established combining several parameters present in a considered area, which are directly or indirectly related to landslide occurrence (lithology, slope angle, previous landslides, groundwater table). For example, in West Virginia, 2416 landslides were studied (LESSING et al. 1983). In each landslide 12 geological factors were tabulated and rated by dividing each factor frequency by its areal extent. Thus, some specific lithologies or slope gradients could be classified considering their susceptibility to failure.

A classical work on landslide susceptibility was performed at San Francisco Bay region (NILSEN et al. 1979). They select different kinds of information which were successively combined. The first step was to overlap the slope gradient map with the landslide inventory map. Analyzing the document obtained, a slope hazardness ranking was established from the density of movements. An additional overlap with the lithological map, produced a new document where lithology and slopes associated to landslide occurrence were easily identified.

Data treatment applied to zoning: Factorial and discriminant analyses have been discussed as powerful hazard assessing techniques. When large amount of landslides are present in a region, parameters relative to landslide geometry, geology, land use, etc. may be processed with the aid of a computer. Multivariate methods have been tested in Calabria, Southern Italy (CARRARA 1983 a, b). Data relative to rock type, vegetation, slope gradient, surface roughness, stream frequency, presence of landslides among others had been stored onto a computer-based data bank. The area was divided into cells of a 200 m grid, and a set of 14 geological and geomorphological variables was selected. After both discriminant and multiple regression analyses, stable and unstable areas were mapped. A test of the accuracy of this method, gave a percentage of cell between 73 to 83 % correctly classified.

Modelling unstable areas: There exists little experience in modelling slope behaviour because soil properties and morphological slope characteristics are not homogeneous. Recently, a three-dimensional groundwater flow model coupled with an infinite slope analysis to predict relative slope stability has been developed (OKIMURA & KAWATANI

1987). A digital landform model was created with a 10 m grid space, and water tables in the soil layer were calculated using a finite difference method based groundwater model. Some assumptions were made such as soil thickness and hydraulic conductivity being constant. The safety factor for each cell was calculated every hour under effective rainfall intensity of 20 mm/h and continuing during 50 hours. The cells predicted to be dangerous in the model, overlapped those at which failure had occured.

Zoning based on monitoring data: Certain areas are found to be clearly unstable. When these areas are inhabited a more precise analysis is needed. Maps performed using data gathered from monitoring slope deformation are very expensive, though the quality of information allows the establishment of accurate models. An example is the DUTI project of the ÉCÔLE POLYTECHNIQUE FÉDÉRALE, LAUSANNE. From photogrammetry and other classical surveying methods, moving areas were detected and their rates of displacement calculated. Thus, a ranking of instability zones within the landslide moving mass based on magnitude of displacements, was established.

Temporal analysis of moving areas showed which parts of the landslide were in process of acceleration or stabilization. From these data, probabilistic analyses of diminishing, constant or increasing velocity of landslide movement have been performed (BONNARD & NOVERRAZ 1984).

Conclusions

Geomorphological analysis and data treatment techniques are increasingly used in landslide risk assessment. The information contained in this review shows several promising lines and topics where further study is required.

In slope instability recognition, statistical techniques have proved to be very useful. More research is needed in characterizing the fabric of landslide deposits in relation to colluvial and glacial sediments. Remote sensing techniques offer new possibilities with SPOT satellite images. The frequent availability of images allows the performance of detailed reconnaisances as for example, before and after an extreme rainfall event.

Hazard assessment is directed towards the prediction of the type of landslide, localization of the potentially failure site, velocity and expected travelled distance and time failure forecasting. Several empirical expressions relate these aspects with specific characteristics of the slope (geometry, lithology, slope use) and climatic conditions (rainfall records, freeze-thaw periods). Much care has to be taken when using these formulae which usually have a limited validity restricted to the regions where they have been deduced. In the mathematical models of failure forecasting the effect of rain variability (extreme rainfall events) or ground shaking, should also be considered. Otherwise, risk may be underestimated.

Several key questions remain still unanswered, specifically those concerning the prediction of the type of movement and its dimensions, the appearance of premonitory signs of instability which could be identified and the long term variability of stability conditions like cohesion decay or regional uplifting.

Hazard zoning and mapping offer a complete possibility only dependent on the available means and time requirements.

Acknowledgements

A. GENS from Geotechnical Engineering Department of the Technical University of Catalunya has reviewed the first draft of this manuscript and offered numerous suggestions for improvement.

References

ALONSO, E. (1987): Riesgos geológicos asociados a las avenidas y su prevision.- In: [BERGA, L. & DOLZ, J.]: Avenidas: Sistemas de previsión y alarma: 37-58, Madrid (Colegio de Ingenieros de Caminos).

ANTOINE, P. (1977): Reflexions sur la cartographie Zermos et bilan des expériences en cours.- Bull. Bur. Rech. géol. min. (2) III, 1/2: 9-20, Orléans.

BONNARD, C. & NOVERRAZ, F. (1984): Instability risk maps: from the detection to the administration of landslide prone areas.- 4th int. Symp. Landslides 1: 511-516, Toronto.

BRAAM, R.R., WEISS, E.E.J. & BURROUGH, P.A. (1987): Spatial and temporal analysis of mass movement using dendrochronology.- Catena 14: 573-584, Cremlingen.

BRABB, E.E. (1984): Innovative approaches to landslide hazard and risk mapping.- 4th int. Symp. Landslides 1: 307-323, Toronto.

BRAND, E.W. & HUDSON, R.R. (1982): CHASE -- an empirical approach to the design of cut slopes in Hong Kong soils.- Proc. 7th Southeast asian geotechn. Conf. 1: 1-16, Hong Kong.

CAINE, N. (1980): The rainfall intensity-duration control of shallow landslides and debris flows.- Geografiska Ann. 62, A: 23-27, Stockholm.

CARRARA, A. (1983a): Multivariate models for landslide hazard evaluation.- Math. Geol. 15: 403-426, New York.

CARRARA, A. (1983b): Geomathematical assessment of regional landslide hazard.- 4th Conf. applic. Stat. Probabil. in Soil a. Estruct. Eng.: 3-27, Firenze.

CARRARA, A. (1984): Landslide hazard mapping: aims and methods. Mouvements de terrain.- Doc. Bur. Rech. géol. min. 83: 141-151, Orléans.

COLIN-ROUSE, W. & FARHAN, Y.I. (1976): Threshold slopes in South Wales.- Q.Journ. eng. Geol. 9: 327-338, London.

COROMINAS, J. (1986): Identificación de taludes inestables.- In: Jornadas de Inv. aplic. en Ing. Geol., Univ. Santander: 90-116, Santander.

COROMINAS, J. (1987): Criterios para la confección de mapas de peligrosidad de movimientos de ladera.- In: Riesgos Geológicos: 193-201, Madrid (IGME).

COROMINAS, J., PEÑARANDA, R. & BAEZA, C. (1988): Identificación de factores que condicionan la formación de movimentos superficiales en los valles altos del Llobregat y Cardener.- II Simp. Taludes y Laderas Inestables: 195-207, Andorra la Vella.

COSTA, J.E. (1984): Physical geomorphology of debris flows. - In: [COSTA, J.E. & FLEISHER, P.J.]: Developments and applications of geomorphology: 268-317, Berlin (Springer).

COSTA, J.E. & JARRET, R.D. (1981): Debris flows in small mountain steam channels of Colorado and their hydrologic implications.- Assoc. eng. Geol. Bull. **18**: 309-322, Lawrence.

CROZIER, M.J. (1984): Field assessment of slope stability.- In: [BRUNSDEN, D. & PRIOR, D.B.]: Slope Instability: 103-142, New York (John Wiley & Sons).

FREDLUND, D.G. (1987): Slope stability analysis incorporating the effect of soil suction.- In: [ANDERSON, M.G. & RICHARDS, K.S.]: Slope stability: 113-144, Chichester (John Wiley & Sons).

FUKUZONO, M.J. & TERASHIMA, H. (1985): Experimental study of slope failure in cohesive soils caused by rainfall.- Proc. Int. Symp. Erosion, Debris Flows and Disaster Prevention: 347-350, Tsukuba.

GARDNER, J.S. (1983): Rockfall frequency and distribution in the Highwood Pass area, Canadian Rocky Mountains.- Z. Geomorph. n.F. **27**: 311-324, Berlin.

GATES, W.C.B. (1987): The fabric of rockslide avalanche deposits.- Bull. Assoc. eng. Geol. **24**: 389-402, Lawrence.

GENS, A., HUTCHINSON, J.N. & CAVOUNIDIS, S. (1988): Three dimensional analysis of slides in cohesive soils.- Geotechn. **38**: 1-23, London.

GONZALEZ DE VALLEJO, L.I., OTEO, C.S. & SERRANO, C. (1988): Reactivación de un deslizamiento provocado por la construcción de un tunel.- II Simp. Taludes y Laderas Inestables: 149-165, Andorra la Vella.

GILI, J.A. (1989): Control de movimentos: auscultación de taludes y laderas inestables.- Soc. Esp. Geomorf., Mon. **3**: 167-214, Zaragoza.

HANSEN, A. (1984): Landslide hazard analysis.- In: [BRUNSDEN, D. & PRIOR, D.B.]: Slope Instability: 523-602, Chichester (John Wiley & Sons).

HARTLEN, J. & VIBERG, L. (1988): Evaluation of landslide hazard.- 5th int. Congr. on Landslides 2: 1037-1057, Lausanne.

HSÜ, K.J. (1975): Catastrophic debris streams (sturzstroms) generated by rockfalls.- Geol. Soc. Am. Bull. 86: 129-140, Boulder.

HUPP, C.P. (1983): Geobotanical evidence of late Quaternary mass wasting in block field areas of Virginia.- Earth Surf. Process. a. Landforms 8: 439-450, Chichester.

HUTCHINSON, J.N. (1988): Morphological and geotechnical parameters of landslides in relation to geology and hydrogeology.- 5th int. Congr. on Landslides 1: 3-35, Lausanne.

INNES, J.L. (1983): Lichenometric dating of debris flow deposits in the Scotish Highlands.- Earth Surf. Process. a. Landforms 8: 579-588, Chichester.

JACKSON, L.E., KOSTASHUK, R.A. & MAC DONALD, G.M. (1987): Identification of debris flow hazard on alluvial fans in the Canadian Rocky Mountains.- Geol. Soc. Amer., Rev. eng. Geol. 7: 115-124, Boulder.

JONES, F.O., EMBODY, D.R. & PETERSON, W.C. (1961): Landslides along the Columbia river valley, Northeastern Washington.- U.S. geol. Surv. prof. Pap. 367: 1-98, Washington.

KOJAN, E., FOGGIN, G.T. & RICE, R.M. (1972): Prediction and analysis of debris slide incidence by photogrammetry, Santa Ynez-San Rafael Moutains, California.- Proc. 24th int. geol. Congr. 13: 124-131, Montreal.

KOVARI, K. (1990): Methods of monitoring landslides.- 5th int. Congr. on Landslides 3: 1421-1433, Lausanne.

LESSING, P., MESSINA, C.P. & FONNER, R. (1983): Landslide risk assessment. - Envi ronm. Geol. 5: 93-99, New York.

LOPEZ, C. (1982): Dinámica de los desplazamientos de materiales sobre pendientes naturales.- VII Simp. Nac. Obras Superf. en Mec. Rocas 1: 2.5.1-2.5.16, Madrid.

LUMB, P. (1975): Slope failures in Hong Kong.- Q. Journ. eng. Geol. 8: 31-65, London.

LLORET, A., GILI, J.A., GENS, A. & ALONSO, E. (1984): Avances recientes en el análisis de la estabilidad de los taludes.- Jornadas Inest. Laderas Pirineo: PG.II.1-PG.II.64, Barcelona.

MILLS, H.H. (1984): Clast orientation in Mount St. Helens debris flow deposits, North Fork, Toutle River, Washington.- Journ. sed. Petrol. **54**: 626-634, Lawrence.

MOSTYN, G.R. & SMALL, J.C. (1987): Methods of stability analysis.- In: [WALKER, B.F. & FELL, R.] : Soil slope instability and stabilization: 71-120, Rotterdam (Balkema).

NELSON, F.E. (1985): Preliminary investigation of solifluction macrofabrics.- Catena **12**: 23-33, Cremlingen.

NEULAND, H. (1976): A prediction model for landslips.- Catena **3**: 215-230, Cremlingen.

NILSEN, T.H., WRIGHT, R.H., VLASTIC, T.C. & SPANGLE, V.E. (1979): Relative slope stability and land-use planning in the San Francisco Bay Region, California.- U.S. geol. Surv. prof. Pap. **944**: 1-96, Washington.

OGAWA, S. & KAWAMURA, K. (1989): Topographical consideration for landslide prediction.- Proc. 12th int. Conf. Soil Mech. a. Found. Eng. **3**: 1587-1590, Rio de Janeiro.

OKIMURA, T. & KAWATANI, T. (1987): Mapping of the potential surface-failure sites on granite mountain slopes.- In: [GARDINER, V.]: International Geomorphology **1**: 121-138, Chichester (John Wiley & Sons).

ONODERA, T., YOSHINAKA, R. & KAZAMA, H. (1974): Slope failures caused by heavy rainfall in Japan.- Proc. 2nd int. Congr. int. Assoc. eng. Geol.: V.11.1-V.11.10, Sao Paolo.

OSTERKAMP, W.R. & HUPP, C.R. (1987): Dating and interpretation of debris flows by geologic and botanical methods at Whitney Creek Gorge, Mount Shasta, California.- Geol. Soc. Amer. Rev. eng. Geol. **7**: 157-163, Boulder.

OYAGI, N. (1984): Landslides in weathered rocks and residual soils in Japan and surrounding areas: a state-of-the-art report.- 4th int. Symp. Landslides **3**: 1-31, Toronto.

PITTS, J. (1979): Morphological mapping in the Axmouth-Lime Regis Undercliffs, Devon.- Q. Journ. eng. Geol. **12**: 205-217, London.

PLAFKER, G. & ERIKSEN, G.E. (1979): Nevados Huascaran avalanches, Peru. - In: [VOIGHT, B.]: Rockslides and avalanches **1**: 277-314, Amsterdam (Elsevier).

RAPP, A. (1960): Recent developments of mountain slopes in Kärkevagge and surroundings.- Geografiska Ann. **42**: 65-200, Stockholm.

RAT, M. (1988): Essai de prevision de la date de rupture d'un grand glissement.- II Simp. sobre Taludes y Laderas Inestables: 419-431, Andorra la Vella.

RENAU, S.L. & DIETRICH, W.E. (1987): The importance of hollows in debris flow studies; examples from Marin County, California.- Geol. Soc. Amer. Rev. eng. Geol. **7**: 165-180, Boulder.

RIB, H.T. & LIANG, T. (1978): Recognition and identification.- Transp. Res. Board Spec. Rep. **176**: 34-80, Washington (National Academy of Sciences).

ROCHET, L. (1987): Application des modèles numeriques de propagation a l'etude des eboulements rocheux.- Bull. Liaison Ponts et Chaus. **150-151**: 84-95, Paris.

SAITO, M. (1965): Forecasting time of occurence of a slope failure.- Proc. 6th int. Conf. Soil Mech. Found. Eng. **2**: 677-683, Mexico.

SHEKO, A.I. (1977): Theoretical principles of a regional temporal prediction of landslide activation.- Bull. int. Assoc. eng. Geol. **16**: 67-69.

SIDLE, R.C., PEARCE, A.J. & O'LOUGHIN, C.O. (1985): Hillslope stability and land use.- Am. geophys. Un. Water Res. Mon. **11**: 1-140, Washington.

SKEMPTON, A.W. (1952): Stability of natural slopes in London Clay.- Proc. 4th int. Conf. Soil Mech. **2**: 378-381, London.

SKEMPTON, A.W. (1964): Long term stability of clay slopes.- Geotechn. **14**: 75-105, London.

SOUTADE, G. (1980): Modèlé et dynamique actuelle des versants supraforestiers des Pyrenées Orientales.- 452 pp., Albi (Coop. Sud-Oest).

SOUTHWORTH, C.S. & SCHULTZ, A.P. (1986): Photogeologic interpretation reveals ancient rockslides in the Appalachian Valley and Ridge Province, Virginia and West Virginia.- Assoc. eng. Geol. Newsl. **29/2**: 31-33, Lawrence.

STEVENSON, P.C. (1977): An empirical method for the evaluation of relative landslip risk.- Bull. int. Assoc. eng. Geol. **16**: 69-72.

VARNES, D.J. (1978): Slope movement types and processes.- Transp. Res. Board spec. Rep. **176**: 11-33, Washington (Nat. Acad. Sciences).

VARNES, D.J. (1984): Landslide hazard zonation: a review of principles and practice.- Natural Hazards **3**: 1-63, Paris (UNESCO).

VOIGHT, B. & KENNEDY, B.A. (1979): Slope failure of 1967-69, Chuquicamata mine, Chile.- Develop. geotech. Eng. **14**, B: 595-632, Amsterdam (Elsevier).

VOIGHT, B. (1989): Materials science law applied to time forecast of slope failure. - Landslide News **3**: 8-11, Tokyo.

WEBER, Ch. (1984): Remote sensing and natural hazards contribution to spatial imaginery to the evaluation and mitigation of geological hazards.- Proc. 27th int. Geol. Congr. **18**: 211-228, Moscow.

WILSON, S.D. & MIKKELSEN, P.E. (1978): Field instrumentation.- Transp. Res. Board spec. Rep. **176**: 112-138, Washington (National Academy Sciences).

YANO, K. & SENOO, K. (1985): How to set a standard rainfalls for debris flow warning and evacuation.- Int. Symp. on Erosion, Debris Flows and Disaster Prevention: 451-455, Tsukuba.

ZANG, X., LIU, T., WANG, Y. & LUO, J. (1985): The main features of debris flows and control structures in Hunshui Gully, Yuannon Province, China.- Int. Symp. on Erosion, Debris Flows and Disaster Prevention: 181-186, Tsukuba.

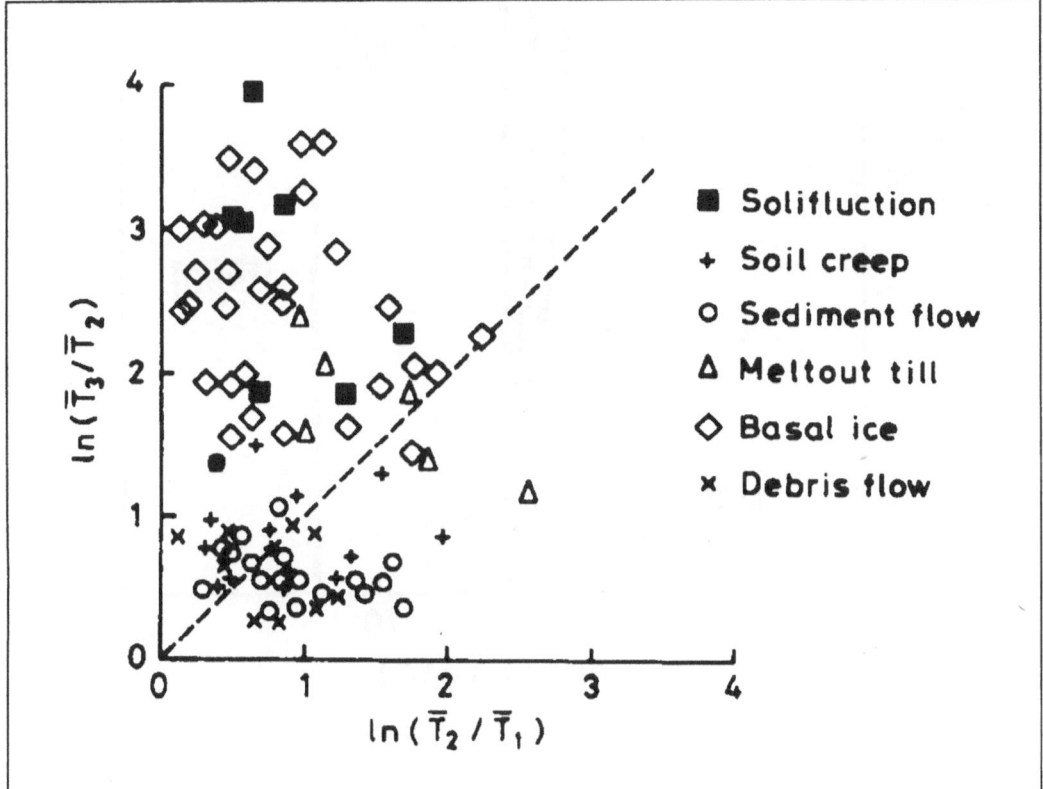

Figure 1: Logarithmic ratio plot of eigenvalues from various types of deposits reported in the literature by NELSON (1985).

Dashed diagonal line indicates position of the cluster-girdle transition zone (ratio ordinate/abscissa is unity). Values greater than unity represent clustered distributions.

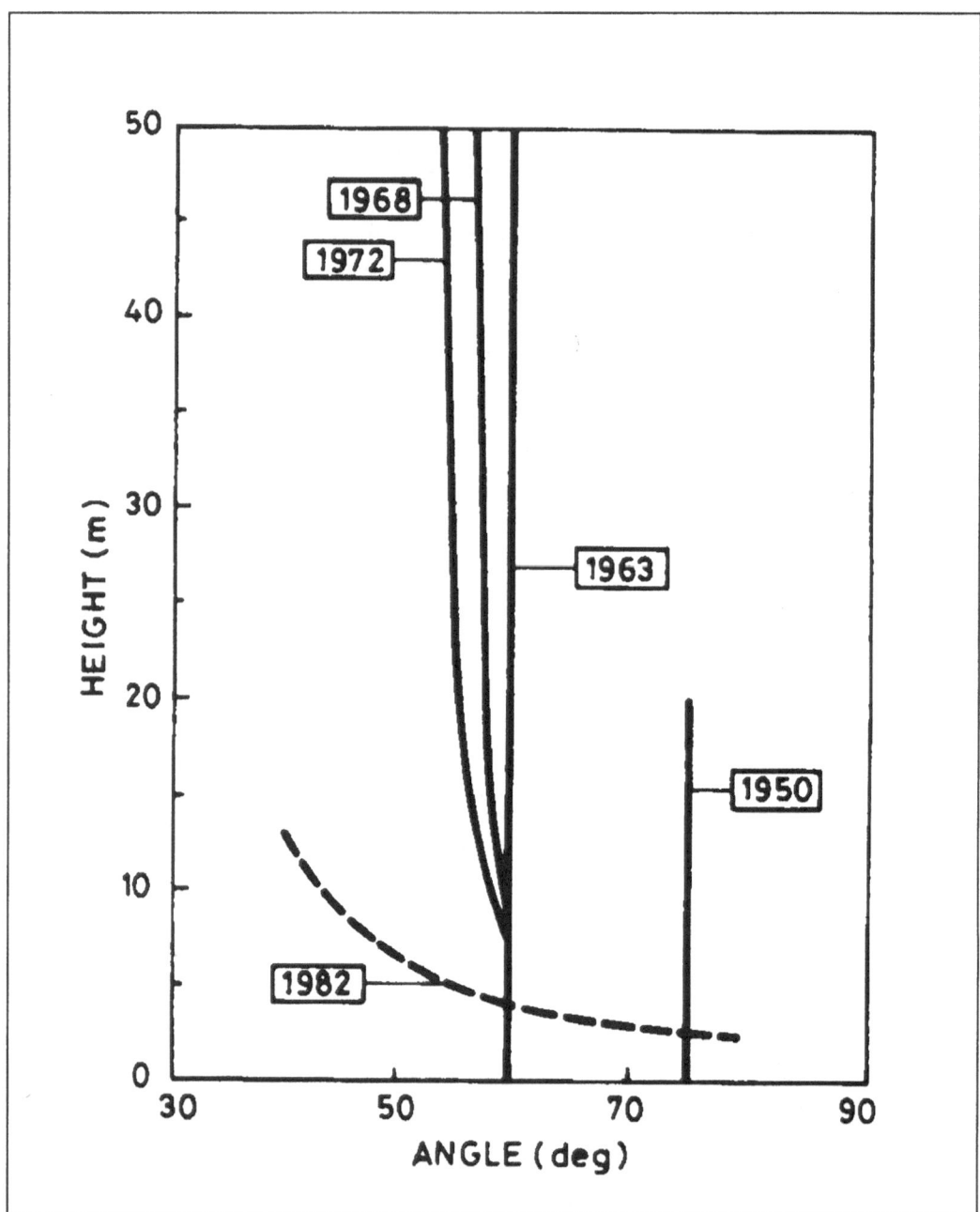

Figure 2: Height versus slope angle plot for 177 failed and unfailed slopes in Hong Kong (modified from BRAND & HUDSON 1982).

Previous rules have been discontinued. 1982 dotted line represents a safety factor of 1.4.

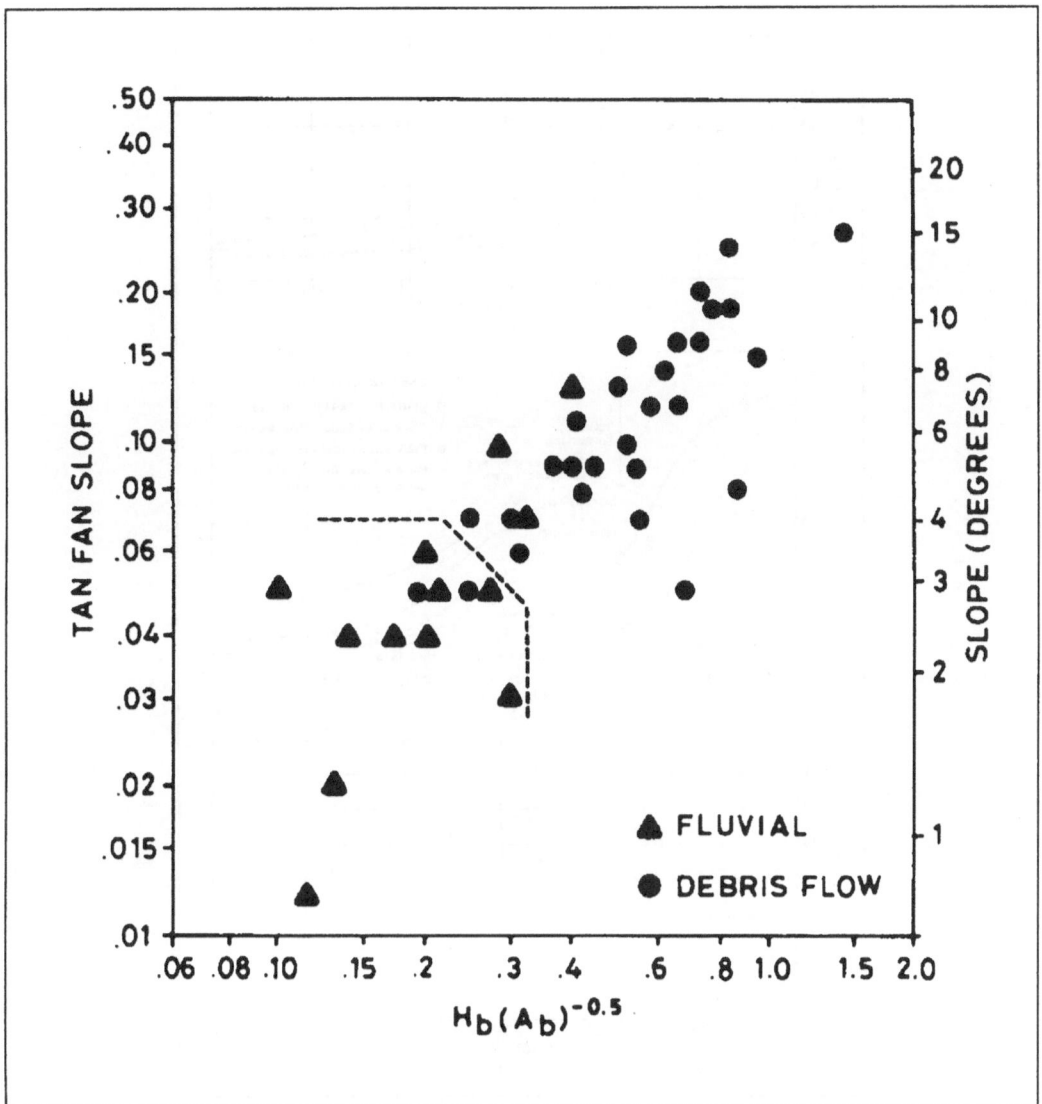

Figure 3:   Plot of tangent of fan slope versus Nelson's ruggedness number ($H_bA_b^{-0.5}$) for 42 fans from Canadian Rocky Mountains.

H_b is basin height measured from fan apex to basin highest point and A_b is basin area above fan apex (JACKSON et al. 1987).

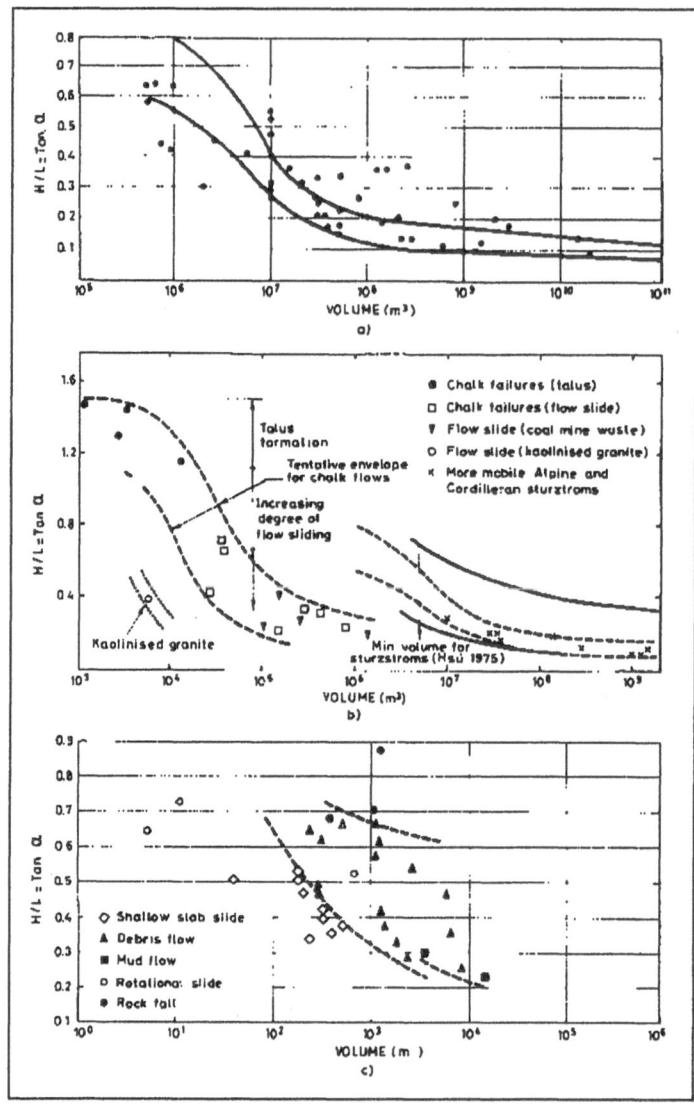

Figure 4: Landslide mobility. Volume of the fallen mass is plotted against drop and horizontal distance travelled ratio.

(a) data corresponding to huge rock avalanches (HSÜ 1975),

(b) data for chalk, coal mine waste and kaolinized granite (HUTCHINSON 1988),

(c) data for small volume movements (COROMINAS et al. 1988).

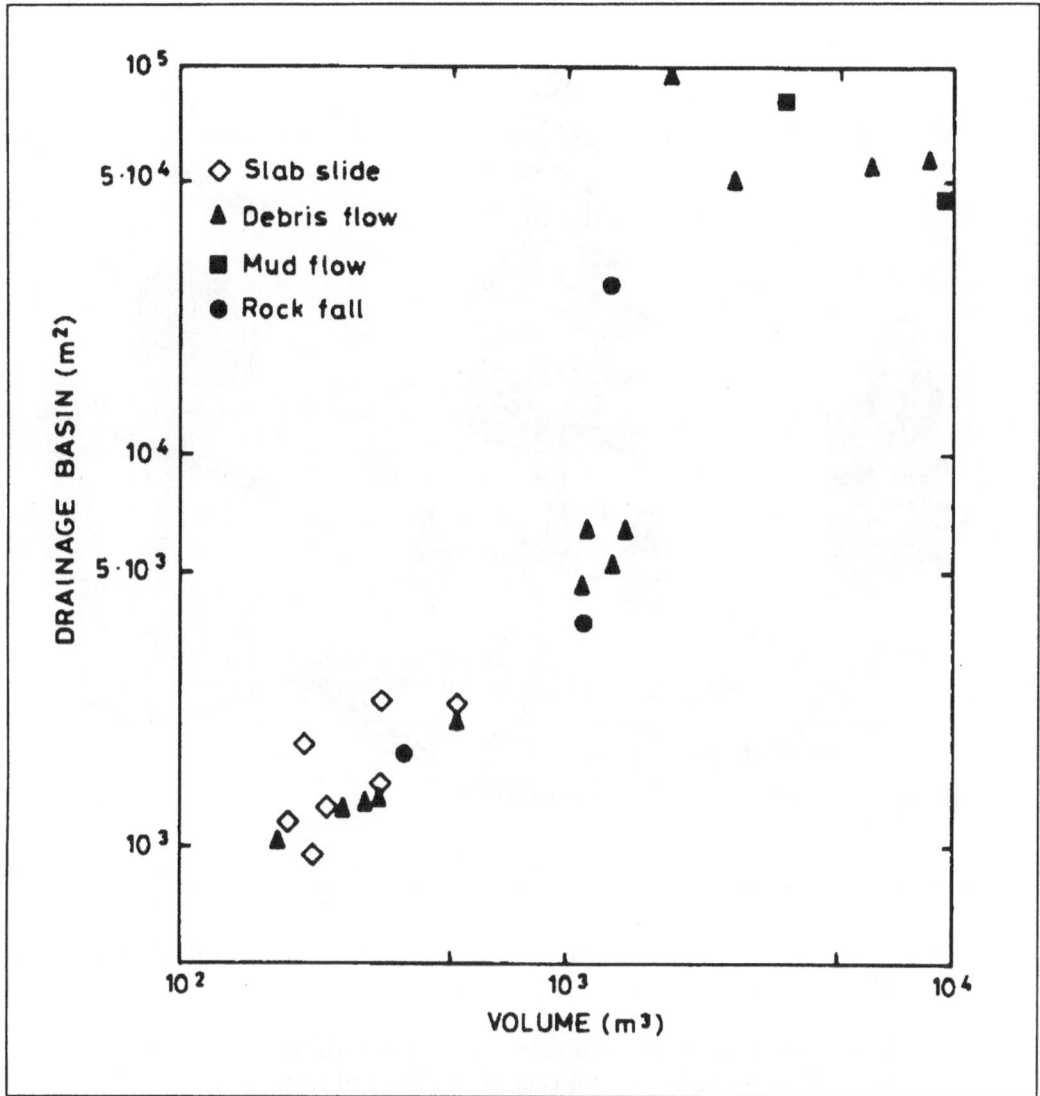

Figure 5: Plot showing relationship between catchment area above failure zone and volume yield by shallow mass movements in Spanish Eastern Pyrenees

(COROMINAS et al. 1988).

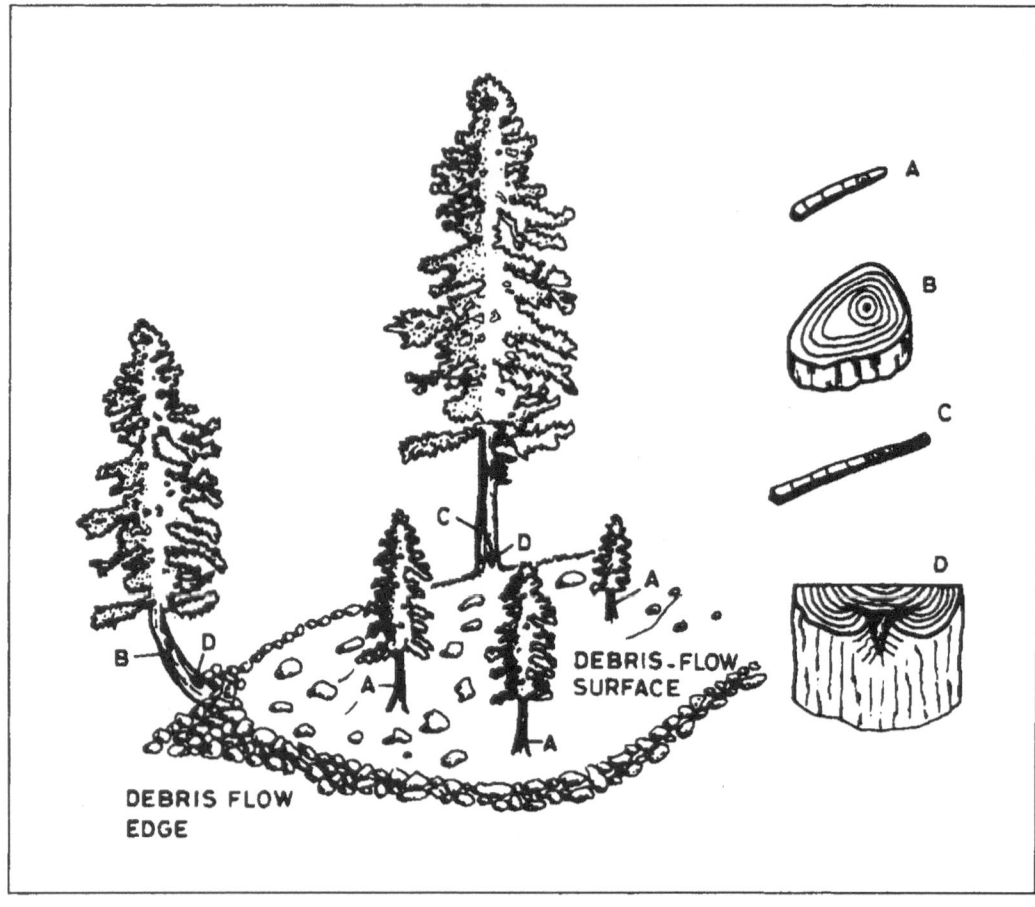

Figure 6:   Dendrochronologic evidence of debris flows (from OSTERKAMP & HUPP 1987). Cores from conhort of young tree

(A)   give similar ages and date debris flow surface.
(B)   Eccentric growth ring caused by tilting of tree.
(C)   Reduction of competition, because of destruction of vegetation, causes suppression-release sequence.
(D)   Scars caused by impacts of debris flow clasts on tree trunks.

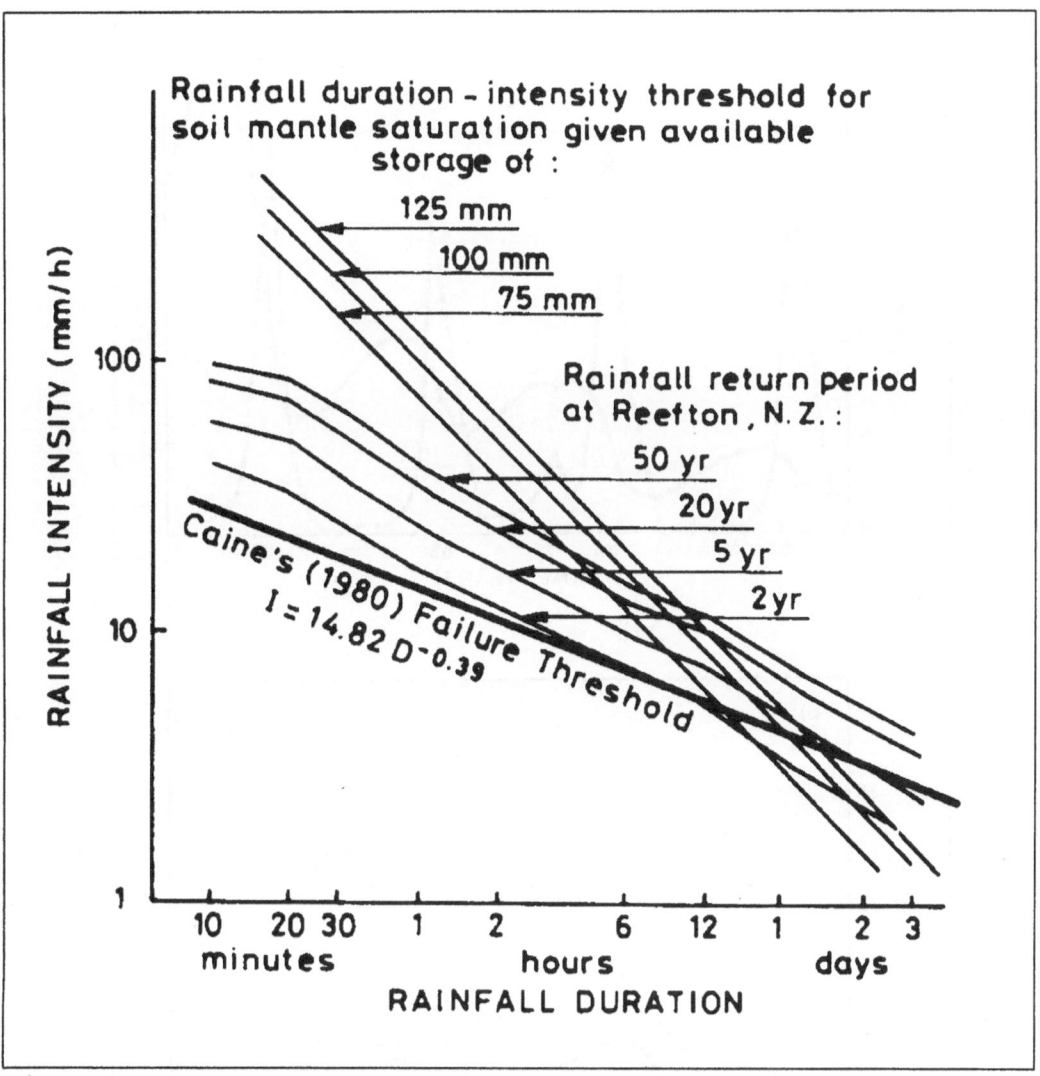

Figure 7:   Rainfall intensity-duration-frequency curves for Reefton, New Zealand.

Compared with intensity-duration combination required for soil mantle saturation and with CAINE'S relationship of rainfall intensity threshold for landslide triggering (SIDLE et al. 1985).

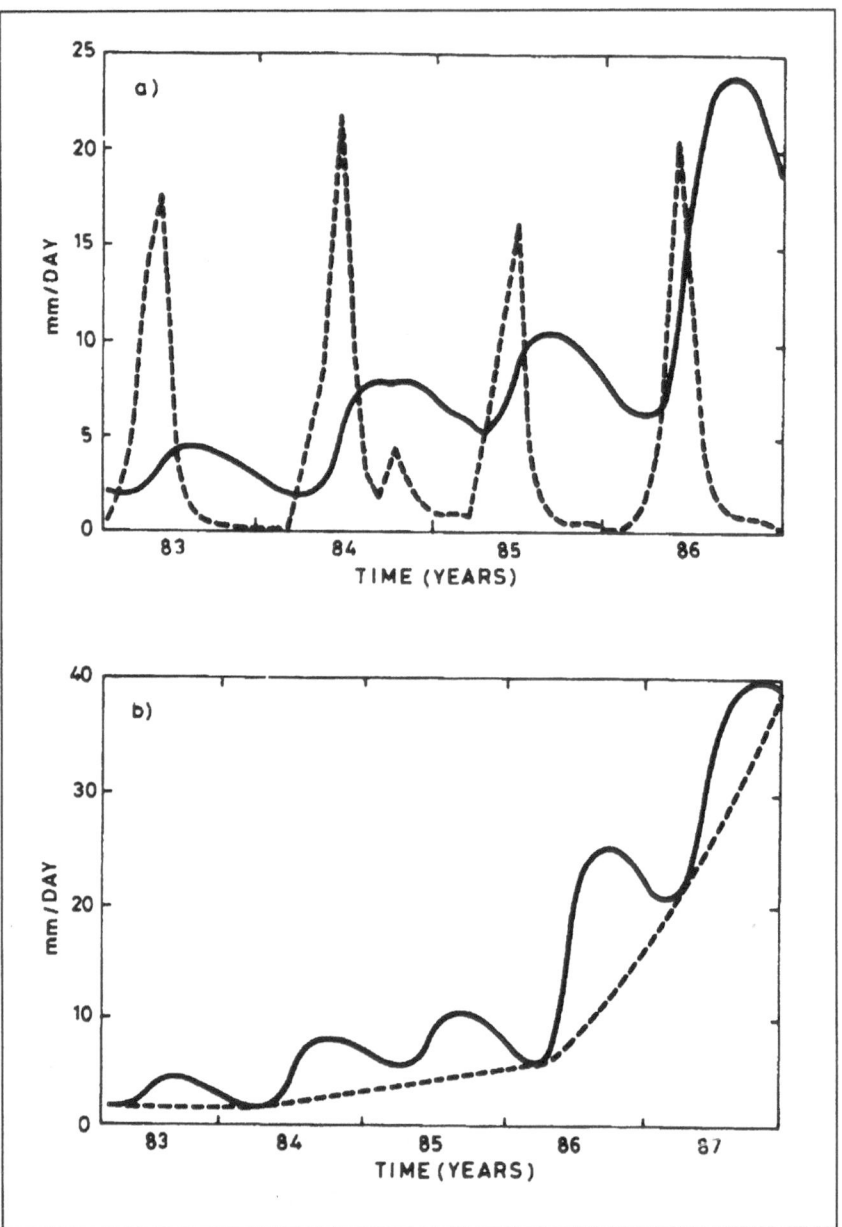

Figure 8: (a) Variability of rate of displacement at La Clapière landslide with discharge of La Tinée river.

(b) Residual trend to acceleration of displacements with time, during 1983-87 period (RAT 1988).

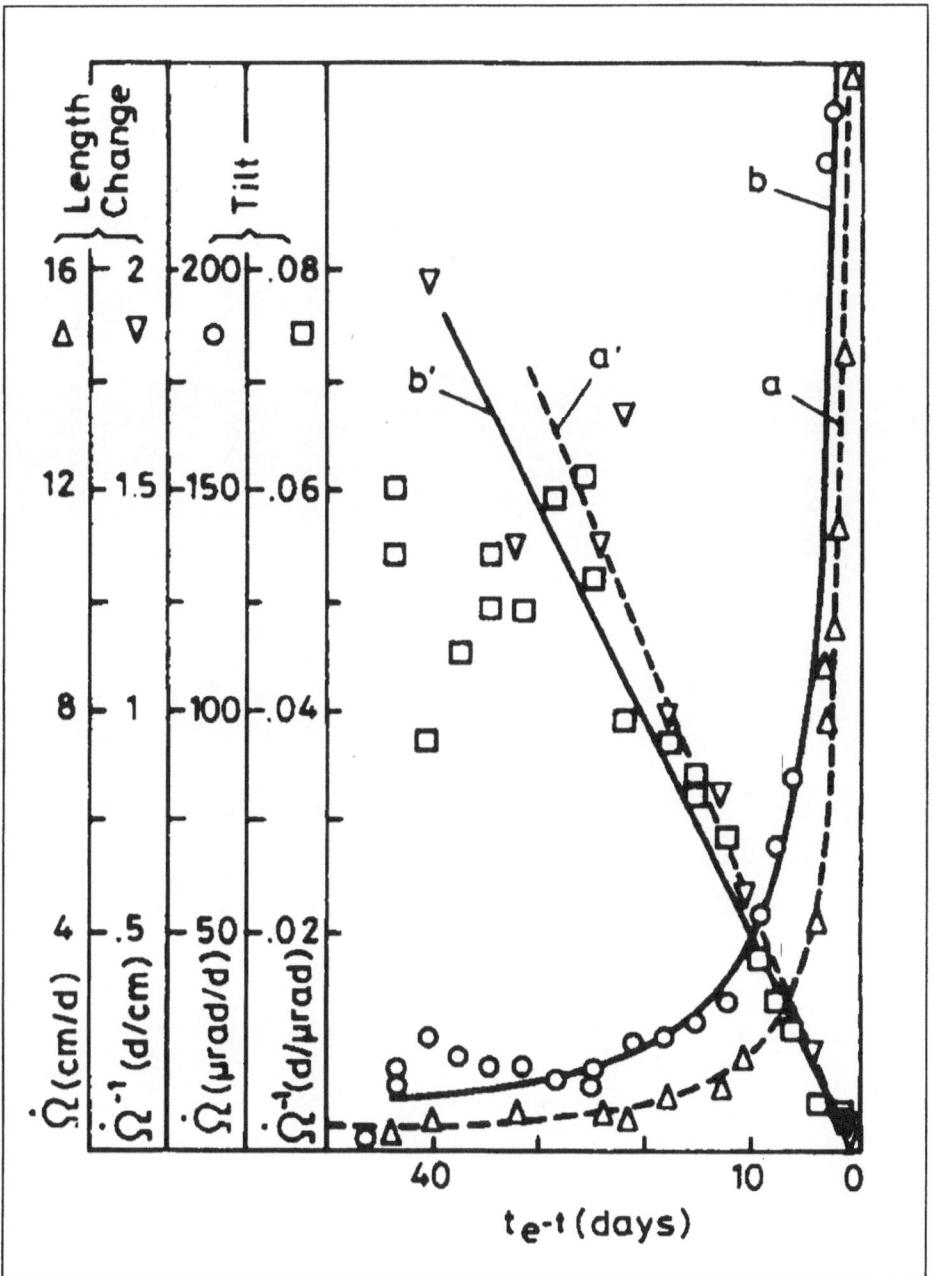

Figure 9:   Mount St. Helens rates and inverse rates against time before 1982
eruption.

Line length changes (curves a, a') or tilt (curves b, b'), with
primes indicating inverse-rate curves (modified from VOIGHT
1989).

Mapping and Modelling Quaternary Depositional Systems to Assess Groundwater Pollution Hazards in the Glaciated Appalachian Region, U.S.A.

KENNETH G. JOHNSON [*]

Introduction

Many communities in the formerly glaciated Appalachian Mountains of the northeastern United States depend entirely on groundwater for their water supply. These communities range in size from places like Schenectady, New York, a city of 150,000 population which relies on munincipal wells producing from thick glacifluviatile gravels to villages like Valley Falls, New York which, with a population of 500, relies entirely on shallow, individual-dwelling wells, most of which produce from sands and gravels of a small Late Pleistocene deltaic complex. The water produced from such glacifluviatile sediments is for the most part of excellent quality and until recently has been more than adequate from the standpoint of volume needed by the various municipalities. This situation is changing due to development pressure connected with increased population in urban centers and expansion of the outer fringes of these centers into nearby rural areas. Ironically, these development pressures are making us painfully aware that the very qualities which make these Pleistocene sedimentary facies such good aquifers are also qualities which make the contained groundwater very susceptible to contamination by landfill leachate, leaking underground storage tanks, agricultural fertilizer infiltration and malfunction of poorly designed and located septic systems.

---

[*] Author's address: Prof. K. G. JOHNSON, Department of Geology, Skidmore College, Saratoga Springs, New York 12866 U.S.A.

The problem of development pressure is especially acute for the rural townships and villages of the glaciated Appalachians inasmuch as they are experiencing a high percentage increase in housing starts, have very low tax bases and as a result are hardpressed to find the money required to cope with the rapid transition from rural to suburban mode. Not the least of their problems, of course, is the threat that uncontrolled development brings to the groundwater on which they are so dependent. They need help, and they need it **now**! One of their most urgent needs is for geological information, information that is basic in the formulation and design of development strategies anywhere, but especially important in regions like the glaciated Appalachians, where subsurface conditions can be agonizingly unpredictable.

The purpose of this paper is to attempt to demonstrate that in the area of development planning an appropriate technique for the less-developed parts of the glaciated Appalachians, as well as for other similar regions throughout the world, is the use of the concept of depositional systems, an idea which evolved in the 1960s among geologists studying the complicated Cenozoic geologic record of the northern Gulf of Mexico. A depositional system is an assemblage of process-related sedimentary facies (FISHER & BROWN 1974). Whether modern or ancient, it is a large-scale, natural, genetic unit that evolved through processes of sedimentation and which can be subdivided into smaller components to the extent required for a particular analysis. This analysis might be exploration for fuel resources of determination of subsurface distributions of surficial sediment and/or bedrock characteristics for purposes of environmental management. Since the 1970s, the concept has been used widely in the petroleum industry and has also been extensively used by the Texas Bureau of Economic Geology in its aggressive program of resource capability mapping and inventory (FISHER & McGOWEN 1967, WERMUD 1974, TEXAS BUREAU OF ECONOMIC GEOLOGY 1988).

Depositional Systems and their Utility in Quaternary Mapping and Modelling

Geologists of the Texas Bureau of Economic Geology have been very successful in using the concept of depositional systems to map and catalog the land, water and environmental resources of non-glacigenic Quaternary Sediments. A good example of their work is the study by R.F. SOLIS I. (1981), in which he reports on his analysis of the Upper Tertiary/Quaternary depositional systems of the central coastal plain of Texas. An important part of his report is an evaluation of the sediments in these systems as aquifers and potential liquid-waste repositories. In the north-central United States, geologists of the Illinois State Geological Survey have had similar success in mapping and modelling Quaternary sediments, in this case, mostly of glacigenic origin. The results of many of these studies have appeared in the Survey's Environmental Geology Notes Series (see bibliography). The first publication in the series appeared in 1965 and since then about 135 of these succinct, widely-distributed, generally high-quality notes have been published at an amazingly low cost. Elsewhere in North America, Canadian scientists have a long history of applying their knowledge of Quaternary glacigenic sediments to planning initiatives and engineering projects (LEGGETT 1939, 1973; CHRISTIANSEN 1970, PREST & HODE KEYSER 1977, COOPER et al. 1989, WHITE 1989).

Glaciated Appalachian Mountains of the Northeastern United States

The Appalachian Highlands dominate the eastern margin of North America and are geographically subdivided from southeast to northwest into the

1)   Piedmont Plateau Province,

2)   Blue Ridge Mountains Province,

3)   Valley and Ridge Province and

4)   Appalachian Plateaus Province.

Pleistocene glaciation affected only the northern half of the Appalachians.

Location

For purposes of this paper, the glaciated Appalachian region extends from northern Pennsylvania northeastward across New York State and the New England States and into the Canadian maritime Provinces. The region is bordered on the northwest by the St, Lawrence River and Lakes Ontario and Erie. On the southeast, the Appalchians are bordered by coastal plain sediments which are for the most part submerged, since the coastal area has not yet fully rebounded from the weight of the Pleistocene Wisconcinan ice sheet. Figure 1 shows the extent of Pleistocene glaciation in the United States and Figure 2 the surface morphology of the northeastern U.S. and southeastern Canada. Accounts of the geomorphology and glacial geology of this region are found in THORNBURY (1965), FLINT (1971) and HUNT (1974).

Historical Context of Use of Geoscience Information in Planning and Management in the Northeastern United States

The earliest contacts between Native North Americans and Western Europeans occured along the seaward margin of the glaciated Appalachians. From the standpoint of U.S. and Canadian early colonial development and penetration into the interior of the continent, and later national emergence, the Hudson and St. Lawrence Rivers made it relatively easy for French, Dutch and English explorers to reach the Great Lakes and Mississippi River Valley and later for individuals and families to migrate westward beyond the Appalachian barrier. For these reasons, among others, some of the oldest and largest cities in Canada and the U.S. are located in the glaciated Appalachians.

In the United States, a number of these munincipalities have expanded along the eastern seaboard to the extent that they now form an urban corridor over 900 km long extending from Portland, Maine to Richmond, Virginia. About half of this corridor lies in the glaciated Appalachians. The rate of urban expansion accelerated significantly in the decades following World War II and it became evident that planning and control were urgently needed. Starting in the 1960s, a significant portion of the financial and manpower resources of the United States Geological Survey were directed to studies focusing on generation of earth-science maps applied to land and water management (e.g. U.S.G.S. 1967, ROBINSON & SPIEKER 1978). These initiatives have played an important part not only in improving the quality of regional planning and management in U.S. urban areas but also in helping public officials, legislators and managers to understand that without such information they simply cannot hope to function effectively. Geoscientists are communicating better and the people in need of geoscience data are starting to listen. However, problems, as always, remain.

One of these problems is the need for generation of low-cost geoscience information which will be of use to rural townships and villages, many of which are located adjacent to expanding urban centers and are feeling heavy development pressure. In a very real sense they are in need of scientific information and appropriate technology not unlike that needed by developing nations around the world.

The Glacial Sedimentary System

A glacier can be considered a sedimentary system in which accumulation, transfer and deposition of mass (snow, ice, water, rock debris) occurs in response to additions and losses of mass and energy (CHORLEY, SCHUMM & SUDGEN 1984). Depositional landforms that result may be characterized as ice-advance forms and stagnation forms. Most attempts to develop a classification scheme for glacial depositional landforms have focused on

1) whether the glacier was actively flowing or stagnant and
2) whether the landforms were formed subglacially or at the glacier surface margin.

It is not the purpose of this paper to develop, or even present in outline, a detailed classification scheme. Good introductions to glacial processes and landform classifications are provided by FLINT (1971), EDWARDS (1978) and CHORLEY, SCHUMM & SUDGEN (1984) and others.

EDWARDS recognizes three categories of glacial sedimentary facies:

1) subglacial deposits (till),
2) supraglacial and proglacial deposits (stratified gravel and sand) and
3) glacimarine and glacilimnic deposits (mud, sometimes with large clasts).

All of these facies are well represented in the formerly glaciated Appalachians with each presenting its own special opportunities and limitations with respect to water resources and land-use.

Use of a Depositional Systems Model to Predict Subsurface Conditions

FETTER (1980) describes the glaciated Appalachians as being mantled by a generally thin cover of glacial drift which in most areas is suitable only for domestic water supplies. The weathered zone at the bedrock surface yields similar very modest volumes of water, except in the Connecticut River Valley and parts of central New York State where bedrock wells are somewhat more productive. As indicated in the introduction to this paper, glacifluviatile deposits form productive aquifers in some areas. This facies is especially well-developed in the upper Hudson River Valley of eastern New York State (Figure 3), a region where rural townships adjacent to the Capital District urban complex are feeling substantial development pressure. Two of these are the Towns of Schaghticoke and Pittstown (Figure 4), which lie at the confluence of the Hudson and Hoosic Rivers. Planning officials of these towns have had the benefit of geoscience data and workshops directed toward helping them to understand the relatively simple depositional system on which their constituents depend for nearly all of their water supply and into which most of their sewage is released. This depositional system is the Hoosic Delta, a complex which evolved along the eastern shore of Lake Albany, a northward-expanding proglacial lake that developed in the Hudson Valley between 18,000 and 13,200 yBP during the final retreat of the Laurentide Ice Sheet. Many other communities in this region, including some cities, rely on similar aquifers. Unfortunately, in more than one of these, serious groundwater pollution has resulted from careless disposal of industrial waste.

The Hoosic Delta Depositional System

The areal extent and sediment distribution of the Hoosic Delta system was outlined many years ago by STOLLER (Figure 4). Recent use of modern delta system process-response paradigms and facies analysis based on surface exposure and water-well data has permitted formulation of the cross-sections shown in Figures 5 and 6. The system consists of five facies:

1)   delta plain,

2)   proximal prodelta,

3)   distal prodelta,

4)   lacustrine and

5)   ice-contact.

Further subdivision into sub-facies, although possible, is not necessary for purposes of general land-use planning and assessment of groundwater pollution hazard.

The ice-contact facies has highly variable texture, ranging from clays to small boulder gravels. Contacts with adjacent facies and underlying bedrock are unconformable and, although the facies is probably present in many places at the base of the delta sequence, prediction of subsurface occurence and extent is difficult. Significant volumes of water are present in this facies, especially in the pre-glacial bedrock channel near the northwest end of Section HRD-I.

The lacustrine facies, since it consists of interlaminated silt and clay, is an aquiclude. It grades upward and laterally into the distal prodelta facies and rests unconformably on bedrock or the ice-contact facies. It is important inasmuch as in many parts of the delta system it protects underlying ice-contact sediments from surface contaminant infiltration.

The distal prodelta facies is composed of interlaminated and interlensed silt and clay with abundant calcareous marlekor and locally abundant dropstones. Like the lacustrine facies, it is an aquiclude and for this reason the two facies are not differentiated on the cross-sections in Figures 5 and 6. The facies grades downward into the lacustrine facies or rests unconformably on bedrock or the ice-contact facies. It grades laterally and upward into the proximal prodelta facies.

The proximal prodelta facies consists of silt which grades upward into very fine-grained sand. Cross-stratification and calcareous nodules and masses ranging up to .3 m thick and several square meters in lateral dimension are abundant in this unit. Sands of the facies grade upward into delta plain sands and gravels, and below and laterally silts of the facies are interbedded and interfingered with distal prodelta clays.

The delta plain facies consists of pebble gravels and fine to medium-grained sands. Basal contact with proximal prodelta sands is gradational; basal contacts within the facies are mostly erosional.

Significance of Hoosic Delta Facies Distribution With Respect to Groundwater Pollution Hazard

The internal structure of the Hoosic Delta System with respect to three-dimensional variations in particle-size, porosity, permeability and homogenity is a function of the depositional processes which were operative as the system evolved. An understanding of the sedimentological response to these processes is a key element in predicting probable patterns of fluid migration within the system -- that is, from a practical standpoint, predicting how surface water will behave as it moves into and within the subsurface.

The morphology and sedimentology of the Hoosic Delta System is typical of a Gilbert-type, or lacustrine, delta. These form when in-flowing (river) water and reservoir water are of the same density. The inflowing river water and reservoir water mix readily and as a consequence the sediment load of the stream is deposited much more quickly than would be the case when river (fresh) water discharges into marine (highly saline, i.e., more dense) water. One result of this very rapid deposition of river-transported sediment in a lake is that, as the delta progrades, there is a vertical development of relatively sharp boundaries between sediment deposited by river processes and that deposited by lacustrine processes. BATES (1953) and JOPLIN & MCDONALD (1975) provide excellent discussions of the hydrodynamics of delta formation.

The following tabulation outlines the general characteristics of facies within the Hoosic Delta System:

Facies	Depositional Processes	Hydrogeologic Character
Delta Plain	Deposition from river traction and suspension loads	Groundwater conductivity high, but characteristically variable both laterally and vertically. Aquifer at risk.
Proximal Prodelta	Rapid deposition from suspension load	Groundwater conductivity moderately high, but decreases progressively towards base of facies. Aquifer at risk.
Distal Prodelta and Lacustrine	Slow deposition from suspension	Groundwater conductivity nil both laterally and vertically. Aquiclude, in some places protecting underlying aquifer(s).
Ice-Contact	Fluviatile and lacustrine deposition against stagnant ice with postmelting slumpage	Groundwater conductivity and retention variable, from excellent to nil.

Conclusion

Subsurface facies distribution and probable groundwater conductivity within a delta system are relatively predictable compared to that within other glacigenic depositional systems. For example, a buried esker system, an outwash system, a valleyfill system or a kame complex are much less straightforward with respect to facies analysis. However, although the challenge is greater, it is possible to understand these systems and to develop predictive models for them by using such mapping and modelling techniques as profile-legend maps (HAGEMAN, 1963), fence diagrams and stack-unit maps (KEMPTON, 1981).

In the petroleum industry, it is an axiom that oil is found in the minds of men. The axiom applies equally well in environmental management and planning, for without a well-founded conceptual model, no matter how much data, funding and technology are available, prediction of fluid behavior in subsurface Quaternary sediments is largely conjecture.

References

AMERICAN PETROLEUM INSTITUTE (1972): The migration of petroleum products in soil and ground water -- Principles and counter-measures.- Am.Petrol.Inst.Publ. **4149**: 1-36, Washington, DC.

ANDERSEN, J.R. & DORNBUSH, J.N. (1967): Influence of sanitary landfill on ground water quality.- J.amer. Water Works Assoc. **59**: 457-470, Denver, CO.

APGAR, A.A. & LANGMUIR, D. (1971): Groundwater pollution potential of a landfill above the water table.- Ground Water **9**: 6, Worthington, OH.

ARCHER, A.A.; LÜTTIG, G.W. & SNEZHKO, I.I. (1987): Man's Dependence on the Earth: 216 pp.- Nairobi, Paris, Stuttgart (UNEP, UNESCO, Schweizerbart)

ASHLEY, G.M. (1987): A facies model for temperate continental glaciers.- J. geol. Educ. **35**: 208-216, Lawrence, KS.

BAEDECKER, M.J. & BACK, W. (1979): Hydrogeological processes & chemical reactions at a landfill.- Ground Water **17**, 5: 429-437, Worthington, OH.

BATES, C.C. (1953): Rational theory of delta formation.- Bull. am. Assoc.Petrol.Geol. **37**: 2119-2162, Tulsa, OK.

BERG, R.C.; KEMPTON, J.P. & STECYK, A.N. (1984): Geology For Planning in Boone and Winnebago Countries.- Illionois State geol.Surv.Circ. **531**: 69 pp, Urbana, IL.

BERGSTROM, R.E. (1968): Disposal of Wastes: Scientific and Administrative Considerations.- Illinios Geol. Surv. Environ. Geol. Not. **20**: 1-12, Urbana, IL.

BERGSTROM, R.E. (1970): Geology for Planning at Crescent City, Illinois. - Illinois State geol. Surv. environ. geol. Note **36**: 1-15, Urbana, IL.

BERGSTROM, R.E.; PILSKIN, K. & FOLLMER, L.R. (1976): Geology for planning in the Springfield-Decatur region, Illionois.- Illinois State geol. Surv. Circ. **497**: 1-76, Urbana, IL.

BIRCH, F.S. (1989): A geophysical study of Quaternary sediments near the late Pleistocene marine limit in Epping, New Hampshire. - Northeast. Geol. **11**, 3: 124-132, Troy, NY.

BURT, E. (1972): The use, abuse, and recovery of a glacial aquifer. - Ground Water **10**, 1: 65-72, Worthington, OH.

CARTWRIGHT, K. & MCCOMAS, M.R. (1968): Geophysical surveys in the vicinity of sanitary landfills in northeastern Illinois. - Ground Water **6**: 23-30, Worthington, OH.

CAVANAGH, T.E. jr. (1973): Sanitary landfill management.- 70 pp., Illinois environ. Prot. Agency, Urbana, IL.

CHERRY, J.A.; GILLHAM, R.W. & PICKENS, J.F. (1975): Contaminant hydrogeology: part 1: physical processes: 76-84, Geoscience Canada, Toronto, Ont.

CHORLEY R.J.; SCHUMM, S.A. & SUDGEN, D.E. (1985): Geomorphology.- 605 pp, New York, Methuen & Co.

[CHRISTIANSEN, E.A.] (1970): Physical environment of Saskatoon, Canada. - 68 pp, Ottawa, Saskatchewan Research Council and National Research Council of Canada.

COOKE, R.U. & DOORNKAMP, J.C. (1977): Geomorphology in Environmental Management - An Introduction.- 413 pp, Oxford, Oxford Univ. Press.

COOPER, A.J.; FUNK, G.H. & ANDERSON, E.G. (1989): Using Quaternary stratigraphy to help locate a hazardous waste treatment site. In: [MULDER, E.F.J. DE & HAGEMAN, B.P.]: Applied Quaternary Research, Rotterdam, Netherlands, A.A.Balkema.

DOONAN, C.J.; HENDRICKSON, G.E. & BYERLAY, J.R. (1970): Ground-Water and Geology of Keweenaw Peninsula, Michigan. - Michigan Dep. Nat. Res. , geol. Surv. Div. Water Invest. **10**: 1-40, Lansing, MI.

DREIMANIS, A. (1977): Correlation of Wisconsian glacial events between the eastern Great lakes and the St. Lawrence lowlands. Geogr. phys,Quatern. **31**: 37-51, Montreal, Que.

EDWARDS, M.G. (1978): Glacial Environments, Ch. 132, In: [READING, H.G.]: Sedimentary Environments and Facies. - 557 pp, New York, Elsevier.

EMBLETON, C. & KING, C.A.M. (1968): Glacial and Periglacial Geomorphology. - 608 pp., New York, St. Martin's Press.

FETTER, C.W. jr. (1980): Applied Hydrogeology. - 488 pp, Columbus, OH, Merill Publ.

FISHER, W.L. & MCGOWEN, J.H. (1967): Depositional systems in the Wilcox Group of Texas and their relationsship to occurence of oil and gas. Trans. Gulf Coast Assoc. geol. Soc. 17: 105-125, Houston, TX.

FISHER, W.L. & BROWN, C.F. jr. (1972): Clastic Depositional Systems. - A Genetic Approach to Facies Analysis (annoated outline and bibliography). - Spec. Publ. Bur. econ. Geol., University of Texas, 211 pp., Austin.

FLINT, R.F. (1971): Glacial and Quaternary Geology. - 892 pp., N.Y., Wiley & Sons.

FOXWORTHY, B.L. (1978): Nassau County, Long Islands, New York, Water Problems in Humid Country. - U.S. geol. Surv. Prof. Paper 950: 55-68, Washington, DC.

FRYE, J.C. (1967): Geological Information For Managing the Environment. - Illinois State geol. Surv. environ. Geol. Notes 18: 1-12, Urbana, IL.

GARLAND, G.A. & MOSHER, D.C. (1975): Leachate effects from improper land disposal.- Waste Age 6: 42-48, Washington, DC.

GERHARDT, R.A. (1977): Leachate Attenuation in the Unsaturated Zone Beneath Three Landfills in Wisconsin. - Wisconsin Geol. and Natural History surv., Information Circ. 35, 93 pp., Madison, WI.

HACKETT, J.E. (1960): Groundwater Geology of Winnebago County. - Illinois State geol. Surv. Rep. Invest. 213: 1-63, Urbana, IL.

HACKETT, J.E. (1968): Geologic Factors in Community Development at Naperville, Illinois. - Illinois State geol. Surv. Environ. Geol. Notes 22: 1-15, Urbana, IL.

HACKETT, J.E. & MCCOMAS, M.R. (1969): Geology for Planning in McHenry County. - Illinois geol. Surv. Circ. 438: 1-29, Urbana, IL.

HAGEMAN, B.P. (1962): The Holocene of Voorne-Putten. - Meded. geol. Sticht., n.S. 15: 85-92, The Netherlands.

HAGEMAN, B.P. (1963): A new method of representation in mapping alluvial areas.- Verh. koninkl. Nederl. geol. mijnbouwk. Genoots., geol. Ser. 21-22, Sub. Conv. pt. 2: 211-219, The Netherlands.

HEATH, R.C.; FOXWORTHY, B.L. & COHEN, P. (1966): The Changing Pattern of Groundwater Development on Long Island, New York. - U.S. geol. Surv. Circ. 524, Washington, DC.

HORNBERG, C.L. (1953): Pleistocene Deposits Below the Wisconsin Drift in Northeastern Illinois. - Illinois State geol. Surv., Rep. Invest. **165**: 1-61, Urbana, IL.

HUGHES, G.M. (1972): Hydrogeologic Considerations in the Siting and Design of Landfills. - Illinois geol. Surv. environ. Geol. Notes **51**: 1-61, Urbana, IL.

HUGHES, G.M.; LANDON, R.A. & FARVOLDEN, R.N. (1969): Hydrogeologic Data From Four Landfills in Northeastern Illinois. - Illinois geol. Surv., Environ. Geol. Notes **26**: 1-43, Urbana, IL.

HUGHES, G.M.; LANDON, R.A. & FARVOLDEN, R.N. (1971): Hydrogeology of Solid Waste Disposal Sites in Northeastern Illinois. - U.S. environ. Prot. Agency, Solid Waste Managem. Ser. Rep. **SW-12d**: 1-145, Washington, DC.

HUGHES, G.M.; SCHLEICHER, J.A. & CARTWRIGHT, K. (1976): Supplement to the Final Report on the Hydrogeology of Solid Waste Disposal Sites in Norteastern Illinois. - Illinois State geol. Surv., Environ. Geol. Notes **80**: 1-24, Urbana, IL.

HUNT, C.B. (1974): Natural Regions of the United States and Canada. - 725 pp., San Francisco, CA, W.H. Freiman & Co.

JOHNSON, T.M. & CARTWRIGHT, H. (1980): The Monitoring of Leachate Migration in the Unsaturated Zone in the Vicinity of Sanitary Landfills. - Illinois geol. Surv. Circ. **514**: 1-82, Urbana, IL.

[JOPLING, A.V. & MCDONALD, B.C.] (1975): Glaciofluvial and Glaciolacustrine Sedimentation. - Soc. Econ. Pal. & Miner., spec. Publ. **23**: 1-320, Tulsa, OK.

KEMPTON, J.P. (1981): Three-Dimensional Geologic Mapping for Environmental Studies in Illinois.- Illinois State geol. Surv., envir. Geol. Notes **100**: 1-43, Champaign, IL.

KEMPTON, J.P.; RINGLER, R.W.; HEIGOLD, B.C.; CARTWRIGHT, K. & POOLE, V.L. (1981): Ground-Water Resources of Northern Vermillion County, Illinois. - Illinois State geol. Surv., environ. Geol. Notes **101**: 1-36, Champaign, IL.

KIMMEL, G. & BRAIDS, O. (1974): Leachate plumes in a highly permeable aquifer. - Ground Water **12**, 6: 388-392, Worthington, OH.

KRUMBEIN, W.C. (1933): Textural and lithological variations in glacial till. - Journ. Geol. **41**, 4:382-408, Univ. of Chicago Press, Chicago IL.

KNILL, J. (1970): Environmental geology. - Proc. geol. Assoc. London **81**: 529-537, London.

LEFLEUR, R.G. (1974): Glacial geology in rural land use planning and zoning. Publ. Geomorphol: 375-388, Binghampton, New York, NY, SUNY.

LANGER, W.H. & JOHNSON, L.H. (1978): Connecticut River Valley, Connecticut, East Granby - a plan of development for a rural community. - U.S. geol. Surv. prof. Paper **950**: 47-51, Washington, DC.

LEGGET, R.F. (1939): Geology and Engineering. - (1st Edition) 650 pp., New York, NY., McGraw Hill.

LEGGET, R.F. (1973): Cities and Geology. - 624 pp., New York, NY, McGraw Hill.

[LEGGET, R.F.] (1976): Glacial Till: An Interdisciplinary Study. - Royal Soc.Canada, spec. Publ. **12**: 1-412, Toronto, Canada.

LINDORFF, D.E. & CARTWRIGHT, K. (1977): Groundwater Contamination: Problems and Remedial Actions. - Illinois geol. Surv., environ. geol. Notes **81**: 1-58, Urbana, IL.

LOBECK, A.K. (1948): Physiographic Map of North America, scale: 1:12,000,000. - Ma plewood, NJ, The Geographical Press.

LÜTTIG, G.W. (1989): Quaternary deposits: suppliers of mineral raw materials and pre requisites for human development. In: [MULDER, E.F.J. DE & HAGEMAN, B.P.]: Applied Quaternary Research. - 83-104, Rotterdam, A.A. Balkema.

MCOMAS, M.R.; HINKLEY, K.C. & KEMPTON, J.P. (1969): Coordinated Mapping of Geology and Soils for Land-Use Planning. - Illinois State geol. Surv., environ. Geology Notes **29**: 1-11, Urbana, IL.

MELVIN, R.L. (1978): Applications in a New England environment: Connecticut River valley, Connecticut, Use of drainage-area maps in evaluating waste-disposal conditions, assessing the impact of highway salting, and designing bridges and culverts. - U.S. geol. Surv. prof. Papers **950**: 52-54, Washington, DC.

MILLER, D. W.; DELUCA, F.A. & TESSIER, T. (1974): Ground-water Contamination in the northeast states.- U.S. environ. Protection Agency Technology Series, 325 pp., Washington, DC.

[MULDER, E.F.J. DE & HAGEMAN, B.P.] (1989): Applied Quaternary Research. - 185 pp., Rotterdam, Netherlands, A.A. Balkema.

NORRIS, S.E. (1963): Permeability of Glacial Till. - U.S. geol. Surv. prof. Paper **450-E**: 150-151, Washington, DC.

NORRIS, S.E. & SPIEKER, A.M. (1966): Ground-water Resources of the Dayton Area, Ohio. - U.S. geol. Surv., Water Supply Paper **1808**: 1-167., Washington, DC.

PARIZEK, R.R. (1971): Impact of highways on the hydrogeologic environment. - In: [COATES, D.R.]: Environmental Geomorphology: 151-199, publ. in Geomorphology, State Univ. of New York, Binghampton, NY.

PETTIJOHN, R.A. (1977): Nature and extent of ground-water quality changes resulting from solid waste disposal, Marion County, Indiana.- U.S. geol. Surv., Water-Res. Invest. 77-40: 1-119, Washington, DC.

PLUHOWSKI, E.J. & KANTROWITZ, I.H. (1964): Hydrology of the Babylon-Islip Area, Suffolk County, Long Island, New York. - U.S. geol. Surv. Water Supply Paper 1768: 1-119, Washington, DC.

PLUMMER, P.M. (1973): An Investigation of the Effects of Waste Disposal Practices on Groundwater Quality of Montgomery County, Ohio. - Miami Conservancy District Report, 68 pp. and appendices, Miami, OH.

POOLE, V.L. & HEIGOLD, P.C. (1981): Geophysical Assessment of Aquifers Supplying Ground Water to Eight Small Communities in Illinois. - Illinois State geol. Surv., environ. Geol. Note 91: 1-61, Champaign, IL.

PREST, V.K. & HODE KEYSER, J. (1977): Geology and Engineering Characteristics of Surficial Deposits, Montreal Island and Vicinity, Quebec.- Geol. Surv. Canada Paper 75-27: 1-29, 2 maps, charts, Ottawa, Ont.

[ROBINSON, G.D. & SPIEKER, A.M.] (1978): Nature to be Commanded ... Earth-science Maps Applied to Land and Water Management. - U.S. geol. Surv. prof. Paper 950: 1-85, Washington, DC.

SCHICHT, R.J.; ADAMS, J.R. & STALL, J.B. (1976): Water Resources Availability, Quality and Cost in Northeastern Illinois.- Illinois State Water Surv. Rep. Invest. 83: 1-90, Champaign, IL.

SINGH, K.P. (1978): Water Supply Alternatives for the City of Danville, prepared for the Division of Water Resources. - 124 pp., Champaign, IL, Illinois Depart. Transport.

SOLIS, I. & RAUL, F. (1981): Upper Tertiary and Quaternary Depositional Systems, Central Coastal Plain, Texas -- Regional Geology of the Coastal Aquifer and Potential Liquid-Waste Repositories. - 89 pp., Bureau econ. geol. Univ. of Texas, Rep. Invest., Austin, TX.

STEPHENSON, D.A. (1967): Hydrogeology of Glacial Deposits of the Mahomet Bedrock Valley in East-Central Illinois. - Illinois State geol. Surv., Circ. 409: 1-51, Champaign, IL.

STOLLER, J.H. (1920): Galcial Geology of the Cohoes Quadrangle. N.Y. State Mus. Bull. Nos. 215/216: 1-48, Albany, NY.

[TEXAS BUREAU OF ECONOMIC GEOLOGY] (1988): Annual Report. - 72 pp., Austin, TX.

THORNHURY, W.D. (1967): Regional Geology of the United States. - 609 pp., New York, NY, Wiley.

[U.S. GEOLOGICAL SURVEY] (1967): Engineering Geology of the North-East corridor. Washington, DC, to Boston, Massachusetts: Bedrock Geology. - U.S. geol. Surv. kiscell. Geol. Invest. Map 1-514-A, Washington, DC.

[WALKER, R.G.] (1984): Facies Models. - 317 pp., Newfoundland, Geol.Ass. of Canada.

[WERMUND, E.G.] (1974): Approaches to Environmental Geology.- 268 pp., Austin, TX, Bureau of Economic Geology.

WHITE, O.W. (1888): Quaternary geology and urban planning in Canada. - In: [MULDER, E.F.J. DE & HAGEMAN, B.P.]: Applied Quaternary Research: 165-175, Rotterdam, A.A. Balkema.

WINTER, T.C. (1973): Hydrogeology of Glacial Drift, Mesabi Iron Range, Northeastern Minnesota. - U.S. geol. Surv. Water Supply Paper 2029A: 1-23, Washington, DC.

WIRENIUS, J.D. & SLOAN, S.L. (1973): Determining the life of a landfill site. Public Works 104: 118-119, Ridgewood, NY.

YOUNG, H. (1976): Digital Computer Model of the Sandstone Aquifer in Southeastern Wisconsin. - Southeastern Wisconsin reg. Plann. Comm., techn. Rep. 16: 1-42, Waukeska, WI.

ZEONONE, C.; DONALDSON, D.E. & GRUNWALDT, J.J. (1975): Ground-water quality beneath solid-waste disposal sites at Anchorage, Alaska. - Ground Water 13, 2: 182-190, Worthington, OH.

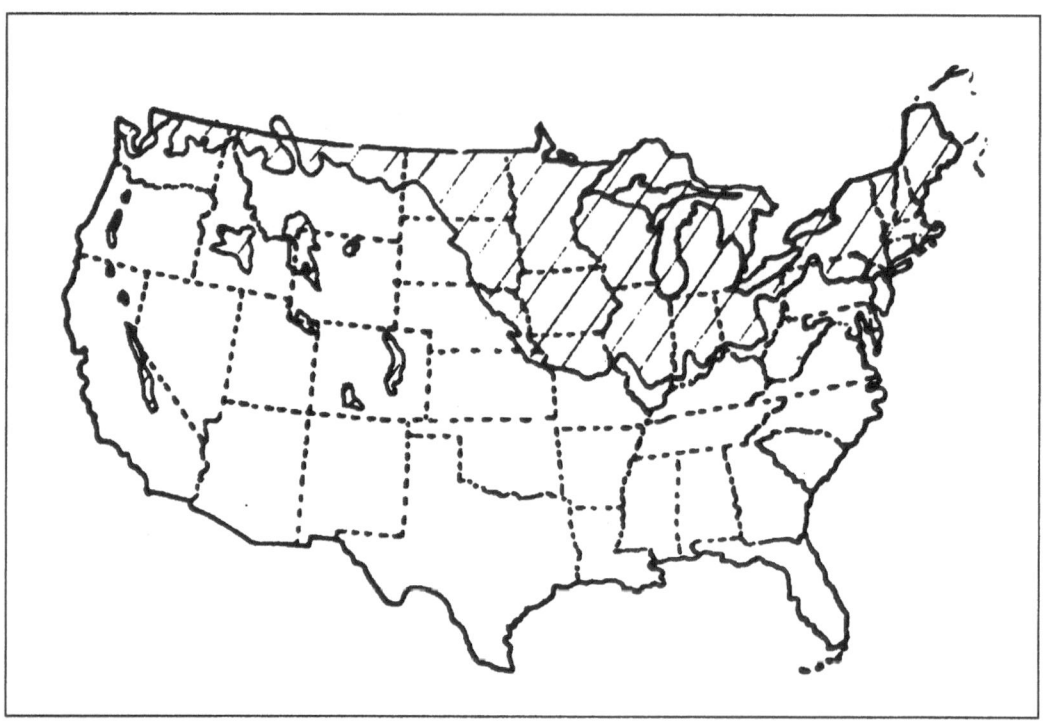

Figure 1: Extent of Pleistocene glaciation in the United States. In the glaciated Appalachian Mountains, maximum ice-advance reached into northern Pennsylvania.

Figure 2: Surface morphology of the northeastern United States and southeastern Canada. The glaciated Appalachians extend northeast from the heavy dashed line.

(after LOBECK, 1948)

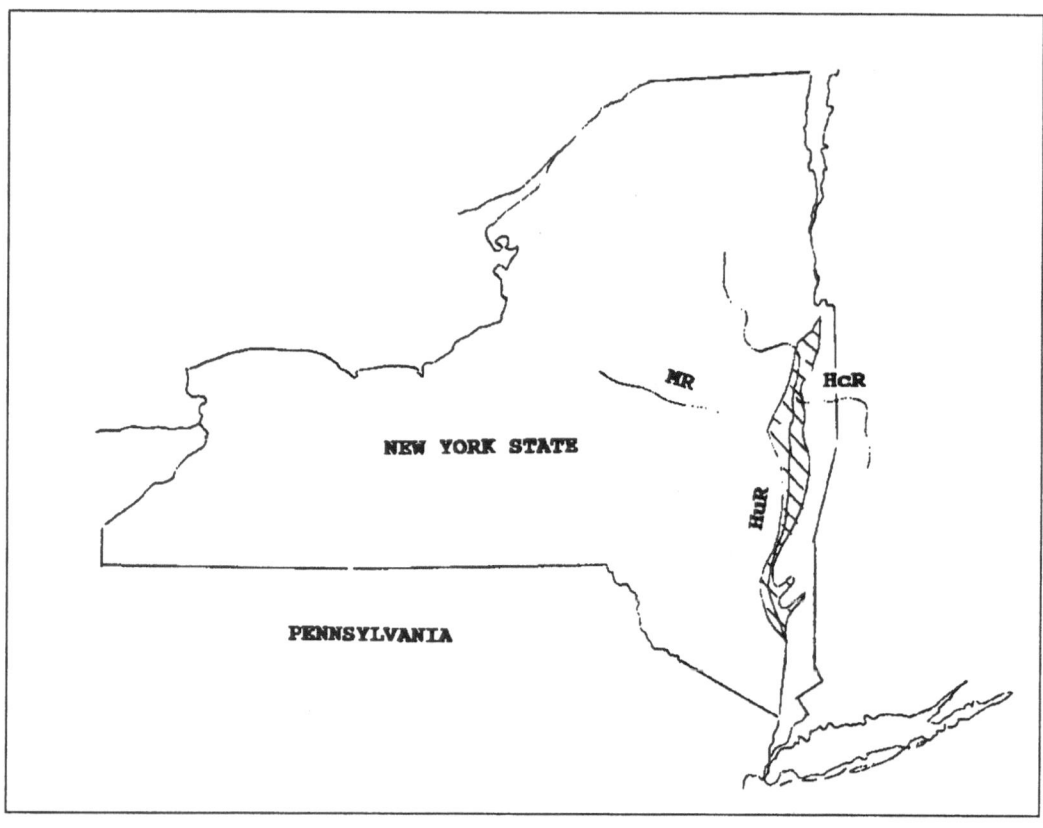

Figure 3: New York State, showing Hudson River (HuR), Mohawk River (MR), Hoosic River (HcR) and Late Pleistocene Lake Albany (diagonal line pattern).

The Capital District urban complex is at the confluence of the Mohawk and Hudson Rivers. The towns of Schaghticoke and Pittstown are at the confluence of the Hudson and Hoosic Rivers.

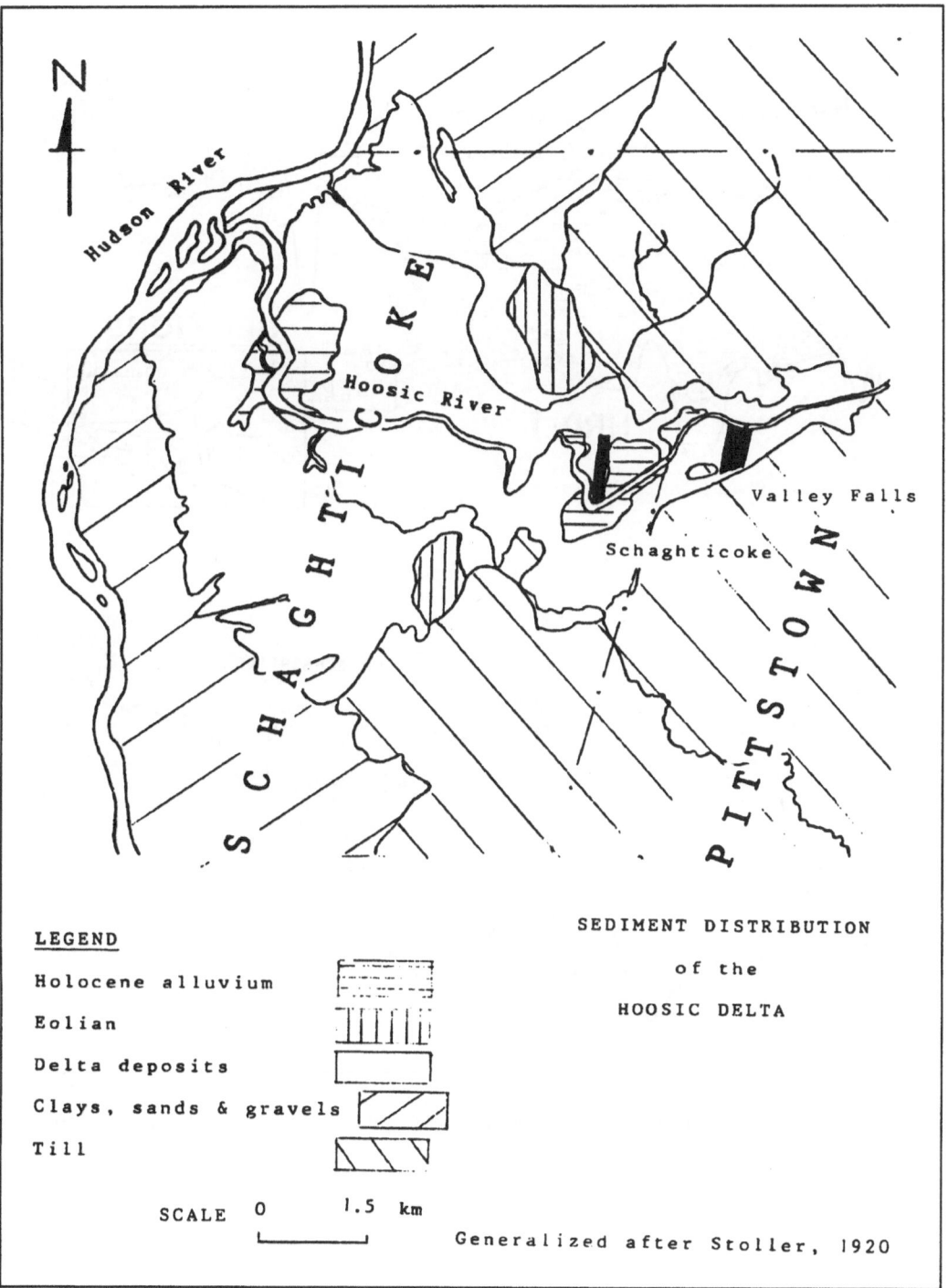

Figure 4: Surficial geology of the Pleistocene Hoosic Delta which underlies northern Schaghticoke and northwestern Pittstown Townships (after STOLLER, 1920).

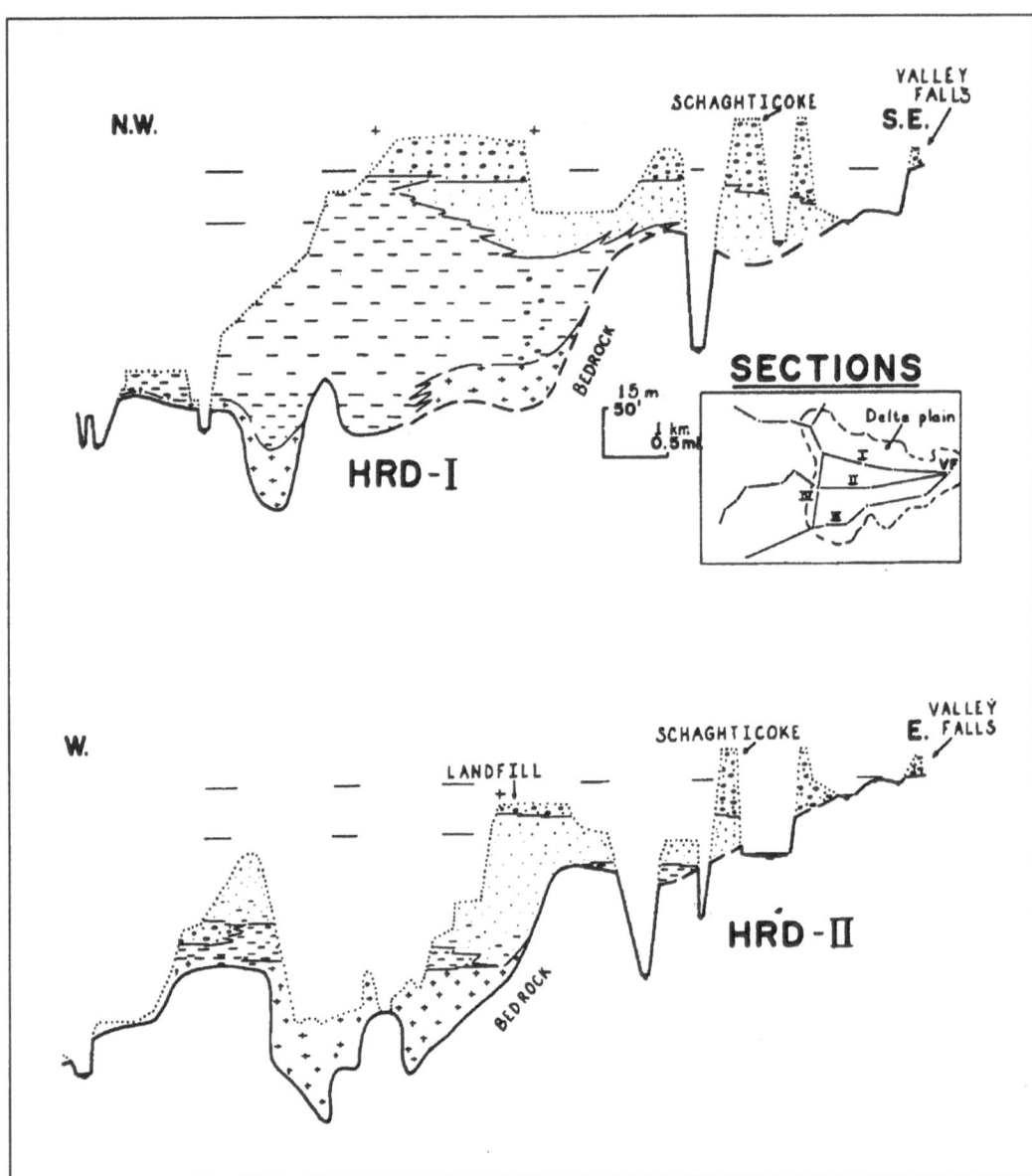

Figure 5:   Proximal-distal stratigraphic cross-sections of the Hoosic Delta System.

Showing sub-surface facies distribution in the northern (HRD-I) and central (HRD-II) parts of the system. Ice-contact facies is indicated by plus-mark pattern; lacustrine and distal prodelta facies by horizontal dashed lines; proximal prodelta facies by stippling; and delta plain facies by stippling/ ellipsoid pattern.

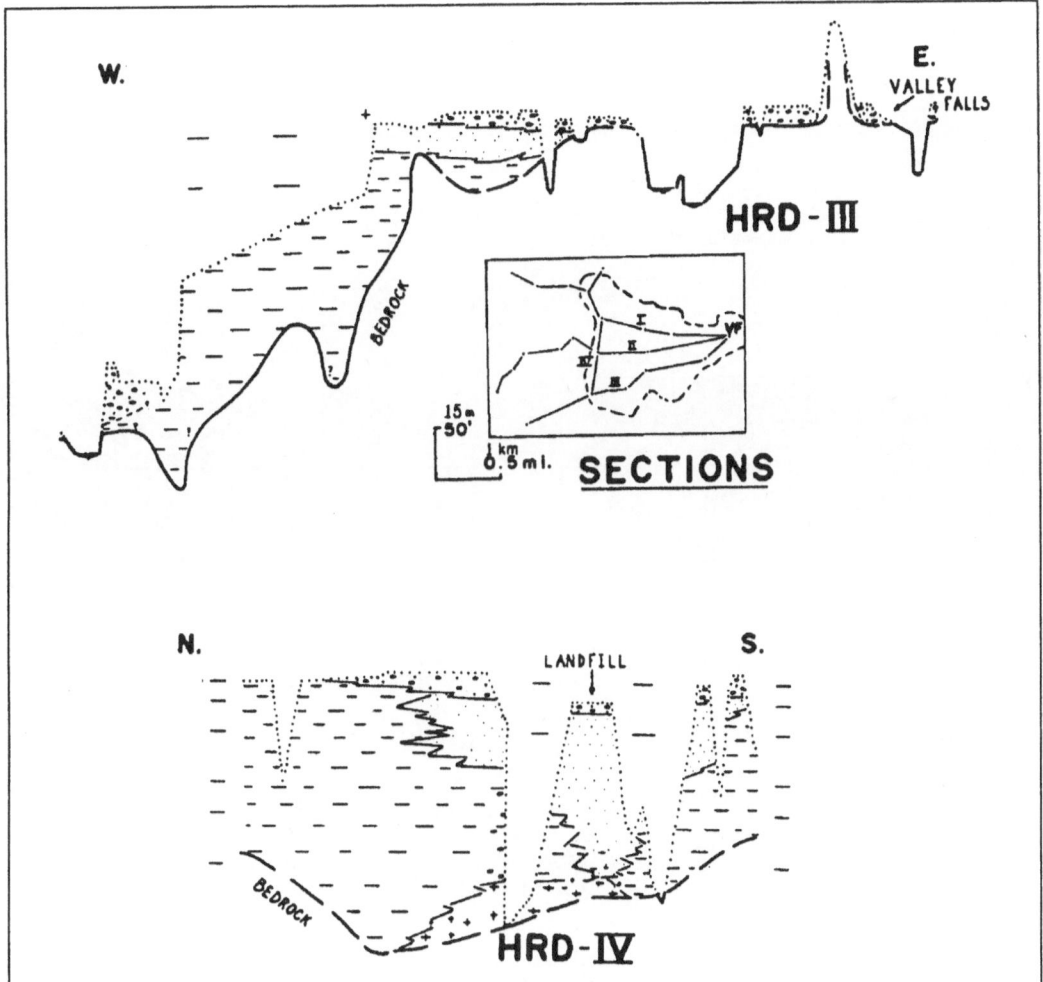

Figure 6: Proximal-distal (HRD-III) and transverse (HRD-IV) stratigraphic cross-sections of the Hoosic Delta System, showing sub-surface facies distribution.

Ice-contact facies is indicated by plus-mark pattern; lacustrine and distal prodelta facies by horizontal dashed lines; proximal prodelta facies by stippling; and delta plain facies by stippling /ellipsoid pattern.

Geoenvironmental Units as a Basis for the Assessment, Regulation and Management of the Earth's Surface

A. CENDRERO, E. FRANCIS & J.R. DÍAZ DE TERÁN *

Abstract

A series of case studies is presented showing how units defined on the basis of geological-geomorphological criteria can be used for summarizing information about a great variety of environmental parameters and for the assessment of many qualities of interest for planning and decision-making. These qualities refer to both geological and non-geological aspects. The utilization of geoenvironmental units for the establishment of land-use regulations, for planning management actions and for evaluation of some economic aspects is also illustrated.

Introduction

It is a well known fact that in the last couple of decades, especially in the last few years, environmental concern has increased worldwide very considerably (WORLD COMMISSION ON ENVIRONMENT AND DEVELOPMENT 1987). The growth in

* Authors' address: Prof. Dr. ANTONIO CENDRERO, Division of Earth Sciences, DCITTYM, University of Cantabria, Santander, Spain.

human population and the development of a more powerful and sophisticated technology, have brought about a much greater degree of intervention in the natural systems, modifying their natural characteristics and dynamics. These modifications have resulted in many instances in a serious level of deterioration, so much so that there is good reason to worry about the future of certain ecosystems of world importance (Mediterranean, Amazonia) or even of some general planetary characteristics (climate and global change).

One of the main reasons for this state of affairs is the careless and ill-planned use of the earth's surface. As more and more space is needed by human beings, the use of marginal areas, with unfavourable conditions, becomes more frequent. Valuable and fragile natural systems are thus affected and deteriorated (coastal areas, mountains, tropical forests), zones subject to natural hazards are occupied (floodplains, unstable slopes), and the impacts on the environment and the risks for the human population become greater. The only way to correct this situation and to prevent future damage, both for human activities and for the environment, is to rationalize the use of the earth's surface, through careful planning and management.

The surface of the earth is the compendium of the human environment, with its resources, potentialities and limitations. Each portion or unit of this surface is underlain by certain types of rocks and unconsolidated surficial deposits, many of which are valuable raw or construction materials; it also has some specific landforms, which might favour some human activities and hinder others, surface and groundwater, soil that can support agriculture. The climate is also different from the one in other areas and this, together with the internal dynamics of the planet, determines the processes which are active in each zone. As a result of the former, abiotic characteristics, certain plant and animal assemblages exist. Finally, human communities and their activities are established in response to the former features. The earth surface in itself is also a very valuable resource, as it provides the space in which to locate the different activities.

Therefore, the earth's surface, our environment, can be conceived as the result of the superposition of three groups of factors, abiotic, biotic and human (CENDRERO

1980). The abiotic features (bedrock and surficial materials, landforms, distribution of continents and oceans, climatic character) determine, on the whole, the distribution of plant and animal communities and are more permanent than these. Both, in turn, have influenced throughout history the distribution and types of activities of the various human groups. Of course, there are also mutual interactions and interdependence, so that apart from the "mainstream dependence" indicated above, there are other less marked relationships between the three groups of factors (Fig.1).

It follows that abiotic features such as surface and subsurface materials, landforms and surficial processes, etc. constitute the framework, the most permanent foundation of the environment. If this is so, it should be possible to describe the environment of a region -- and to understand the way it functions -- through the identification and representation of the abiotic units that make it up and of the processes which operate in them, together with the representation and description of their biotic and human characteristics and processes. The advantage of this approach is that abiotic features are normally not subject to great changes within time spans comparable to human life -- or even several generations -- and can be identified and represented objectively. Thus, they constitute a "permanent" basis for the representation, evaluation, planning and management of the environment.

The study, identification, description and interpretation of the materials, landforms and processes on the earth's surface, their origin, evolution and present condition, makes it possible to define a series of "geoenvironmental units" and to represent them in map form. Using these units as a basis, other environmental features such as soils, vegetation, fauna, land use and land occupation, etc. can be incorporated, thus transforming the former units into "integrated, homogeneous environmental units". Each one of these units has similar environmental features in any point and, therefore, should have a similar response to human actions; that is, a certain action should produce the same impacts in any part of a given homogeneous unit and all points of the unit should have the same level of risk with respect to natural processes. Also, as geoenvironmental units summarize the characteristics, potentialities and limitations of the territory, it is possible to use them as a basis for the assessment of a great variety of qualities -- some directly related to geoscientific aspects and others only vaguely connected with them -- significant for planning and management. This assessment, in turn, can provide the grounds for decision-making, including the establishment of legal norms, the design of master plans and of management actions or the evaluation of costs of such actions.

This presentation will illustrate the concepts briefly explained above, presenting a series of examples showing how geoenvironmental units can be applied to the assessment of hazards, risks, impacts, quality for conservation (including visual landscape evaluation), etc., to the setting-up of environmental education activities, to the establishment of land-use regulations, or to the design of management plans and the economic evaluation of conservation and management actions.

Needs regarding environmental conservation and management

The conservation and careful use of the environment and its resources require:

a) scientific knowledge of the different elements or components that make up the environment in each part of a region or a country,

b) an understanding of the natural processes which constitute the dynamics of the different systems,

c) the comprehension and appreciation, by decision-makers and general public, of the value of the various environmental components/processes,

d) the establishment of legal norms to regulate the use and protection of the environment,

e) the technological capabilities and the financial resources that make intervention in the environment possible,

f) an institutional organization that enables the control of human activities affecting the environment.

The concept of "integrated geoenvironmental units, as I will attempt to show with the examples below, is very useful to summarize the knowledge referred to in points a) and b), to serve as a means of communication for the purpose indicated in c), to facilitate the establishment of the norms mentioned in d), and also application of the technical and institutional instruments included in e) and f).

Case studies
Assessment of quality for conservation in Valencia

One of the aims of a general study carried out in the province of Valencia (about 10,000 km and 1,700,000 inhabitants) for environmental planning was the classification of the territory according to its quality for conservation, in order to define where future developments could have a greater impact on the natural environment.

To do this, a hierarchy of terrain subdivisions was defined and mapped at the 1:200,000 scale (CENDRERO et al., 1986, 1990). These subdivisions were called: "morphodynamic environments" (defined on the basis of climate and large-scale morphostructural features), "morphodynamic systems" (established according to lithology and geomorphology) and "morphodynamic units (defined on the basis of soils, active processes, vegetation and fauna). Thus, each "morphodynamic unit", although defined essentially using geological and geomorphological criteria, was homogeneous from the point of view of bedrock, landform, climate, soil, active processes and biologic assemblages.

To define categories of quality for conservation and to establish a rank among the units, the following features or "elements" of the environment were considered: geological-geomorphological factors, vegetation, fauna, aquifer vulnerability, visual landscape quality and fragility. For each one of those "elements", the different "types" (units or categories) existing in the area were ranked on a scale of five terms, according to their interest for conservation. The criteria used to establish the rank were: scarcity, scientific or educational value, diversity and abundance of species, accessibility, etc. In the case of visual landscape a different procedure was followed.

Landscape was considered as the superposition of a series of "descriptive parameters", the integration of which results in a certain preception by viewers. These parameters were: type of relief and topographic complexity, difference in altitude, land-use and vegetation, presence of water bodies or rivers, presence of human structures or modifications, accessibility, visual incidence. The different existing situations for each one of these parameters were identified and ranked in a hierarchy of "quality" and an other of "fragility" with regards to visual perception. Some of the parameters mentioned (accessibility, visual incidence) do not have an influence on the quality of the landscape, but only on its vulnerability as a consequence of human intervention (fragility).

Weights between 3 and 1 were assigned to the parameters just indicated and the following indices were obtained:

$$lq = \frac{\sum Pqi \times Vqij}{\sum Pqi}; \quad lf = \frac{\sum Pfi \times Vfij}{\sum Pfi}; \quad LV = \frac{2\ lq + lf}{3}$$

where there are: lq = quality index,

Pqi = weight of parameter i for quality,

Vqij = value of the type j of parameter i for quality,

If = fragility index,

Pfi = weight of parameter i for fragility,

Vfij = value of type j of parameter i for fragility, and

LV = landscape value index.

The individual values of LV thus obtained for each morphodynamic unit were then grouped into five classes, with values 1 to 5, using a statistical procedure.

To obtain the class of value for conservation of each morphodynamic unit a similar approach was used. A weight was assigned to each one of the five elements indicated above (geology-geomorphology, vegetation, fauna, aquifer vulnerability and visual landscape), and the overall value for conservation was derived using the expression:

$$VC = \Sigma \, p_i * v_i$$

where there are: VC = value for conservation of the morphodynamic unit,

p_i = weight of element i,

v_i = value of the type of element i in the unit.

With the method described, units defined on the basis of geological-geomorphological criteria can be used to integrate the rest of the information about the environment and to establish a system for the assessment of qualities such as landscape value or value for conservation. These qualities are significant for decision-making in environmental planning and management, and yet difficult to evaluate, because such an evaluation has a certain element of subjectivity. This procedure enables, once a series of criteria have been defined and accepted, to obtain similar results by any operator carrying out the evaluation.

It is interesting to point out that with the method described, all the areas which had previously been declared of interest for conservation by several government and private organizations -- some of them legally protected already -- happened to be made up of units with very high or high value for conservation. Apart from these, other units not previously identified as valuable, also appeared in the same high classes of the rank, because they had similar constituent features.

Preparation of a general master plan in the Region of Cantabria, Spain

In the drawing-up of the general master plan of the Autonomous Region of Cantabria (about 5,000 km and 530,000 inhabitants), integrated environmental units were defined, at the beginning of the process, according to those features considered to be most relevant from the point of view of environmental protection and natural resource use and conservation (FRANCÉS et al., 1990 a). The criteria used for the definition of units were:

a) Landscape. Corresponded to large units of high visual landscape quality and/or fragility, or to small areas of special visual interest.

b) Agricultural productivity. Determined by the type of soil and the kind of use it is subject to.

c) Ecological character. Defined on the basis of proximity to the climax, diversity and rarity of plant and animal species.

d) Scientific-cultural interest. Presence of sites or points with especially valuable natural or historical-cultural elements.

e) Resource use. Possibility to improve the exploitation of certain resources or to restore degraded areas.

The integrated or synthesis units obtained (Fig. 2a) included, among others: morphological units of high visual landscape value; units of high soil quality and productivity; units of medium soil quality and productivity; well preserved fluvial complexes; wetlands; estuaries; beach-dune complexes; autochthonous climax woods; units with vegetation of special interest; habitats of vertebrate species in danger of extinction; degraded ecosystems; units with under-exploited resources. Although the basis for diagnosis was, in each case, one of the factors indicated above, units were described with regards to all other environmental factors.

The limitations affecting the use of the territory were considered and mapped separately. These limitations were:

a)   aquifer recharge zones vulnerable to pollution,

b)   areas of slope instability hazard,

c)   areas vulnerable to erosion,

d)   flood hazard zones,

e)   coastal hazard zones.

The units thus defined and described were the basis to propose general legal regulations regarding the use of land and the siting of activities in the region. Each diagnosis unit, considered as a "planning category", was evaluated from the point of view of ecological value, productivity value, landscape value and scientific-cultural value, using a qualitative rank. A global value was obtained as well, by simple aggregation of the four aspects indicated.

The "carrying capacity" of diagnosis units for different activities was also assessed. This was done using a matrix (Table 1) in which each unit was confronted with the activities considered, a total of 51, grouped in the following categories: a) nature conservation and restoration, b) recreation and outdoor sports, c) agrarian activities, d) urban, e) industrial, f) infrastructures, g) waste disposal. For each unit it was established whether the activity was: optimum for the unit and coincident with present use; optimum for the unit but not coinciding with present use; compatible with the unit; compatible but subject to special permit; compatible after a possitive EIA has been accepted; non compatible.

Finally, proposals were made for the establishment of legal norms regulating land-use. These included the activities that should be forbidden or limited, on the basis of the impact they could produce or the existing hazards, the activities that should be promoted, because they represent a rational use of natural environmental potential and resources, and the actions to undertake for improving the quality of the environment. Figure 2b shows non-compatible, compatible and vocational or optimum activities for the area represented in Figure 2a.

Planning and management of natural park areas in Northern Spain

Planning of natural parks usually includes the establishment of levels of protection for different zones of the area considered, the setting-up of land-use regulations and the definition of specific management actions to be undertaken for the conservation, improvement or use of the environment. The identification and mapping of integrated environmental units based on geological and geomorphological features can facilitate those tasks.

In the Natural Park of Oyambre (Cantabria, Northern Spain) a detailed map (at the 1:10,000 scale) of integrated, morphodynamic units, defined using a method very similar to the one described above, was made (FRANCÉS et al. 1990 b). The environmental characteristics of these units, including bedrock and surficial deposits, landform, soils, active processes, vegetation, fauna, landscape, and human activities, were described and a diagnosis of the most relevant features from the point of view of conservation and management was made for each one of them (Fig. 3a).

Integrated morphodynamic units were transformed into "diagnosis categories". These "categories" served as the basis to define the legal norms that should regulate the use of the area under consideration, and also to establish a series of actions that should be part of a management plan for the park. For each unit represented on the map, the forbidden activities, acceptable activities and activities to promote were indicated. Details were also given about management actions such as substitution of vegetation, re-landscaping, protection of valuable sites, establishment of visitors' areas, subsidizing of traditional agriculture, etc. (Fig. 3b). As the different actions were referred to units clearly defined and represented in the maps, it was relatively simple to make an evaluation of their costs, as well as the cost of the whole management plan.

A very similar approach was adopted in the planning of a future natural park in the Jaizkibel massif, by the Spanish-French border, in the Basque Country (FRANCÉS et al.

1990c). In this case the basic aim was to create, close to a fairly important urban-industrial agglomeration, a public space for conservation and recreation, although maintaining, if possible, traditional agricultural activities.

A map of integrated environmental units (Fig. 4a) was prepared, again according to a sequential subdivision of the area. The first subdivision corresponded to a series of "visual incidence" units; that is, units defined on the basis of morphological features, which determine the visibility of actions carried out on them and, therefore, the visual impact of those actions. The next subdivisions were morphodynamic systems, units and elements, as described above. For each integrated unit thus obtained information was provided on the following items: surface, slope, altitude, orientation, accessibility, property status, present use, degree of human occupation, bedrock, surficial deposits, landform, active processes, soil, limitations for construction, surficial water, water quality, aquifer vulnerability, vegetation, fauna, visual landscape, microclimatic conditions, sites of scientific or cultural interest.

Landform, vegetation, fauna, sites of scientific or cultural interest and visual landscape were the parameters selected as relevant for establishing a rank of quality for conservation in the massif.

The zoning of the area and the definition of management actions to be implemented in each integrated unit were carried out using the following criteria:

1. All units containing at least one type of any parameter with the highest value of quality for conservation should be devoted to conservation, first order.

2. All sites and points of scientific or cultural interest were assigned to the same category.

3. Second order conservation was proposed for units covered by *Pinus pinaster* and for units formed by small steep valleys adjacent to first order conservation units.

4. Protective reforestation was proposed for units subject to erosion or to surficial landslides.

5. In some units, of high slope instability, it was recommended to take no specific action and to avoid the siting of any activity. In a few cases the need for protective works was indicated.

6. Reforestation with autochthonous species was recommended for units surrounding existing forested areas, in order to increase their surface. It was also recommended in some units of interest for conservation but with bush vegetation.

7. In the case of units with high visual incidence and strong visual impacts (mainly forest tracks) reforestation with species of rapid growth was recommended.

8. In units with good soils, presently devoted to cultivation, traditional agricultural activities should be maintained.

9. Intensive recreation areas should be established on units with the following conditions: presence of some services and facilities constituting a nucleus which will facilitate implantation; easy road access; wide surfaces with gentle slopes; good panoramic views and/or good microclimatic conditions; far from units of high quality for conservation, which could be subject to undesirable impacts.

10. Recreation without installations was recommended for other units with gentle slopes, adjacent to the coast or near conservation areas but without road access.

As a result of the applications of those criteria, the map of integrated environmental units, defined essentially according to geomorphological characteristics, was transformed into a map of land-use recommendations and management actions (Fig. 4b). The first correspond to activities which could be accepted in the massif with a minimum environmental impact; the second indicate the actions to be carried out to improve the environmental quality of the zone.

Hazard and impact assessment for urban planning

When master plans regulating construction and other activities in a town have to be drawn, it is important that the regulations affecting each individual property are made clear and can be applied without ambiguity. Integrated environmental units or morphodynamic units, although defined on the basis of objective features, such as type of materials, landform, soil, presence of hazardous processes, etc., are not always clearly visible in the terrain by non-specialists. These units are very useful to describe the area subject to planning and to make a diagnosis about its qualities and limitations, but not for establishing legal regulations for very detailed plans, which can be directly applied on cadastral maps, because the limits between natural terrain units do not normally coincide with clearly defined landmarks.

In the elaboration of the master plan of the coastal town of Suances, near Santander, detailed maps of morphodynamic units, at the 1:5,000 scale, were prepared (CENDRERO 1989). The natural hazards affecting each unit and the impacts that could be produced by different activities were assessed. Morphodynamic units were then classified according to the categories contemplated within the usual rural and urban land-use regulations in Spain. This was done identifying in each unit the most limiting factor from the point of view of land-use.

The categories established were: 1) areas where high density building is allowed, 2) areas where low density building could be allowed but only after the elaboration of a detailed special plan, 3) areas where no building is allowed, 4) areas of ecological and/or landscape protection, 5) areas with building restrictions due to natural hazards, 6) areas reserved for agriculture, 7) areas with building restrictions due to the protection of aquifers, 8) areas with building restrictions due to high visual incidence, but where building could be allowed after the preparation of a detailed project.

The great variety of morphodynamic units was thus reduced to eight "planning categories". In order to make the use of these categories simple and straight forward from the point of view of the application of municipal planning regulation, the natural boundaries of the units were transformed and adapted to existing roads, tracks and property limits (Fig. 5). The general criterion followed in this transformation was to increase the area of protection of units vulnerable to impacts or the area of limitation in the case of hazards. These planning categories with the new boundaries constituted the spatial reference for the promulgation of building and land-use norms.

Hazard and risk evaluation

Morphodynamic units are particularly suitable for the assessment of natural hazards, as most of them are of a geodynamic character. By their very definition, morphodynamic units are homogeneous from the point of view to many environmental factors, among them enables the delimitation of zones subject to various processes (flooding, landslides, coastal erosion, volcanism, etc.) with different levels of intensity and/or periodicity. Of course, the precision in the determination of periodicity varies very much for different processes and regions, depending on the frequency of the event and on the quality of the historical and geological records.

Thus, morphodynamic units can easily be transformed into hazard units. Such units are usually qualitative, and are expressed as a rank with regards to each specific hazard (high, medium or low flood hazard; classes 1 to 5 of landslide hazard; etc.). However, if the available data are sufficiently precise, hazard units can be more quantitative; for instance, flood hazard level can be expressed as: areas subject to floods every 1-5 years; areas flooded with a periodicity between 5 and 30 years; areas flooded every 50-100 years. In the case of landslides, the quantification can be of the form: units which will be affected by landslides in more than 20% of their surface in a period of 30 years; units affected in 10-20 % of the surface; units affected in less than 10% of their surface.

In the mapping and assessment of natural hazards carried out in several munincipalities in the Basque Country at the 1:5,000 scale (CENDRERO et al. 1987), several types of hazards units were defined (Table 2). Units of flood hazard were established combining the observation of fluvial geomorphology and historical records of floods. Slope instability hazard units were defined by multiple correlation between existing landslide deposits and landforms and factors determining slope instability, such as bedrock, structure, topographic slope, type and thickness of the cover material, vegetation. The combination of macro-, meso-, and micro-orientation with frequency and intensity of the prevailing winds was used to map wind hazard units. Coastal hazard units were defined through the combination of coastal geomorphology and existing records of damages. Most of these units were expressed as a qualitative rank, although in some cases it was possible to indicate the probable periodicity. A synthesis map of hazards plus vulnerability was also made (Table 2).

A more recent analysis of some of those hazard maps (DUQUE et al. 1990), specifically referred to landslide hazard units, has indicated that the systematic observation and mapping of the landslides produced during a sufficiently long (for instance, 30 years) and well-recorded period, through the use of air photographs, could be used to transform the qualitative rank of "landslide hazard classes" into quantitative or semiquantitative units, expressed as percentage of each hazard unit which would be affected by landslides within a period of 30 years, if the existing conditions remain unchanged.

Hazard units, defined on the basis of geomorphological criteria, can also be used as a reference to express the value of buildings, infrastructures, utilities etc., using data from cadastral maps and other sources. So, the total value of human property and structures in a morphodynamic or a hazard unit can be known and relected on the map. This, in turn, can be used for the assessment of risks, expressed in monetary terms.

For the assessment of risks, it is possible to use an expression of the type:

$$Ir = V * v/P$$

where there are: I_r = risk index for the unit ($/year),

V = value of structures and properties in the unit,

v = vulnerability (% of the value which will be damaged in one event, such as a flood, a storm etc.),

P = average periodicity between dangerous events (years).

Vulnerability can be estimated on the basis of past damages produced in the zone by similar events, or by comparison with damages produced in other areas by similar processes. Normally this parameter can be estimated only approximately. The determination of periodicity can be very precise for frequent events, and only roughly approximate for events happening with intervals of decades or centuries.

In the case of landslides, the factor v/P can be substituted by "% of the value which will be destroyed by landslides in the unit within the next 30 years".

In the end, the use of geoenvironmental units and the application of expressions like the one proposed above, make it possible to assess natural risks in monetary terms, and to prepare risk maps depicting zones or units with different levels of losses to be expected, irrespective of the type of hazard affecting them. This type of map is obviously very useful for the establishment of legal norms or for making decisions regarding public investments for risk prevention or correction.

In other cases, risk assessment can be much simpler and straight forward, because it might refer to single events that could happen in the future and for which a certain strategy must be prepared. This is, for instance, the case of sea-level rise.

The definition of hazard and risk units becomes easier here. Lowland areas that could be affected by a rise in sea level can be identified for different scenarios, basically on the basis of their elevation with respect to present level. Zones that could be affected by coastal erosion can be defined on the basis of lithology, structure, sediment type, morphology and present evolutionary state. Again, cadastral and other data can be used to estimate the value of properties and structures in the different zones. As in this case there is no need to define periodicity but to provide an answer to questions such as: if sea level rises 1 m, what will be the capital value at risk in different parts of the coastal areas?

This kind of approach has been followed for the assessment of risks associated to sea-level rise along the eastern part of the Bay of Biscay (RIVAS & CENDRERO 1990). Low-lying areas in this coastal area correspond mainly to reclaimed lands. They were identified and mapped according to position in relation to present sea level, kind of process used for reclamation (filling, draining, isolation) and present land-use; that is, homogeneous units were defined according to "geomorphological" factors which are due essentially to human processes.

In each type of area the surface potentially subject to the effects of sea-level rise (Fig. 6) was measured and the value of land as well as that of existing permanent structures was calculated. The area under risk covers 73 km and the total capital value in amounts to US$ 5,000 million. The impact of sea-level rise on the biological productivity of these areas was also calculated. A comparison between the present productivity of reclaimed agricultural land devoted to agriculture (about 60 %) and the productivity of natural intertidal and wetland areas, both of them expressed in terms of the market value of produce obtained by the primary sector, showed that the latter produce, on the average, about US$ 266,000/year more than the former. In this respect the impact of sea-level rise appears to be positive, although this factor is indeed small compared with the losses that could be expected in built-up areas. Nevertheless, the evaluation described shows that the advisable strategy to cope with the effects of an eventual rise in sea level should only contemplate, whereas it should "let nature follow its course" on zones devoted mainly to agrarian uses.

Assessment and zoning of a Biosphere Reserve in the Andes

The concept of morphodynamic units or integrated environmental units can be very useful in the assessment of territories for which there is limited information and where detailed surveys are not possible within a reasonable period of time. This has been the case in the study of a lacustrine basin in the High Central Andes, the Laguna de Pozuelos, at the Argentinian-Bolivian border (TECCHI et al., in prep.).

The basin of the Laguna de Pozuelos, with about 4,000 km , altitude between 3600 and 4800 m, annual rainfall in the order of 350 mm and a population of 3,600 inhabitants, has been studied during 1989-1990 with the aim of creating a Biosphere Reserve within the program Man and Biosphere of UNESCO. The aims in the establishment of these reserves (Anonymus, 1984) are to ensure the protection and conservation of relatively un-altered, representative ecosystems, making this protection compatible with the improvement of traditional forms of land and resource use, the development of new forms of resource exploitation that respect the environment, and with the preservation of local cultures and traditions.

In the basin of Pozuelos, the population consists almost exclusively of Coya Indians, whose main productive activities are sheep and llama shepherding, with some goats and cows, and a few small mining operations. The study of the proposed reserve includes the identification and mapping of units of high priority for conservation, and the zoning of the areas most suitable for the following activities: conservation, pasture cultivation, intensive shepherding, moderately intensive shepherding, collection of firewood, reforestation with "queñoa" (*Polylepis tomentella*), tourism without installations, restoration of degraded zones, tourism with permanent buildings and facilities, human settlements. In particular it was intended to define the areas devoted to total protection, constituting the nucleus of the reserve, and the surrounding buffer zones, with a more limited degree of protection.

A method of definition and mapping of integrated morphodynamic units is being used. As in other case-studies described, climatic and geological-geomorphological criteria are the basis for the establishment of a hierarchy of morphodynamic "environments", "systems" and "units", but many other environmental factors are included in the description and diagnosis of those subdivisions. Each morphodynamic unit is described using a form containing information about 90 items, grouped in the following categories: climate, topography, geology and geomorphology, active processes, hydrology-hydrogeology, vegetation, fauna, points/sites of scientific, cultural or historical interest, visual landscape, present land-use, social structure, human settlements, accessibility, resources, degradations. A diagnosis is included as well about: quality for conservation, land-use recommendations, land-use limitations.

Using a weighing/scaling method somewhat similar to the one described for the province of Valencia, the quality for conservation of each unit and its aptitude for the different activities proposed have been assessed.

Finally, a sequence of criteria, of the type described for the natural park of the Jaizkibel massif, have been applied to obtain a proposed zoning of the reserve. The method has been tested so far only in a pilot strip representing about 10 % of the basin. The distribution of activities obtained is: 16.3 %, total protection (the water body, also devoted to full protection, is not computed here); 6.1 %, limited site protection; 19.6 %, pasture cultivation; 8.3 %, intensive shepherding; 27.7 %, moderately intensive shepherding; 10.3 %, extraction of firewood; 3 %, tourism without permanent facilities; 14.8 %, no specific use.

The methodology described will be extended to the whole basin, in order to provide a series of recommendations for the establishment of regulations by the Council responsible for the management of the reserve. These recommendations will include not only the zoning for nature conservation and various activities, but also indications about carrying capacity of units suitable for shepherding, soil conservation practices in units prone to erosion, tourism management of sites of cultural or scenic interest, exploitation of autochthonous flora for the obtention of wood for construction or for medicinal plant production, use of local building materials for the improvement of construction design using low-cost, appropriate technology, etc.

Assessment of capacity and impact for tourism development

Geological and geomorphological factors, together with other abiotic and biotic factors, can also be considered separately for the environmental assessment of human activities. This is, for example, the case of site selection for specific projects. A site should have the maximum conditions favourable for the profitable development of the project considered (capacity), and the minimum conditions that would represent a damage for the environment (impact). These factors can be reflected in a series of separate thematic maps and considered independently form each other in the initial process of evaluation, arriving at an integration of all factors at the end of the process.

This kind of method was adopted for the selection of an optimum site, from the point of view of the natural environment, for a proposed tourism project in the north coast of Spain (CENDRERO et al. 1980). The project contemplated the construction of individual bungalows, a hotel, a sports complex and a park zone, with a total of 260,000 m .

Once the project had been defined, a list of general "attractiveness conditions" for the area were defined. These conditions were: proximity to a beach; proximity to a bay or estuary where a marina could be established; presence, within a radius of 15 km, of a variety of centers of interest for recreation, such as fishing and hunting, cultural or artistic sites, nature trails, etc.; far from pollution sources; a population center of at least 1,000 inhabitants within 5 km; no towns with more than 20,000 inhabitants within 10 km; the total population within a radius equivalent to 15 minutes travel time should number at least 10,000; a big town with airport, railway, a big hospital, commercial centers, etc. should be less than 60 minutes away; the area must have a water supply system or a good source that can be used immediately. By means of generally available maps at the scale 1:50,000, nine potentially suitable zones were identified and one was selected.

The capacity and impact requirements for the project were then defined with greater detail. The requirements for capacity were: absence of geomorphological hazards; slope less than 20 %; orientation towards the S, SE or SW; smooth terrain surface; bedrock with good bearing capacity, easy to excavate and with moderate permeability; thickness of soil and cover material greater than 50 cm; inmediate accessibility through the existing road network; diversity of vegetation, including trees; high landscape quality. Impact requirements included: high productivity soils should be avoided; units with sites of elements of interest for conservation should not be used; low visual impact; low potential erodibility; high permeability zones vulnerable to groundwater pollution must be avoided; the development should not be close to rivers or streams; areas with potential resources of construction materials must not be used; the development should not be next to already built-up areas.

Separate thematic maps were prepared for the followong features: bedrock and surficial deposits, active processes and hazards, topographic slope, orientation, landform, thickness of the regolith, soil type and capacity, vegetation, fauna, mineral resources. Standard topographic maps were also used. By means of a simple manual superposition method, all those units of the thematic maps not fulfilling any one of the capacity or impact requirements defined were eliminated. In the end the areas left were only those were **all** capacity and impact requirements were present. Final site selection was then relatively easy, and also the definition of precautionary measures to prevent impacts, which were minimum due to the fact that the process of site selection followed eliminated all sensitive or vulnerable areas.

Final Comments

We have tried to show how the identification, definition and mapping of integrated environmental units based on geological and geomorphological features can be used for impact, hazard and risk assessment in a variety of problems. Thse units also represent a very good basis for the establishment of norms and regulations for planning and management at different levels and scales.

Going back to the requirements mentioned at the beginning, it must be mentioned that the use of integrated environmental units, or morphodynamic units, is also appropriate as a means of communication with the general public and with decision makers (requirement c). Different types of environmental maps can be derived from the initial maps of morphodynamic units, depicting those features of particular interest and showing their types and variability throughout the area of concern. The values and qualities of interest to the public or significant for authorities (value for conservation, landscape quality, hazard or risk level, soil erosion vulnerability, biological productivity, impact of certain activities, etc.) can thus be shown clearly and simply, so that non-specialists can have an appropriate comprehension of the environmental problems affecting the various parts of a region.

Thus, morphodynamic units are very useful as a tool for summarizing scientific knowledge about the constitution and dynamics of natural systems, as a vehicle for environmental education and as an instrument for the application of technological capabilities and institutional framework to the regulation and management of the environment and its resources.

Planning the use of the earth's surface implies making decisions about the activities that must be promoted, limited or forbidden in different areas. The use of the kind of approach described here can provide a sound geoscientific basis for such decisions.

References

ANONYMUS (1984): Plan de acción para las Reservas de la Biosfera.- La Naturaleza y sus recursos **20**, 4: 1-12, Paris (UNESCO).

CENDRERO, A. (1980): Bases doctrinales y metodológicas. - Actas 1 Reunión Nacional de Geología Ambiental y Ordenación del Territorio, Vol. Ponencias, GEGAOT, Santander: 2-62 Santander.

CENDRERO, A., DÍAZ DE TERÁN, J.R., FRANCÉS, E., GONZÁLEZ LASTRA, J.R. & ORTEGA, J. (1980): Ejemplo de valoración de un proyecto Turístico en relación con los aspectos ambientales del territorio.- Seminario sobre Turismo y Medio Ambiente, Doc. CIFCA/TUMA-80/P.8. CIFCA, 64 pp., Madrid.

CENDRERO, A. (1989): Mapping and evaluation of coastal areas for planning.- Ocean and Shoreline Manag. **12** (5-6): 427-462, Barking.

CENDRERO, A., DÍAZ DE TERÁN, J.R., FERNÁNDEZ, O., GARROTE, R., GONZÁLEZ LASTRA, J.R., INORIZA, I., LÜTTIG, G., OTAMENDI, J., PÉREZ, M., SERRANO, A. & 'Grupo Ikerlana' (1987): Detailed geological hazards mapping for urban and rural planning in Vizcaya (Northern Spain).- Spec.Publ.geol. Survey Norway **2**: 25-41, Trondheim.

CENDRERO, A. et al. (1986): Mapa geocientífico de la provincia de Valencia.- 71+350 pp., Valencia (Diputatión Provinvial).

CENDRERO, A. et al. (1990): Geoscientific maps for planning in semi-arid regions: Valencia and Gran Canaria, Spain.- Engin. Geol. (in press).

DUQUE, A., ECHEVERRÍA, G., FERNÁNDES DE LIENCRES, E., KEREJETA, A., CENDRERO, A. & TAMÉS, P. (1990): Ensayo de un modelo empírico de caracter predictivo para la evalucación de la inestabilidad de laderas.- Act. IV Reun. nac. de Geol. ambient. y Ordenac. del Territ., Vol. Comunicac.: 105-113, Gijon (GEGAOT).

FRANCÉS, E., CENDRERO, A., DÍAZ DE TERÁN, J.R., CASTIEN, E. & SAYAGO, J.M. (1990c): A procedure for environmental analysis and diagnosis of a coastal area for planning natural park zones; application to the Jaizkibel massif, Guipúzcoa, Basque Country.- Proc. litt., Assoc. EUROCOAST **1990**: 380-384, Marseille.

FRANCÉS, E., DÍAZ DE TERÁN, J.R., CENDRERO, A., GÓMEZ OREA, D. & VILLARINO, M.T. (1990b): Integrated study of the natural park of Oyambre, Cantabrian Coast, Spain, for land-use planning.- Proc. litt., Assoc. EUROCOAST **1990**: 370-374, Marseille.

FRANCÉS, E., CENDRERO, A., GÓMEZ OREA, D., DÍAZ DE TERÁN, J.R., FERNÁ-NDEZ, P.R., ECHEVERRÍA, G., ESCOBAR, G. & VILLARINO, M.T. (1990a): Una metodología para la definición de unidades de diagnostico en la elaboración de directrices de ordenación territorial a escala regional: el modelo de Cantabria.- Actas IV Reun. nac. de Geol. ambient. y Ordenac. del Territ., Vol. Comunicac.: 213-223, Gijon (GEGAOT).

RIVAS, V. & CENDRERO, A. (1990): Use of natural and artificial accretion on the north coast of Spain: historical trends and assessment of some environmental and economic consequences.- Journ. coast. Res. 6/4: 2-18, Lawrence.

TECCHI, R., CENDRERO, A., DÍAZ DE TERÁN, J.R., GONZÁLEZ, D., MASCITTI, V. & ROTONDARO, R. (in preparation): Environmental diagnosis for planning and management in the High Andean Region; the Biosphere Reserve of the Laguna de Pozuelos, Argentinia.

WORLD COMMISSION ON ENVIRONMENT AND DEVELOPMENT (1987): Our common future.- 400 pp., Oxford (Univ. Press).

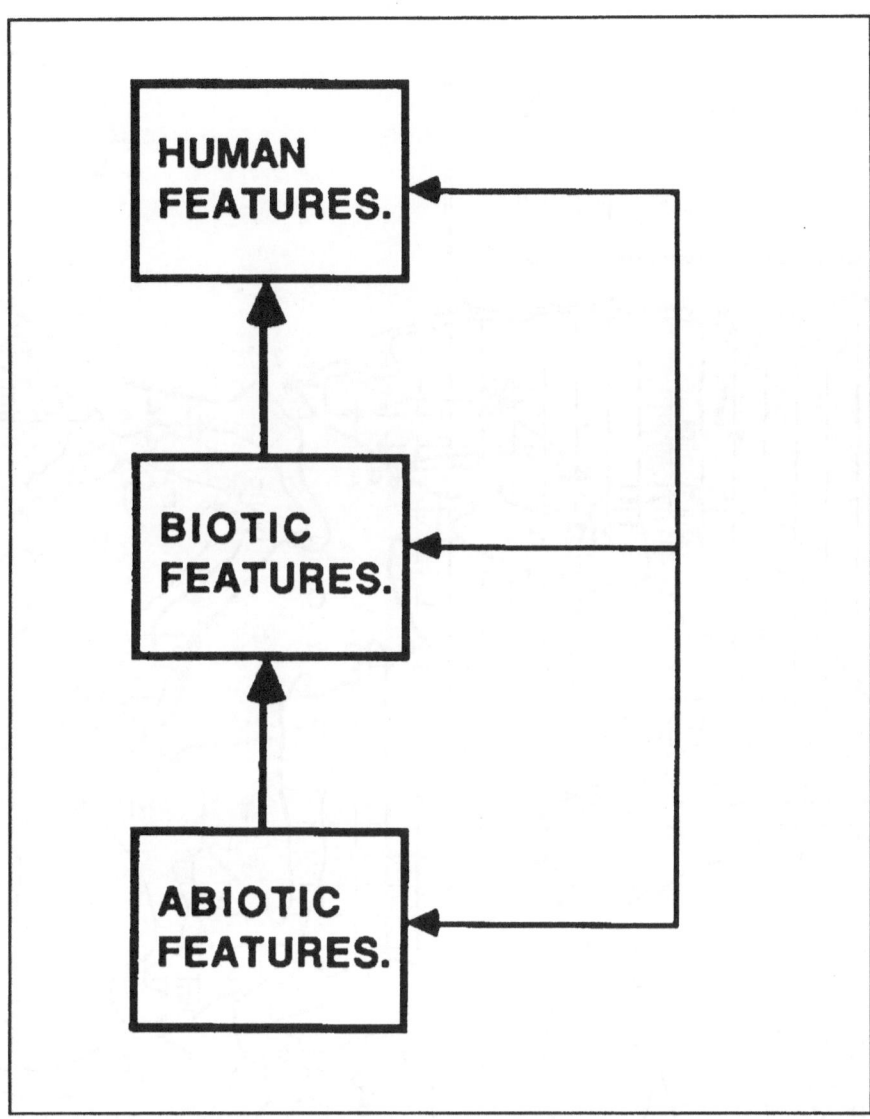

Figure 1: Dependence relationships between the different factors making up the earth's surface. Thick arrows: main dependence; thin arrows: secondary relationships.

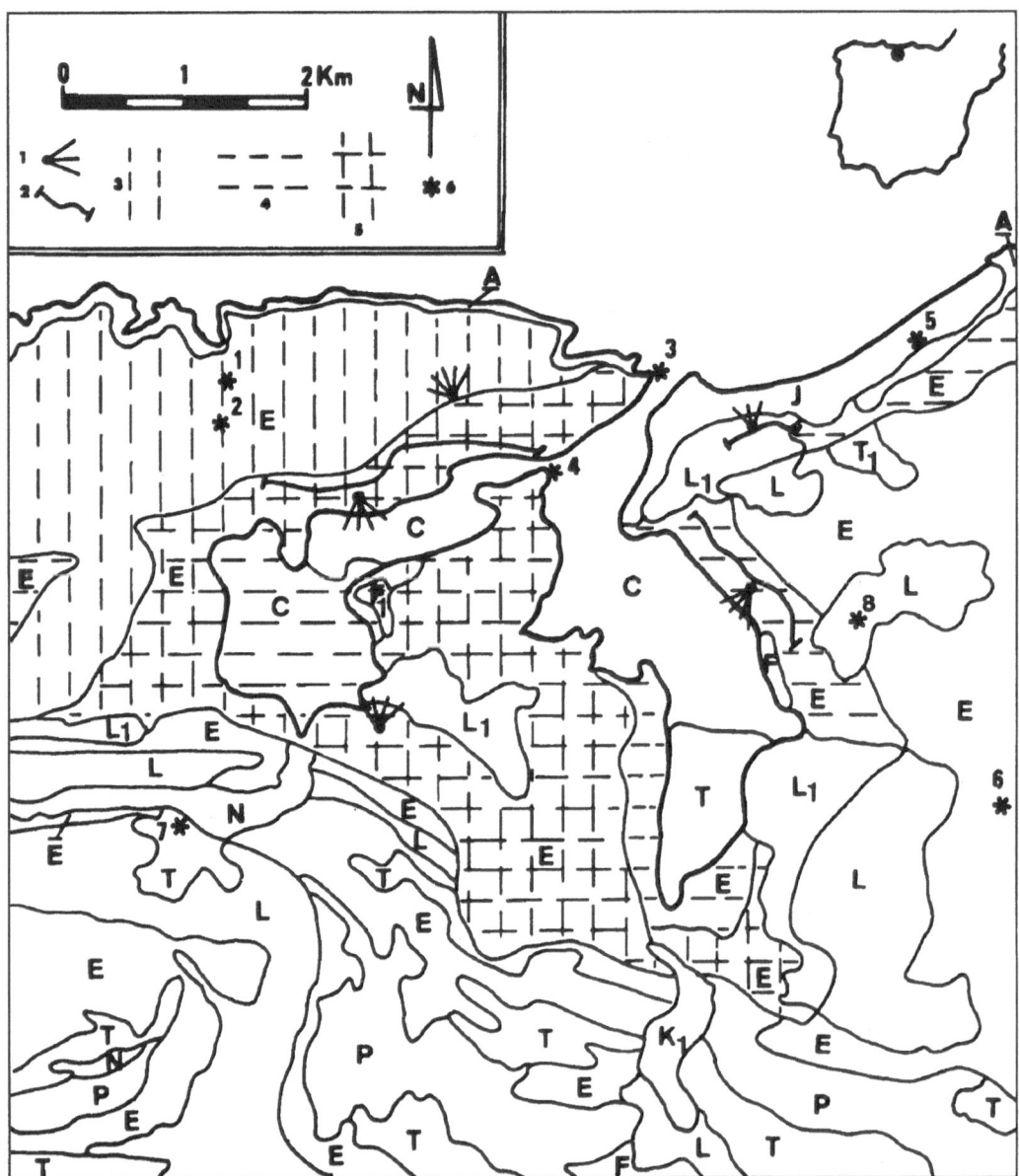

Figure 2.a: Part of the map of synthesis units for the General Master Plan of Cantabria.

F: climax forests; F1: id within valuable landscape; A: coastal cliffs; C: estuaries; N: vegetation singularities; K1: landscape singularities; L: units of high agricultural productivity; L1: id within valuable landscape; E: areas with medium agricultural productivity; T: area reforested with rapid growth species; T1: degraded autochthonous forsts; P: under-exploited units. Visual landscape features are indicated by the following numbers: 1: scenic viewpoint; 2: scenic road; 3: landscape of high quality; 4: landscape of high fragility; 5: landscape of high quality and fragility; 6: site of scientific interest.

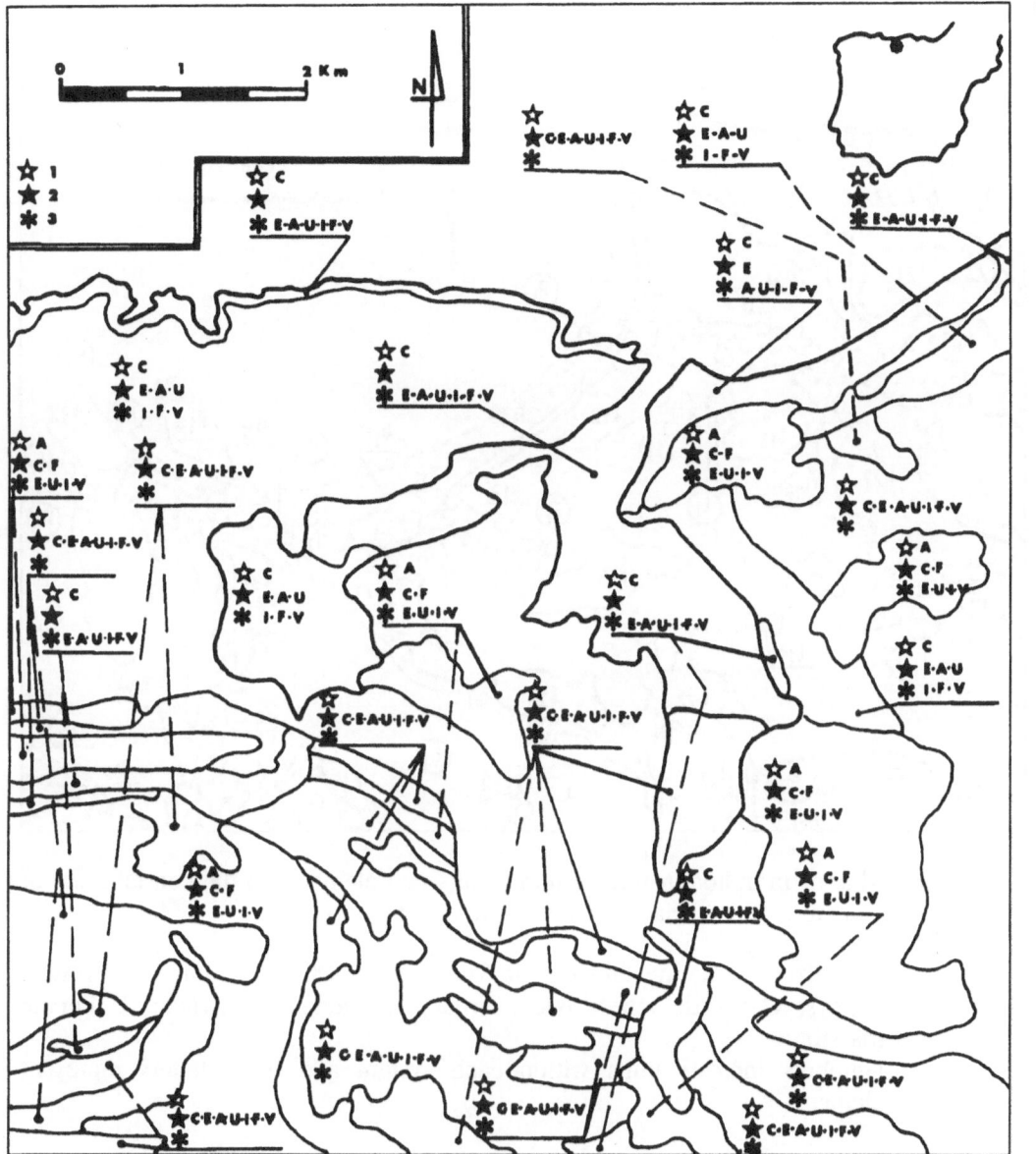

Figure 2.b: Map of main land-use recommendations for the same area.

1: vocational uses; 2: compatible uses; 3: non-compatible uses.
C: conservation; E: recreation under outdoor sports; A: agricultural activities;
U: urban uses; I: industrial activities; F: infrastructures; V: waste disposal.

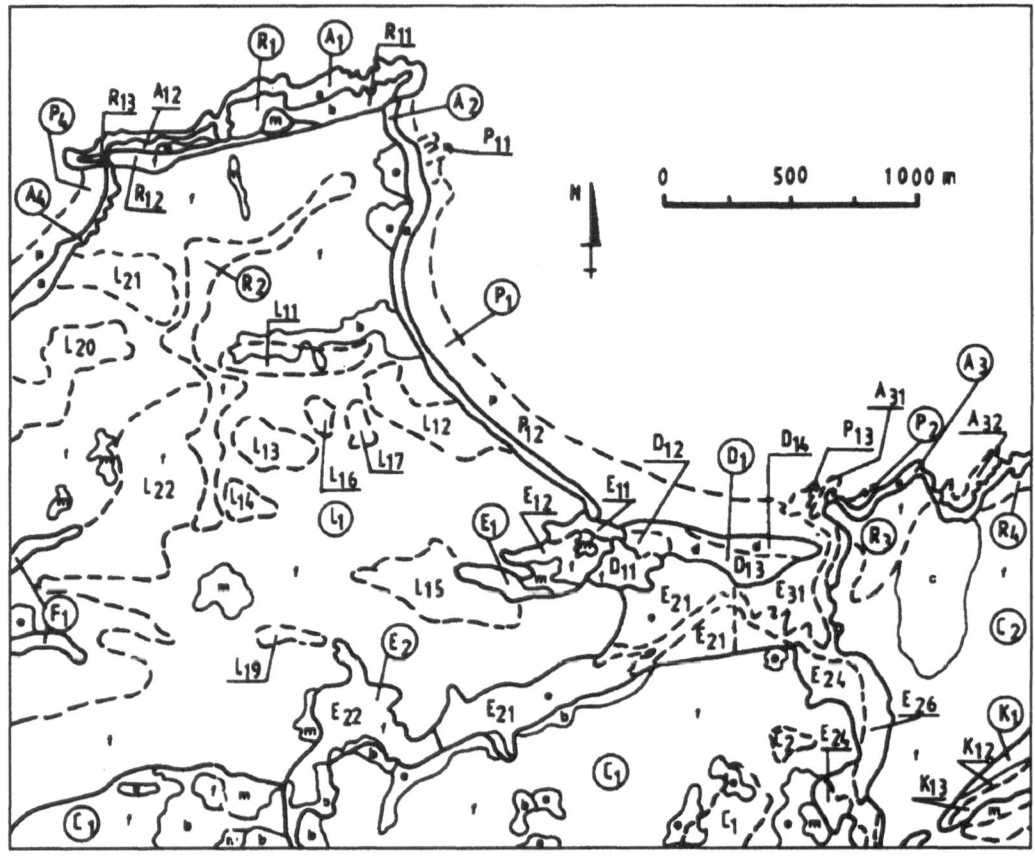

Figure 3.a: Map of morphodynamic systems, units and elements in the coastal zone of Oyambre (Cantabria). Capital letters denote systems.

A: cliffs; R: marine erosion platforms; P: beaches; D: dunes; F: fluvial valleys; C: gentle reliefs over Tertiary terrigenous materials; K: karstic massifs.

Numbers indicate units within each system and small letters represent elements.

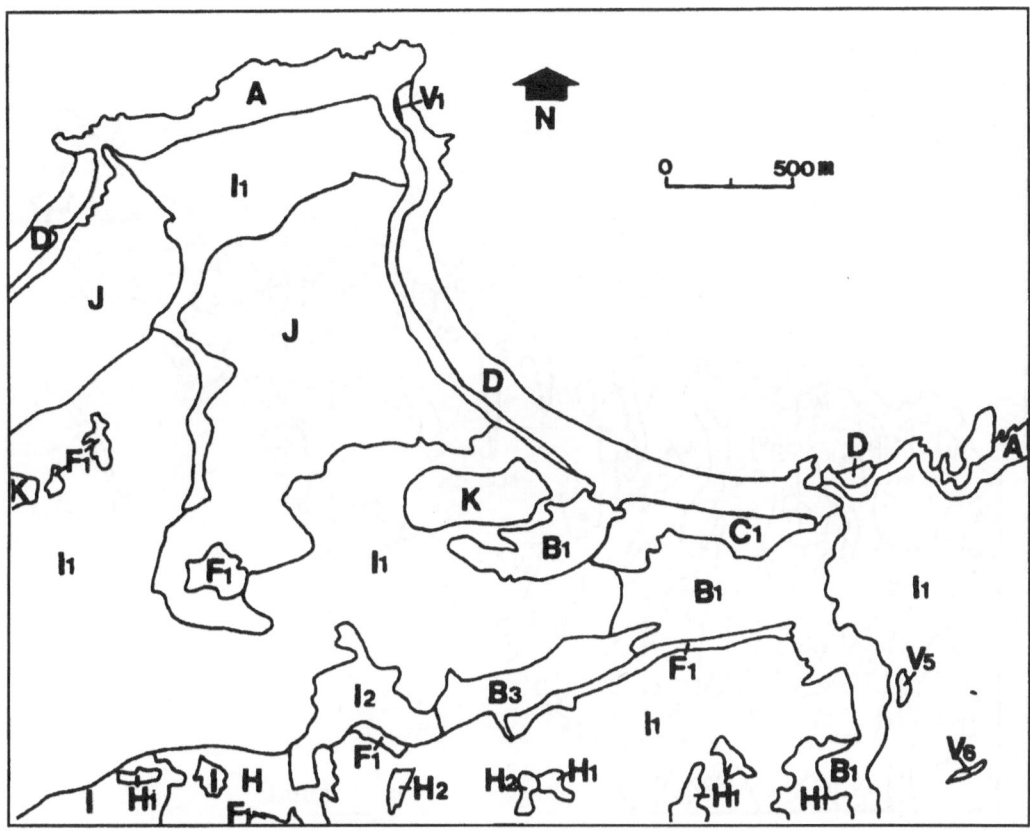

Figure 3.b: Map of management actions for the same zone.

A: protection of cliffs and adjacent marine terraces; B1: control of fishing and shellfish harvesting in estuaries; B3: elimination of barriers to tidal circulation and regeneration of vegetation; C1: cleaning and re-vegetation of dunes; D: control of recreation in beaches; F1: conservation and improvement of autochthonous vegetation; H1-H2: reforestation with autochthonous trees; I1: improvement of grass and fodder cultivations; I2: drainage and improvement of grass production; J: limitation of structures of high visibility; K: re-location of campsites; V: protection of sites of scientific interest.

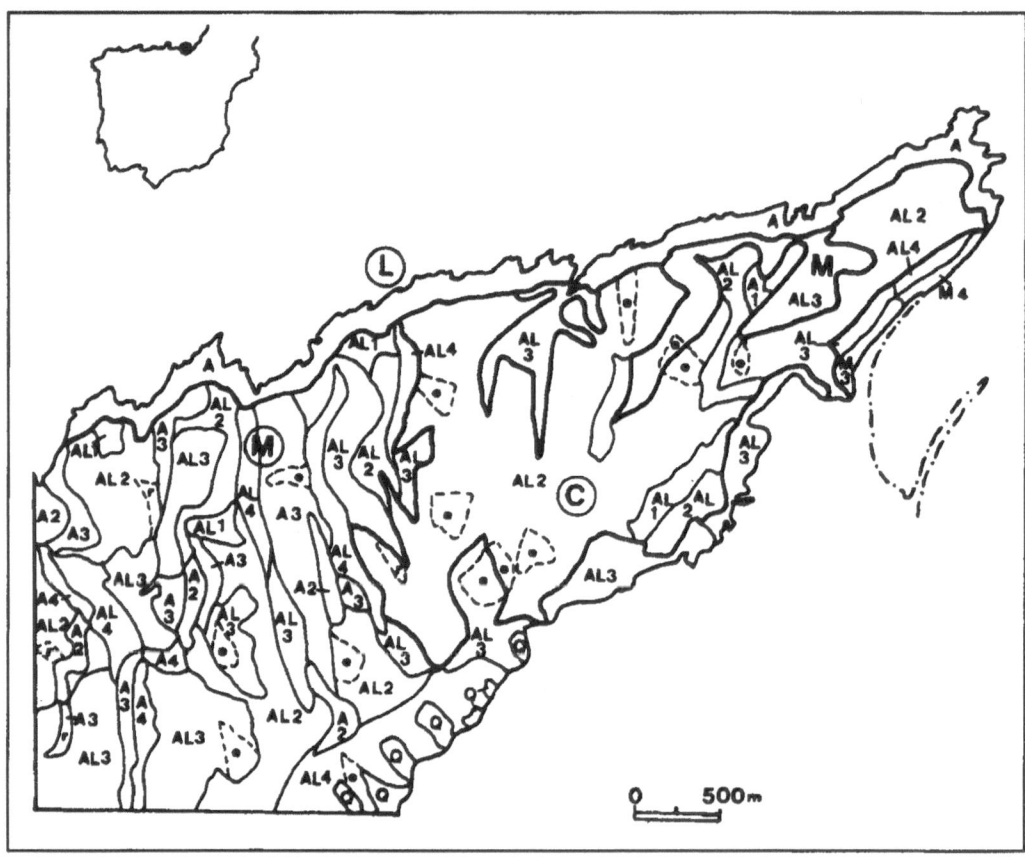

Figure 4.a: Integrated morphodynamic systems, units and elements in the Jaizkibel
Massif (Basque Country).

Large capital letters in circles indicate systems: L: littoral; A: cliffs; C: gentle
reliefs; M: strong reliefs.

Smaller capital letters represent rock types: AL: sandstones and siltstones; A:
sandstones and conglomerates; M: marls; Q: colluvial deposits.

Numbers indicate slope gradient: 1: < 10 %; 2: 10-30 %; 3: 30-50 %;
4: > 50 %.

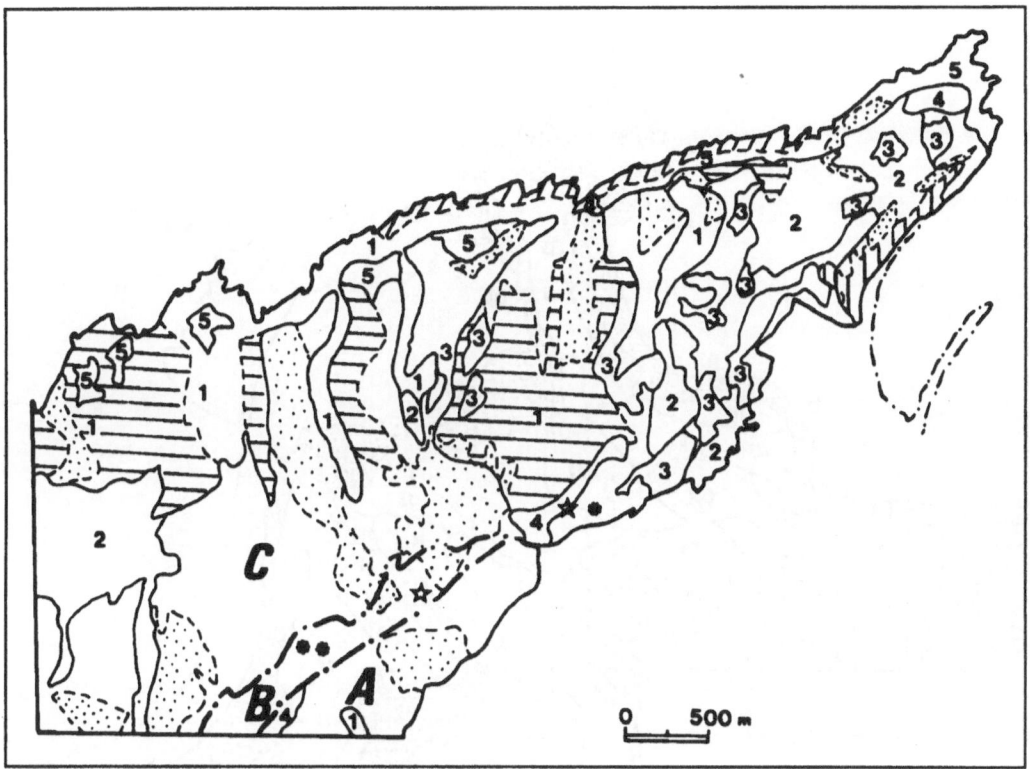

Figure 4.b: Map of land-use recommendations and actions for the same area. Continous lines enclose zones with a certain recommended use:

1: first order conservation; 2: second order conservation; 3: maintenance of traditional agriculture; 4: intensive recreation; 5: recreation without installations.

Dashed line denotes management actions: dotted: reforestation for protection against erosion and surficial landslides; vertical lines: protection works against slope instability; horizontal lines: re-implantation of autochthonous woods; oblique lines: unstable areas where no protection works should be undertaken.

Visual landscape features are indicated by capital letters and symbols: A: high visual incidence; B: medium; C: low. Asterisks: sites of scientific-cultural interest. Stars: panoramic views.

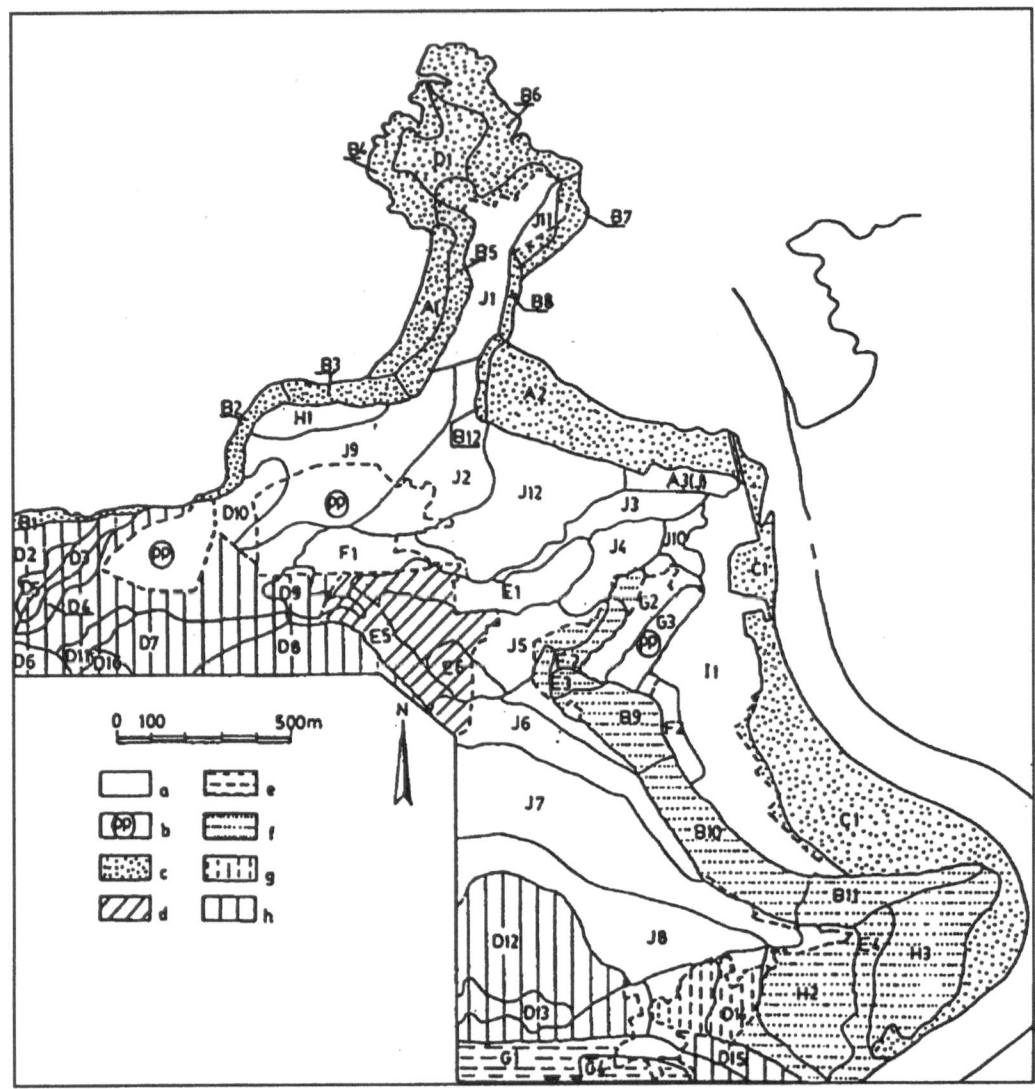

Figure 5: Morphodynamic units (continuous lines) and planning categories (dashed lines) in a part of the coastal munincipality of Suances (Cantabria).

Morphodynamic units indicated by capital letters: A: beaches and dunes; B: cliffs; C: estuaries and wetlands; D: karstic areas; E: unstable slopes; F: flood-prone areas; G: units with high productivity soils; H: units with vegetation or landscape valuable for conservation; I: reclaimed intertidal areas; J: units partially or totally built-up.

Planning categories are shown with shading symbols: a: urban areas; b: areas where building can be allowed after preparation of a special plan; c: building not allowed due to the protection of ecosystems; d: building restricted due to natural hazards; e: area devoted to agriculture; f: building restricted due to landscape protection; g: building not allowed; h: building restricted due to aquifer protection.

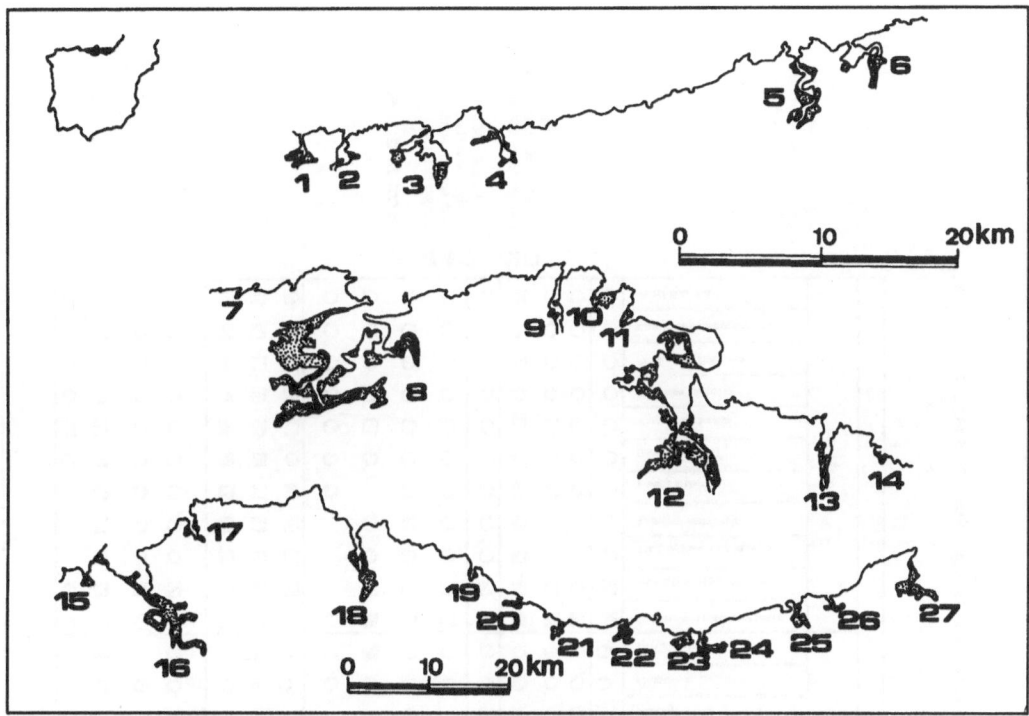

Figure 6:   Reclaimed intertidal and wetland areas in the north coast of Spain that would be under risk in the event of sea-level rise.

The numbers indicate estuaries/zones. 1: Tina Mayor; 2: Tina Menor; 3: S. Vicente; 4: La Rabia; 5: Suances; 6: Pas; 7: S. Juan de la Canal; 8: Bay of Santander; 9: Ajo; 10: Soano; 11: Trengandín; 12: Bay of Santoña; 13: Oriñón, 14: Castro Urdiales; 15: Somorrostro; 16: Bilbao; 17: Plencia; 18: Guernica; 19: Lequeitio; 20: Ondárroa; 21: Deva; 22: Zumaya; 23: Zarauz; 24: Orio; 25: S. Sebastián; 26: Pasajes.

Table 1: Matrix for the definition of carrying capacity of "synthesis units" for different activities.

VOCATIONAL USES
☐ Coincident with present use
▣ Non-coincident with present use

COMPATIBLE USES
▨ Without Limitations
▣ With special permit
▨ Subject to E.I.A.

○ NON-COMPATIBLE USE

TABLE 2 a

Legend of present hazards map.

FLOOD HAZARD.
    High.
    Medium.
    Low.

SLOPE INSTABILITY HAZARD.
    Active instability areas.
    Slide.
    Old slide.
    Deep slide.
    Creeping.
    High hazard areas.
    Medium hazard areas.
    Hazardless or low hazard areas.

COLLAPSE HAZARD.
    Low hazard areas.

WIND HAZARD.
    High.
    Medium.
    Low.

COASTAL HAZARD.
    High.
    Medium.
    Low.

TABLE 2 b

Legend of hazards + vulnerability map.

UTILITIES AND INFRASTRUCTURE.
- Power line.
- Telephone line.
- Sewage.
- Water supply.
- Pipe line.
- Road.
- Protection works along roadcuts.

MINING AND DISPOSAL OPERATIONS.
- Active quarry.
- Abandoned quary.
- Rubble and mining waste deposits.
- Solid waste disposal sites.

BUILDING DENSITY.
High.
Medium.
Low.

INSTITUTIONAL-PUBLIC SERVICE BUILDINGS.

WORKS ALONG RIVERS.
Dedging of river channel.
Margin protection.
Channelisation.

LAND-USE PLANNING.
Reserved area for urban use.
Reserved area for industrial use.

HAZARD KINDS.
High flood hazard.
Medium flood hazard.
High slope instability hazard.
Medium slope instibility hazard.
High wind hazard.
Medium wind hazard.
Areas with no medium or high hazard.

Chapter 3:

Specific Geoenvironmental Topics

Geochemistry and the Environment

BEDŘICH MOLDAN [*]

Abstract

Geochemistry is one of the most important environmental disciplines. Indeed, a more proper term for this scientific field is biogeochemistry. The connection between the crucial aspects of the environment appears clearly from this term. It was coined by V.I. VERNADSKY in 1923 (MOCHALOV 1982). Later he incorporated this idea into the famous concept of the biosphere (VERNADSKY 1926) adopted by UNESCO in 1968 (UNESCO 1970) and is today one of the frequently used terms not only in scientific but also in general vocabulary.

Introduction

Biogeochemical studies have been the cornerstone of the work of the environmental commission created by ICSU in 1969 under the name SCOPE (Scientific Committee on Problems of Environment). Of 36 reports published so far, seven contain this term in the title and at least ten others deal mainly with problems of biogeochemistry. The importance of these studies was stressed by the panel of the U.S.Academy of Sciences which used the

[*] **Author's address:** RN Dr. B. MOLDAN, Ustredni Ustav Geologicny, Praha, ČSFR.

terms "geochemistry of the exogenous zone" and "metabolism of the Earth". JEROME K. NRIAGU preferred the title "environmental biogeochemistry" (NRIAGU 1976). According to him, this means the chemical study of the interaction between the biological life forms and their surroundings. In the narrower, more anthropocentric meaning, it refers to the inquiry about chemical changes in the atmosphere, biosphere, hydrosphere and lithosphere resulting not only from natural forces but primarily from the activities of man. The newest major international scientific project "Global Change"; International Geosphere - Biosphere Programme (IGBP 1988) is unthinkable without the fundamental contribution of biogeochemistry.

Indeed, problems of distinctly biochemical character are among the most urgent environmental concerns of today. Examples of local-scale issues are the pollution of surface waters and groundwaters, especially by non-point sources, and the harmful fumes of toxic and other hazardous waste dumps. On the regional scale we face the dangers of acidic atmospheric deposition with all its consequences. The prime global environmental threat is probably the rise in concentration of greenhouse gases: a biogeochemical problem par excellence.

In what respect can biogeochemistry as a science contribute to the solution of these and other environmental problems? First, it offers fundamental ideas which may serve as the conceptual framework for our better understanding of their nature. Second, it has a highly elaborated methodology.

Biogeochemical studies in various fields have resulted in many interesting and important findings. In a short paper it is not possible to give any systematic treatment of such a broad and complex discipline. The author will use a few insights only to show the possibilities which biogeochemistry may offer to environmental sciences.

Data

The data relevant to environmental biogeochemistry result from a very broad set of field observations and measurements, laboratory results and model computations. Traditionally, the greatest care has been paid to the analytical techniques for determining concentrations of diverse chemical compounds. Dramatic improvements have been made in this respect in the last decades through the development of clean sampling and sample preparation techniques and the improvement of precision and sensitivity of analytical procedures. In general, due to the enormously quick pace of instrumental development, the main burden of problems to be solved lies in the operations outside the laboratory. Much attention still needs to be devoted to the sampling process in order to avoid any sources of contamination, and the sampling scheme must be carefully designed to obtain truly representative samples. To assure the integrity of a sample between collection and analysis is also an important and sometimes difficult task. Often the simplest operations performed by local observers cause the most difficult problems.

Nevertheless, in general, the overall quality of the available data is already rather good and is still improving. This is partly due to the traditionally high level of the geoscientific data base which also serves environmental geochemistry. The use of standard reference materials has a long-established tradition. The same can be stated about interlaboratory comparisons and many other quality control procedures.

Without much exaggeration it may be stated that the complex task of acquisition of high-quality data an ever-broadening set of environmental variables is being pursued successfully. This highlights the more difficult task of the further elaboration of the data. That means first publishing them in a compact and clear way and making them easily available to the whole community of environmental scientists. Of course, the most difficult problem, and therefore often missing, is the creative interpretation of the data. Only then do they make really sense and contribute significantly to better understanding of environmental processes and tackling of environmental problems.

Biogeochemical reservoirs

While working on a local, regional or global scale, biogeochemistry sees the physical environment as a sum of several compartments or biogeochemical reservoirs. For example, the global biosphere or exogenous zone may be considered a system formed by four reservoirs: the continents, the oceans, the atmosphere and the biota, i.e. the sum of living organisms. Of course, for most purposes this scheme is too rough and the basic reservoirs must be further divided into smaller units according to different aspects. Thus, the atmosphere may comprise the troposphere and the upper atmosphere, or southern and northern hemisphere, or continental and oceanic atmosphere, or gaseous and particulate compartments, or various combinations of these and other units. Usually, the compartments are assumed to be of constant size and in most cases well-mixed, i.e. geographically homogenous.

Reservoirs should be clearly defined and their borders unequivocally outlined. Each reservoir has a set of "state variables", the most important being its volume, mass and concentrations of chemical components. In most cases environmental biogeochemistry doesn't study individual compartments as stable units even if such tasks are also performed, e.g. the investigations of concentration distributions within the soil of a given region. The goal is usually to characterize the migration and chemical changes of various components by the processes of mobilization, transport, transformation and deposition of chemical elements and/or compounds. These processes take place within a compartment, at the interface between two of them or within the whole system studied.

An example of major carbon reservoirs is given in Table 1 (BERNER, LASAGA 1989).

Concentrations and fluxes

Concentrations of chemical elements are the most studied "state variables" of biogeochemical reservoirs. In most studies where the concept of reservoirs is employed, the concentration of a chemical element is given by a sole number, perhaps with an indication of the degree of its certainty or reliability by which the "average" is indicated, (e.g., see Table 2). However, it is very well known that in the real world the concentration of any given species is distributed in space and time; it fluctuates; it may have gradients; at least it has a natural, mostly log-normal or other distribution. Moreover, in most works the concentration of chemical elements is given, e.g. S, N or Pb, which may or may not mean the sum of or all chemical forms, often with extremely different chemical properties. For example, when we refer to "nitrogen", we may tacitly disregard the ever-present virtually inert N molecules. For detailed biogeochemical studies not only individual chemical compounds, ions, radicals etc. are of importance but also isotopes whose study is presently one of the most quickly growing fields of biogeochemistry. New types of mass spectrometers, e.g. in combination with inductive-coupled plasma, offer great opportunities here (VÖLLKOPF 1989, personal communication).

The concept of material flow is the specific and important contribution of biogeochemistry to environmental studies. The flow is expressed (by some authors the term flux is used) in the units of mass by time. For example, the flow of matter through the global sedimentary (exogenic) cycle is $0.6 - 1 \times 10^{16}$ g a^{-1} (GARRELS, LERMAN 1977). Human society creates flow three to five times higher, i.e. 3.3×10^{16} g a^{-1} (MOLDAN 1983) of materials mostly taken from geological deposits.

The units of material or mass flux (or mass flux density) are mass by surface area times time. Commonly used units are g m^{-2} a^{-1} or kg ha a^{-1}. Indeed, mass fluxes are fundamental environmental parameters which can be used for characterizing various both natural and man-made processes on local, regional and global scale. For example, erosion rates, velocity or intensity of weathering, specific or elemental runoff, atmospheric deposition and various "pollution phenomena" can be quantified in mass flux units and in this way easily compared. In Table 3 some biogeochemically important fluxes are summarized.

Biogeochemical study of small catchments

The study of biogeochemical processes in small catchments was pioneered in Hubbard Brook Experimental Forest where complex investigation started in the sixties (LIKENS et al. 1977). This approach was later worked out by several groups of scientists in Scandinavian countries, in the U.S.A., in Canada and elsewhere. At present, it represents a rapidly expanding field of research attracting growing international interest. This was reflected by the surprisingly large attendance of 160 scientists from 18 countries at the the first international meeting of this kind, the Geomon Workshop (Prague, Czechoslovakia, April 1987), devoted to geochemistry and monitoring in small drainage basins. In 1988, a SCOPE project "Biogeochemistry of small catchments" was approved (MOLDAN 1988).

A small catchment in this context is a well-defined drainage basin or watershed with the surface area usually of tens to hundreds of hectares but sometimes substancially smaller or larger. The set of investigated processes can be greater if the basin contains one ore more small lakes. Most often the catchment is situated in a comparatively undisturbed landscape, i.e. in areas not affected by local environmental impacts. These areas could be typical for a larger region of a given biome. The ecosystems in the catchment are preferably in a late stage of ecological succession, e.g. forests.

There the "biogeochemical metabolism", i.e. the sum of the processes of mobilization, transport, transformation, and deposition of energy and matter within a small catchment area and across its boundaries, is studied.

Human-caused effects participate in a complex metabolic process; in many cases, as in areas affected by long-range transport of atmospheric pollutants, they are very important.

Studied in the small catchments are the inputs and outputs of energy and matter: the atmospheric deposition of rain, snow, fog, particles or gases, evaporation and degassing, subsurface and surface runoff of water with dissolved and undissolved material, incoming solar radiation and outgoing infrared radiation fluxes. Biological processes comprise the exchange of energy, water, and gases by photosythetic and respirative processes and water transpiration. It is important to investigate any human caused flows of matter and energy.

These observations are complemented by the investigation of processes within the catchment, such as rock weathering, chemical and mechanical erosion, soil processes including soil biology, limnological and hydrobiological processes, internal biogeochemical cycles within the forest and other ecosystems, the dynamics of living organisms and their metabolic role. Attention is paid not only to the more or less regular processes such as the accumulation of a certain compound in the soil but also to all sorts of seasonal or random episodes such as thaws, storms, floods, and remarkable biological events.

To improve our understanding of the functioning of the catchment's metabolism, experiments could be performed. These include liming, fertilization, irrigation, biological manipulation, forest management, artificial acidification, and controlled release of chemicals including radioisotopes. Such studies are also helpful in establishing countermeasures against anthropogenic influences.

Mathematical models are frequently used to enable a more complex elucidation of various metabolic (namely hydrochemical) processes. Models exist, e.g. for short-term episodic simulation of stream water chemistry and for long-term prediction of future response to changes in environmental parameters such as atmospheric deposition patterns.

Long-term integrated research and monitoring of small catchments proved to be a valuable tool for assessing both natural and man-made changes in the environment. The list of environmental impacts reflected by processes in small catchments includes -- among others -- acidification of soils, surface and ground water, depletion of nutrients and accumulation of harmful substances in the soil, transport of atmospheric pollution, changes in physical and chemical parameters of climate, e.g. growing concentration of CO_2 or depletion of stratospheric O_3, effects of land-use practices on the geochemical balance, e.g. autrophication. Monitoring can reveal the effects of the reduction of emissions of sulphur and nitrogen agreed on by European countries. Examples here are the pioneering Norwegian study (OVERREIN et al. 1981), Lake Gårdsjön Study (ANDERSSON, OLSSON 1985), the North America's ILWAYS study established to help to elucidate the "acid rain dispute" between the U.S.A. and Canada (GOLDSTEIN et al. 1985) or the Swedish PMK system for monitored watersheds (NATIONAL SWEDISH ENVIRONMEN-TAL PROTECTION BOARD, 1985). Recently, the biogeochemical monitoring of small catchments has been adopted as the basis of the Intergrated Monitoring Programme by the Economic Commission for Europe in the framework of the "Convention on long range transboundary transport". This project is now in the pilot phase (International Cooperative Programme on Intergrated Monitoring 1989).

Atmospheric deposition

Interactions of the atmosphere with land and oceans are some of the most important biogeochemical processes. First, there are many processes of mobilization of matter (both gaseous and particulate) into the atmosphere: production of sea salt particles, suspension of dust, biological and geological degassing, evapotranspiration, and anthropogenic emissions. The exchange of water vapor, CO_2 and O_2 is the most massive. The latter two gases are related to photosynthesis and respiration in green plants. Other important gaseous inputs result from geological processes (volcanic activity and emissions from the Earth's crust and mantle), microbial activity and fossil fuel combustion.

The average life-time of material in the atmosphere varies widely from several dozen minutes for the largest particles (20 to 100 m) and the most reactive gases to about 10 days for water vapor and sub-micron particles to thousands or tens of thousands of years for such gases as CO_2, O_2, CH_4 and N_2O. Ultimately the rate of physical and chemical transformation will determine the lifetime of atmospheric matter. Physical tansformation includes condensation of water vapor, coagulation of particles, solution of gases or soluble salts by water droplets and adsorption onto aerosols. Chemical transformation includes both homogeneous and heterogeneous photochemically induced oxidation/reduction reactions.

For relatively soluble gases and accumulation mode particles the phenomenon of atmospheric deposition is the most important sink process: the transfer of materials from the atmosphere onto the land and/or water surface.

We differentiate the following:

a)   Wet deposition connected with the atmospheric water cycle. The most important is precipitation which can take the form of rain, snow, drizzle and hail. The annual flux of water onto an average square meter of land surface is about 750 litres with about 10 g of dissolved and undissolved material. Among deposited chemical elements, the most aboundant are: C, S, N, Ca, Cl, Na, K, Mg, Si, Al, Fe, and, of course, O and H. The yearly global flux of water is 5×10^{20} g a^{-1} with 3 to 5×10^{15} g of dissolved solids.

Horizontal deposition, fog and rime-ice, is a quantitatively less important kind of wet deposition. Sometimes, genetically different dew, grey frost, etc., may be significant, too. In localities of higher altitude (mountains) and/or large incidence of fog this type of precipitation may in part be ecologically important. The average ionic strength of the cloud, fog and/or rime water is generally substantially higher compared to rain or snow water and therefore the biogeochemical significance may be enhanced (WEATHERS et al., 1988). Very high values of acidity are encountered in fog water; an extreme pH=1.68 was measured in California (JACOB et al., 1985).

b)   Dry deposition is defined as a downward flux of gases and particles from the atmosphere not associated with atmospheric hydrometeors. Nearly all the atmospheric gases participate in the air-surface exchange. However, only the most reactive are generally considered as having exchange as a net sink from the atmosphere. For example, sulphur dioxide, nitric acid, nitrogen oxides and oxidants (such as hydrogen peroxide and ozone) are all dry-deposited.

Of atmospheric particles, only those with diameters > 10 m settle under the influence of gravity. For fine particles a wide variety of processes that are generally not very effective contribute to the total deposition (e.g. impaction, thermoprecipitation).

For a given chemical element or compound all the deposition mechanisms mentioned may apply. However, they contribute to the total deposition at very different rates depending on the substance's chemical nature and on many local meteorological and topographical parameters. Vertical wet deposition connected with rain and snow is the most important process for soluble gases and particles. In mountainous and other sites frequently covered by clouds, the horizontal deposition may be pronounced. Dry deposition of particles (mostly terrigenous) is often dominant in polluted urban regions and in dry areas. Direct absorption of gases has been studied extensively for compounds of sulphur and nitrogen in respect to acid deposition.

The atmospheric deposition of ten chemical elements into a small catchment called Jezeří, studied by the Geological Survey, Prague, is given in Table 4. The catchment is located in the Krušné Mts. with extremely high pollution.

Biological production and its use by man

The biological productivity of terrestrial and marine ecosystems is expressed by the material flux called annual net primary production (NPP). Energy flux may also by used. It is defined as the energy (almost exclusively solar) biologically fixed in a year by a given ecosystem minus the respiration of primary producers (mostly green plants). NPP is the remaining biological production -- or energy -- left over for consumers and decomposers. Consumer species also include humans. The following data are based on the work of VITOUSEK et al. (1986, see Table 5) and the interpretation given by DIAMOND (1987).

The global NPP is about 225 Pg (1 Pg = 10^{15}g) of organic matter. The proportion of this NPP directly used by humans for food, and of wood used for timber, paper and firewood is estimated as 7.2 Pg, i.e. 3% of the total global NPP. However, the human acquisition of food, wood and other organic products diverts a far larger proportion of NPP. For example, an average less than 10 % of the total organic matter fixed on agricultural land is edible. When a natural forest or a tree plantation is harvested, less than 50 % of its accumulated NPP becomes timber. The total amount of NPP diverted and consumed is 42.6 Pg or 19 % of the Earth's total NPP. To this figure we must add the changes human activity creates by modifying the natural biomes. The croplands, pastures, urbanized areas and man-made deserts do not have the same NPP as the original natural ecosystems. These changes represent the overall reduction of global NPP by 17.5 Pg. Hence our total appropriation is 60.1 Pg which falls far more heavily on the terrestrial NPP (39 %) than on the marine NPP (2 %).

Only 61 % of the global terrestrial NPP is left to all other terrestrial consumer species. This means almost certainly the largest appropriation of NPP by one species and its servant species since life appeared on the Earth. The fraction left to sustain all our fellow species has shrunk dramatically.

Global biogeochemical cycles

Most studies of global cycles have been focused on the elements carbon, nitrogen, sulphur, and phosphorus: the major biogenic elements. The availability, circulation and interaction of these four chemical elements in nature has been decisive for the development of life on Earth and for the maintenance of the global biosphere. Man today perturbs the major biogeochemical cycles significantly. The annual emissions of carbon dioxide into the atmosphere by the combustion of fossil fuels amounts to about 10 % of the amount of CO_2 used by green plants by the process of photosynthesis.

The fixation of nitrogen by combustion and fertilizer production is about half of what is produced naturally. Land transformation, particularly deforestation, urbanization and the conversion of about 10 % of the total land surface into arable land have resulted in a major translocation of nutrients, particularly nitrogen compounds. For example, the anthropogenic contribution to the contemporary flux of nitrogen in the drainage basin of the river Labe in Bohemia is 86 % (PAČES, 1982). Emissions of sulphur oxides into the atmosphere, primarily by fossil fuel combustion, probably exceed the natural gaseous sulphur emissions. The major features of global biogeochemical cycles of C, N, S and P and their interactions were summarized in the SCOPE Report 21 (BOLIN, COOK 1983). The study of the carbon cycle is critical for the understanding of the expected climatic change due to the enhanced greenhouse effect of the Earth's atmosphere. According to BOLIN (1989), the average global warming of 3.5 C within the next 50 years is virtually certain. The main threat is carbon dioxide causing about half of the temperature increase. Before industrialization, during the first half of the 19th century, levels of CO_2 in the atmosphere are thought to have been about 270 ppm whereas the present concentration is 344 ppm. However, also other gases have a marked effect on the greenhouse warming (see Table 6).

In recent years the investigation of biogeochemical cycles has broadened and today global data exist for most of the chemical elements. Information on metals was summarized in the Dahlem Workshop Report edited by NRIAGU (1984). The authors of this volume concluded that the cycles of almost all metals are anthopogenically perturbed at least on the local scale. NRIAGU (1989) assessed the global natural and anthopogenic sources of atmospheric trace metals. His figures appear in Table 7.

A great environmental concern is devoted to persistent, biologically active man-made compounds. A typical example is DDT; its global cycle was studied by CRAMER (1973). According to this author, the time required to return to equilibrium after the termination of DDT usage cannot be precisely determined, but lies somewhere between about 25 and 110 years.

References

[ANDERSSON, F & OLSSON, B.] (1985): Lake Gårdsjön. An acid forest lake and its catchment. Ecol.Bull. **37**: 1-336, Stockholm.

BERNER, R.A., LASAGA, A.C. (1989): Modelling the geochemical carbon cycle. - Scientif. Amer., **1989**/3, 54-61, New York.

BOLIN, B. (1989): Man and climate. Lecture, Ecology 89 Conference, August 1989, Göteborg.

[BOLIN, B. & COOK, R.B.] (1983): The major biogeochemical cycles and their interactions. SCOPE **21**: 1-532, Chichester (John Wiley).

CRAMER, J. (1973): Model of the circualtion of DDT on Earth. Atmosph. Environm. **7**, 241-256, Oxford etc.

DIAMOND, J.M. (1989): Human use of world resources. Nature **328**: 479-480, London.

GARRELS, R.M. & LERMAN, A. (1977): The exogenic cycle: Reservoirs, fluxes, and problems. In [STUMM, W.]: Global chemical cycles and their alterations by Man. Dahlem Workshop Rep., Berlin (Abakon).

GOLDSTEIN, R.A. et al. (1985): Integrated Lake - Watershed Acidification. - Dordrecht (D. Reidel).

JACOB, D.J. et al. (1984): Chemical composition of fog water collected along the California coast. - Environm. Sci. Technol. **18**, 11: 827-833, Washington.

JACOB, D.J., RUEEN-FANG, T.W. & FLAGAN, R.C. (1984): Fogwater collector design and characterization. - Environm. Sci. Technol. **18**: 827-833, Washington.

[KARPE, H.-J., OTTEN, D. & TRINIDADE, S.C.] (1990): Climate and development. - 477 p., 95 Fig., Berlin etc. (Springer).

LIKENS, G.E. et al. (1977): Biogeochemistry of a forested ecosystem. - 146 p., 31 Fig., 22 Tab., New York (Springer).

MOCHALOV, I. (1982): Vladimir Ivanovich Vernadskij (1863-1945). - 488 S., Moskva (Izdatelstvo Nauka).

MOLDAN, B. (1983): Koloběh hmoty v přírodě [The matter cycle of nature]. - 171 p., Praha (Acad.).

MOLDAN, B. (1988): Why study small catchments? Czechoslovak experience. - Lecture, SCOPE General Assembly, Budapest, June 1988.

MOLDAN, B. (1989): Atmospheric deposition: a biogeochemical process. - Praha (Academia), in press.

MOLDAN, B. & DVOŘÁKOVÁ, M. (1987): Atmospheric deposition into small drainage basins studied by Geological Survey. - In: [MOLDAN, B. & PAČES, T.]: Geomon International Workshop on geochemistry and monitoring in representative basins, Prague (Geol. Surv.).

[MOLDAN, B. & PAČES, T.] (1987): Geomon International Workshop on geochemistry and monitoring in representative basins. - Prague (Geol. Surv.).

[NATIONAL BOARD OF WATER AND ENVIRONMENT] (1989): International Co-operative programme on integrated monitoring ECE, 1989, field and laboratory manual, Programme Centre EDC. - Helsinki.

[NRIAGU, J.O.] (1976): Environmental biogechemistry **1**: Carbon, nitrogen, phosphorous, sulfur and selenium cycles: V-XI, 1-423, (Am. Arbor Science).

[NRIAGU, J.O.] (1976): Environmental biogechemistry **2**: Metals transfer and ecological mass balances: XIII-XV, 426-797, (Am. Arbor Science).

[NRIAGU, J.O.] (1984): Changing metal cycles and human health. - 445 p., Berlin (Springer).

OVERREIN, L.N. et al. (1981): Acid precipitation -- effects on forest and fish. - 2nd ed., 175 p., Final Rep. S. N. F. Proj. 1972-1980, Oslo-Ås.

PAČES, T. (1982): Natural and atmospheric flux of major elements from Central Europe. - Ambio **11**: 206-208, Oslo.

[THE NATIONAL SWEDISH ENVIRONMENTAL PROTECTION BOARD] (1985): Monitor 85. - Stockholm.

[UNITED NATIONS EDUCATION, SCIENCE AND COOPERATION OFFICE] (1970): Use and conservation of the biosphere. Proceedings of the intergovernmental conference, Paris, 4-13 Sept. 1968. - 272 p., Paris (UNESCO).

VERNADSKY, V.I. (1926): Biosphere. - Leningrad (Nauchnoe klinikotekhnicheskoe Izdatelstvo).

VERNADSKY, V.I. (1929): Le Biosphère. - 232 p., Paris (Félix Alcer).

VITOUSEK, P.M., EHRLICH, P.R., EHRLICH, A.H. & MATSON, P.A. (1986): Human appropriation of the products of photosythesis. - Bioscience **36**: 368-373, Washington.

WEATHERS, K.C. et al. (1988): Cloudwater chemistry from ten sites in North America. - Envir. Sci. Technol. **22**: 1018-1026, Washington.

Form	Carbon mass (10^{18} grams)	Relative to life
Calcium carbonate (mostly in sedimentary rocks)	35,000	62,500
Ca-Mg carbonate (mostly in sedimentary rocks)	25,000	44,600
Sedimentary organic matter (such as kerogen)	15,000	26,800
Oceanic dissolved biocarbonate and carbonate	42	75
Recoverable fossil fuels (coal and oil)	4.0	7.1
Dead surficial carbon (humus, caliche, etc.)	3.0	5.4
Atmospheric carbon dioxide	0.72	1.3
All life (plants and animals)	0.56	1

Table 1: Global carbon cycle - main reservoirs of carbon (Source: BERNER & LASAGA, 1989)

Element	Earth's crust	Soil	Sea water	Fresh water	Precipitation water
Al	83600	71000	0.002	0.3	0.08
As	1.8	6	0.004	0.0011	0.005
Be	2	0.3	0.0000006	0.00005	0.0001
Ca	46600	15000	412	32	0.45
Cd	0.16	0.35	0.0001	0.00009	0.0002
Cl	120	100	1935,000	13	0.29
Co	29	8	0.00002	0.0002	0.0002
Cr	120	70	0.0003	0.001	0.0015
Cu	68	30	0.00025	0.001	0.008
F	540	200	1.3	0.00018	0.03
Fe	62000	40000	0.002	0.29	0.05
K	18400	14000	339	2.2	0.07
Mg	27600	5000	129000	4	0.06
Mn	1100	1000	0.00002	0.062	0.02
N	19	2000	16	1.6	1.0
Na	22700	5000	1077000	6	0.1
Ni	99	50	0.0006	0.0005	0.001
P (PO$_4$)	1120	800	0.06	0.020	0.015
Pb	13	16	0.00003	0.0008	0.015
S	780	700	905	17.7	1.6
Zn	76	90	0.005	0.01	0.04

Table 2:   Concentrations of chemical elemants in selected biogeochemical reservoirs, mg kg^{-1} (Source: MOLDAN, 1989)

Element	Rock weathering	World river output	River Labe	Global emission into atmo- sphere	Emission in Bohemia and Moravia	Atmo- spheric deposition
Al	2200	73	-	140	-	400
As	0.04	0.13	0.7	0.07	1.9	5
Be	0.07	0.07	0.006	0.0006	0.1	0.05
Ca	1100	3700	9700	70	-	1400
Cd	0.003	0.02	0.02	0.016	0.1	0.2
Cl	3.5	1700	8600	1400	-	380
Co	0.5	0.05	-	0.02	0.2	0.1
Cr	2.7	0.25	-	0.14	0.6	0.5
Cu	1.3	0.73	0.84	0.16	3.9	5
F	25	25	70	0.4	195	140
Fe	1100	130	53	80	-	200
H⁺	-	-	-	-	-	320
K	560	540	1500	30	-	250
Mg	610	100	3400	110	-	140
Mn	25	2	19	2	1.9	12
N	0.7	12.7	1000	1000	-	2600
Na	610	1500	5800	800	-	180
Ni	2.1	0.13	-	0.2	1	0.6
P	27	4.9	-	4	-	10
Pb	0.37	0.7	0.5	1	1	10
S	6.7	930	6600	440	16000	5700
Zn	2	3.7	4.9	0.8	25	30

Table 3: Biogeochemically important fluxes of selected chemical elements
Units: mg m^{-2} a^{-1}

Deposition	Hydrol. year	Water mm	H⁺	Na	K	Mg	Ca	$N^{a)}$	$N^{b)}$	F	Cl	$S^{c)}$
Wet — Vertical	1985	598	37	226	198	133	900	730	540	183	540	1960
	1986	812	55	184	198	119	880	940	620	230	500	2480
Wet — Horizontal	1985	120	22	136	119	80	540	440	330	110	330	1170
	1986	160	33	110	119	72	530	570	370	138	300	1490
Dry — Gases	1985-6	-	870	-	-	-	-	-	2600	-	-	11000
Dry — Particles	1985	-	-	40	189	61	440	274	97	102	77	730
	1986	-	-	21	141	51	300	186	52	102	39	440
Total	1985	720	930	400	510	270	1880	1440	3600	400	950	14900
	1986	970	960	320	460	240	1710	1700	3600	470	840	15400

Table 4: Atmospheric deposition to the Jezeří catchment, situated in area high in air pollution level. Units: $mg/m^2/a$ (Source: MOLDAN & DVOŘÁKOVÁ, 1987)

<u>World NPP</u>

Terrestrial	132.1
Fresh water	0.8
Marine	91.6
Total	224.5

<u>NPP used or diverted by humans</u>

Cropland	15.0
Converted pastures	9.8
Consumed by livestock or	
burnt in natural pastures	1.8
Human-occupied areas	0.4
Wood used for timber, paper	
or firewood	2.2
Forest destruction without use:	
during harvesting of forest	1.3
during harvesting of tree	
plantations	1.6
during clearing for shifting	
cultivation	5.9
during clearing for permanent use	2.0
Fish used for food	2.0
Sub-total	42.6

<u>NPP directly used by humans and domestic animals</u>

Plants eaten by humans	0.8
Plants eaten by domestic animals	2.2
Fish eaten by humans and domestic animals	2.0
Wood used for timber or paper	1.2
Wood used for firewood	1.0
Sub-total	7.2

<u>NPP used, diverted or permanently reduced by humans</u>

NPP used or diverted	42.6
NPP reduced by conversion:	
to cropland	9.0
to pasture	1.4
to desert	4.5
to human-occupied areas and roads	2.6
Total	60.1

Table 5:   Human appropriation of the world's net primary productivity (NPP, in units of 10^{15} g organic matter per year (Source: VITOUSEK et al., 1986)

	Atmospheric concentration (ppbv)	annual rate of increase (%)
carbon dioxide	344,000	0.4
methane	1,650	1.0
nitrous oxide	304	0.25
methyl chloroform	0.13	7.0
ozone	variable	-
CFC 11	0.23	5.0
CFC 12	0.4	5.0
carbon tetrachloride	0.125	1.0
carbon monoxide	variable	0-2

Table 6:   Greenhouse gases in the atmosphere (Source: UNEP, 1987)

Trace metal	Anthropogenic source	Natural source	Total emission
As	19	12	31
Cd	7.6	1.3	8.9
Cr	30	44	74
Cu	35	28	63
Hg	3.6	2.5	6.1
Mn	38	317	355
Mo	3.3	3.0	6.3
Ni	56	30	86
Pb	332	12	344
Sb	3.5	2.4	5.9
Se	6.3	9.3	16
V	86	28	114
Zn	132	45	177

Table 7:   Natural versus anthropogenic emissions of trace metals into the atmosphere in 1983 (Source: NRIAGU, 1989)
All figures in units of 10^9 g a^{-1}

Trace Element Maps and the Regional Distribution of Human Diseases

IAIN THORNTON [*]

Abstract

Deficiency, excess or imbalance of trace elements have been attributed to some human diseases for some time. Maps showing the regional distribution of trace elements are often useful in defining possible problem areas, particularly in several parts of the world in less developed countries where foodstuffs are locally grown. National geochemical atlasses of geochemistry - health connection based on stream sediment and soil surveys are supporting research in that respect. The advantages and limitations of such surveys to human health studies are reviewed. Many of the relations established between the abundance of chemical elements in the environment and human disease are empirical rather than causal.

[*] Author's address: Prof. I. THORNTON, Applied Geochemistry Research Group, Department of Geology, Imperial College of Science, Technology & Medicine, London SW7 2PE, England.

Introduction

There is a common understanding that man's health and wellbeing is influenced by the amounts and proportions of many of the trace elements he is taking in by nowishment, drinking water and the atmosphere. Direct ingestion and inhalation of soil and dust is also sometimes important. The essentiallity and/or toxicity of the elements of the periodic table have been reviewed in detail by UNDERWOOD (1977), CROUNSE et al. (1983), MOYNAHAN (1979) and many others, listing the elements Co, Cr, Cu, F, I, Fe, Mn, Mo, Se, Zn as essential, Ni, Sn, Si, V as possibly essential and Al, As, Cd, Pb, Hg as toxic. Most of the essential elements may also be toxic of present in large concentrations.

Attempts to relate trace element deficiencies and toxicities in man to the geochemical environment have only been partly successful. UNDERWOOD (1979) attributes this to four main reasons:

-- areas of origin to human foods are continually widening so that the overall diet usually contains foodstuffs grown or produced on a range of soils with different geochemical characteristics;

-- modern diets particularly in the western world contain a wide variety of foods, so that trace element imbalance in one type of food may be offset by another;

-- man is at the end of food chain so that gross soil deficiencies or toxicities affecting trace element levels in plant and animal tissues will generally have been modified in order to maintain crop and livestock yields and

-- technical development in food production and food processing, such as the refining of sugar, milling of wheat, canning and use of food additives, can result in gains and losses in trace elements in food, which can erode the direct relationship between man and his natural geochemical environment.

Nevertheless, regional variations in the prevalent human diseases are recognized in many parts of the world, and both epidemiologists on the one hand and geochemists on the other have attempted to seek relationships between these geographical patterns of disease and the geochemical environment. This rapidly advancing research area of **geochemistry and health** has primarily to do with the recognition and understanding of relations between the composition of rocks, soils, dusts, air-borne materials, ground and surface waters and the health of plants, livestock and man (WEBB 1964; WEBB, NICHOL & THORNTON 1968).

Geological Maps and their Limitations

Knowledge of the regional distribution of the chemical elements is, obviously, a prerequisite for studies into the relations between geochemical parameters and health. The regional distribution of soils derived from different parent rocks will generally govern the levels of element abundance in the natural environment. Concentrations of the trace elements in rocks may vary widely (CANNON et al. 1978, Table 1), and there are many geological factors of affecting the health and production of agricultural crops and livestock. For example in rails underlain by ultrabasic rocks levels of Cr and Ni may be elevated, giving rise to toxicity in cereal crops in Scotland (MITCHELL 1964). Black shales often contain high levels of Mo that may induce clinical or subclinical copper defiency in cattle (KUBOTA et al. 1961; THORNSON, THORNTON & WEBB 1972). Areas underlain by carbonate rocks, such as limestone and chalk, some calcareous sandstones, and acid igneous rocks, such as granite, may contain naturally low levels of the essential elements Cu, Co, Zn, Mn and Se, causing clinical effects in crops and grazing farm animals (MITCHELL 1974; DAVIS 1983; THORNTON 1983). Very large concentrations of the one or more of the elements As, Cd, Cu, Cr, F, Mo, Ni, Pb, U and Zn are frequently associated with areas of metalliferous ores and old or present mining activity which may lead to poisoning of vegetation of livestock (THORNTON 1983; DAVIES 1983). It is evident that in many cases geological maps will provide useful information on the likelihood of natural deficiency or excess. However, there are severe limitations to this approach (HOWARTH & THORNTON 1983; PLANT & STEVENSON 1985). In many cases geological maps are too small a scale (i.e. 1:250,000 - 1:1,000,000) to pinpoint localities in which problems might arise, or may only show general geological information without details on lithology. Often map units are based on stratigraphical information rather than detailed lithology. There are also difficulties in applying tabulated information on the average chemical abundance (CLARKE values) of different rock types, as specific discrepancies clearly occur. For example some Scottish granites exceed 36 ppm U compared to the accepted CLARKE value of 4.8 ppm U[16].

The Need for Trace Element Maps

Systematic mapping of element levels in soil or vegetation would provide information closer to man's food chain and remove some of the inconsistencies of geological maps. However in many countries there is little information available on either total or "available" levels of trace elements in soils and the establishment of systematic sampling networks on a grid basis though highly desirable are often pecluded due to cost and time. There are even fewer data on regional variations in trace elements in pasture herbage and food crops.

However, some exceptions exist. For example low density regional soil sampling was carried out by the Branch of Regional Geochemistry of the United States Geological Survey between 1969 and 1973, and regional maps published for the State of Missouri based on 1140 agricultural soils over 180,500 km², giving 10 samples from each of the 114 counties in the state (MIESCH 1976). More detailed soil inventories are currently being prepared for England and Wales based on sampling topsoils (0-15 cm) on a 5 km grid network (DAVIES 1986; MCGRATH, CUNCLIFF & POPE 1986). In Scotland maps have been based on the analysis of B horizon soils from 1000 soils profiles representing the various soil types (BERROW & URE 1976).

Geochemical Surveys

The methodics of compiling multielement geochemical atlasses have been previously reviewed (HOWARD & THORNTON 1983). There is no general concensus on the best methods available as different approaches may be suited to particular geographical areas and climatic conditions. Possible sampling media include rock, soil, stream sediment, lake sediment, surface waters and vegetation. Soils and stream sediments are most widely used. However a soil sample is usually only representative for a small area, and the use of soil to produce trace element maps requires a high-density sampling regime. Active stream sediment however represents natures closest approximation to a composite sample of the weathered products of rock and soil upstream from the sampling site, and the sample is representative of the catchment area. Stream sediment surveys heve been widely used for mineral exploration in many parts of the world and are now also accepted as a means of producing trace element maps for application to agriculture, environmental questions and human health.

Sampling density used to produce geochemical maps largely depend on the size of the target area. For studies related to human health, the requirement would normally be for maps showing regional patterns of trace element distribution rather than point source information which is necessary to locate a problem on a farm or field basis. HOWARD and THORNTON (1983) describe three approaches for regional studies:
-- extrapolation of results obtained for type localities to surrounding regions,
-- very low density random sampling of well defined "homogenous" mapping units and
-- semi-systematic sampling of the entire survey region.

Extrapolation has inherent difficulties and is subject to factors such as changes in geochemical facies within mapped geological units, the presence of transported materials such as fluviatile and aeolian deposits which may mask the geochemical nature of the underlying material, and different weathering and climatic regimes.

In the Soviet Union, KOVALSKY (1970, 1979) applied the concept of biogeochemical mapping, in which he recognized zones characterized by relatively uniform soil-forming processes, climate and comparable biogeochemical behavior of the elements. These were intended to show regions in which endemic diseases and specific metabolic disorders in man and animals due to geochemical influences may occur. Such an approach does not lead to the compilation of trace element maps but rather to the definition of large areas of geochemical similarity within which particular diseases are recognized.

The multielement regional or national geochemical survey is a more widely accepted approach. Such surveys are now undertaken in many parts of the world routinely. The development of systematic sampling procedures and multielement analytical techniques such as direct reading and ICP emission spectrography have resulted in multi-purpose maps with useful application to human health.

Examples of Geochemical Maps Applied to Human Health

1.)   United Kingdom

Relations between geochemistry and health in the U.K. have been the subject of research for the past 20 or more years and have been successively reviewed by THORNTON & WEBB (1979), THORNTON & PLANT (1980) and THORNTON (1985). The subject was considered by a Royal Society Working Party over the period 1979-1981, and their published findings (ROYAL SOCIETY 1983, BOWIE & THORNTON 1985) included the recommendation that current geochemical mapping programmes should by continued as a matter of priority, with the aim of providing systematic data over the whole of Britain on the geographical distribution and chemical and biological forms of a wide range of major and trace elements. They added that research into the interpretation of geochemical maps is of equal inportance in order to ensure their maximum contribution to the understanding of plant, animal, and human health and disease.

The requirement for systematic data on trace element levels in the U.K. has been met in part by geochemical reconnaissance surveys carried out by the Applied Geochemistry Research Group (AGRG) of Imperial College, London and the British Geological Survey (previously known as the Institute of Geological Sciences, IGS). The data obtained by these surveys are available in published atlasses and in machine readable form. They are based mainly on the systematic collection and multielement analysis of stream sediment samples taken at densities ranging from 1 sample per 1 to 2.5 km . A summary of the chemical elements available in these published atlasses is given in table 2. The first national geochemical atlas to be published anywhere in the world was the Wolfson Geochemical Atlas of England and Wales (WEBB et al. 1977), the brainchild of Professor JOHN WEBB who saw such atlasses as an important cartographic requirement for any country. This was based on a survey carried out in 1969 in which 50,000 stream sediment samples were taken over 64,000 sq.mi. over a 10 week period. The atlas published at a scale of 1:2 million is presented as computer smoothed maps for 21 elements, many of potential significance to human health. The IGS atlasses present maps as point source data and as geochemical 'landscape' maps at a scale of 1:625,000.

Research to data has been mainly concerned with the applications of these maps to crop and animal nutrition and to pollution studies, and human health has received less attention. This is possibly due to the fact that to date, there is little evidence directly linking the geochemical environment and the presence of human disease in the U.K. with the exception of

a)   the extremely well documented enverse relationship between the fluorine content of the water supply and the incidence of dental caries, and

b)   the much diminished prevalence of endemic goitre associated primarily with a deficiency of iodine in water supplies and food.

Empirical relationships are more common and include

a)   a negative correlation between water hardness and the prevalence of cardiovascular disease,

b)   relationships between the distribution of cancers and heavy metals in the soils and

c)   variations in the prevalence of dental caries and concentrations of Mo and Pb.

In these latter examples no causal factor linked to geochemistry has yet been established.

However, there have been two major studies into the possible effects of increased exposure to metals in "geochemical hotspot" situations, delineated by the Wolfson Atlas maps, both related to widescale metal pollution from historical mining and smelting activities:

--   The map for lead (not illustrated) highlights anomalously high levels of this metal in several discrete areas totalling some 4,000 km . In an area of past lead-zinc mining in Derbyshire, a collaborative study by the Paediatric Unit, St. Mary's Hospital Medical School and AGRG, showed, when households were grouped according to the Pb content of their garden soils, Pb in the blood and hair of the 2-4 year old children increased with that in soil and housedust (Table 3). However, none of the Pb values in children were high enough to be considered hazardous at that time, even though the amounts in soil and dust were extremly high, peaking at 2.8 and 2.5 % Pb respectively (BARLTROP et al. 1975). The children's Pb status does, however, reflect that of the environment, and in the absence of evidence linking the human burden with Pb in foodstuffs of water supplies, it was suggested that the major pathway of Pb into the children was through inhalation and involuntary ingestion of dust particles (BARLTROP et al. 1975).

--   The map for cadmium (figure 1) again mainly focuses attention on areas contaminated by mining and smelting operations.

A comprehensive environmental health study was conducted around the village of Shipham, Somerset where Zn was mined as the ore smithsonite ($ZnCO_3$) over the period between 1700 and 1850. The ore, with appreciable amounts of Cd as a guest element, was washed in ponds and calcined within the village. Collaborative investigations involving national and local government, Westminster Medical School and AGRG showed:

-- 60 % of household garden soils with over 60 ppm Cd, greatly exceeded the levels
in polluted paddy soils associated with the well documented 'itai-itai'disease in
Japan; these soils also contained large amounts of Zn with Zn:Cd rations of around
90:1;

-- housedust samples averaged 26 ppm Cd and 2300 ppm Zn;

-- from studies on metals in locally grown vegetables and diets, an average human
intake of 200 ug Cd per week was calculated, compared with the average U.K.
intake of 140 ug Cd per week; individual intakes rarely exceeded the World Health
Organization's provisional tolerable weekly intake of 450-500 ug Cd (SHERLOCK
et al. 1983);

-- from health inventories and biochemical tests on 548 residents of Shipham and 543
controls from a nearby incontaminated village only slight differences attributable to
Cd were found (BARLTROP & STREHLOW 1982).

In this unique situation it is possible that the large amounts of Zn and Ca (Cd?)
present in the environment are affording some degree of protection against possible adverse
effects of cadmium.

These two examples clearly show the potential for trace element maps to define
areas with high levels of potentially toxic elements. These anomalous areas may be due
to the presence of naturally metal-rich rocks or man-made pollution. Potential applications
of geochemical maps are however far wider than this, as they present a complete
geochemical landscape ranging from very low to very high concentrations of a wide range
of element relations with disease and multielement associations and interactions.

To date in the U.K. there has been only a limited attempt to relate this large
geochemical data bank to epidemiology. This is not due to a shortage of reliable
epidemiological data. The development of mortality mapping in the U.K. has been
reviewed by GARDNER (1988), culminating in the publication of two mortality atlasses

for England and Wales respectively for cancers and other causes of death based on records between 1964 and 1978 (GARDNER et al. 1978, 1984). Regional information on mortality from cardiovascular diseases is also available and in a study of 234 British towns has been negatively associated with water hardness (POCOCK et al. 1980).

Problems of relating information, even at an empirical level, between geochemical and epidemiological data bases relate to two principal factors:

-- the respective data and the maps compiled from them, are originally obtained for clearly defined but different reasons, using a different sampling strategies and different criteria for representivity. Mortality data are usually based on administrative units, which in U.K. may be at Local Authority district or County level. Thus the finest detail mapped in the two above atlasses is based on 1366 separate districts. Less common causes of death are mapped on a larger geographical scale. Conversely geochemical maps are based on a much more detailed sampling network and the patterns of element abundance produced frequently reflect chemical variations in the surface environment bearing no relationship to administrative areas or hospital catchment areas. The use of postal codes to provide a more rigorous location of residence may in the future lead to better definition of mortality, but the release of such point source information to map compilers may raise problems of medical ethics.

-- The majority of the U.K. population lives in large cities and towns in which the "chemical environment" may bear little relationship to "baseline geochemistry". In particular inputs of metals from industrial and urban sources may greatly modify the home environment. A national survey of metals in U.K. urban dusts and soils based on 5300 households in 53 localities has confirmed elevated concentration of lead, zinc, cadmium and copper in urban compared to rural areas, with significant differences between towns depending on their industrial history and age of urban development (THORNTON et al. 1985). This and other studies have demonstrated the need for the development of sampling and analytical strategies to "fingerprint" the chemistry of the urban environment to which the majority of the population is exposed. Such an approach coupled with traditional geochemical surveys will provide a more robust trace element data base for comparison with human health.

2.) United States of America

Perhaps the largest multielement stream sediment programme undertaken to date is the National Uranium Resource Evaluation programme of the U.S. Department of Energy, with a nominal sampling density of 1 per 10 km over much of the U.S. and analysis of 16 to 48 elements. However, only data for limited areas are available in map form, including those compiled into the Geochemical Atlas of Alaska (LOS ALAMOS NAT. LAB. 1983). A recent review of regional scale geochemical mapping in the United States (MCNEAL 1986) discusses the procedures used in this and earlier programmes including the systematic sampling of surficial materials by SHACKLETTE and co-workers (1971, 1984).

Considerable effort has also gone into mapping human morbidity and mortality in the U.S., and it is not possible to review the extensive literature in this article. Atlasses showing cancer mortality by county were based on mortality data for the period 1950-1969 (MASON et al. 1975, 1976). Particular attention has also been focused on cardiovascular diseases (COMSTOCK 1986).

Interest in relations between geochemistry and health was stimulated by National Academy of Sciences Subcommittee on the Geochemical Environment in Relation to Health and Disease which organized several multi-disciplinary workshops extensively reviewing the many U.S. studies (NATIONAL ACADEMY OF SCIENCES 1974, 1977, 1978). The proceedings of these workshops and those of the continuing series of conferences on Trace Substances in Environmental Health organized at the University of Missouri, Columbia from 1966 to the present (HEMPHILL 1966, 1985) provide a substantial literature base. However, it is perhaps fair comment to conclude that there is still little evidence conclusively linking human disease and the geochemical environment in the U.S.A. Several empirical relationships have been established, including, as in Britain, that between cardiovascular disease mortality and water quality (COMSROCK 1986). Comparisons between the NURE geochemical data bank and the cancer maps and those for other diseases provide future possibilities, but success will be subject to the same limitations as in the U.K. and any other industrialized country.

3.) China

Within the last decade the most significant advances in geochemistry and health have taken place in China, where the occurrence of both Keshan Disease, an endemic cardiomyopathy, and Kaschin-Beck Disease, an endemic osteoarthropathy, have been shown to be related to the selenium status of the environment (TAN et al. 1982, ZHU et al. 1984, XU & JIANG 1986). However as yet there are no national maps of Se. These causal relationships have been shown as a result of studies with Se and other elements on soils, foodstuffs and drinking waters in problem and control areas coupled with positive responses to Se supplementation in the population. Systematic studies on trace elements in soils, waters and crops are being planned by the Institute of Geochemistry, Academia Sinica, and trace element maps will be compiled for some provinces. A national atlas of cancer mortality has been published (EDIT. COMM. 1979), and the occurence of goitre and human fluorosis is also being mapped. This latter problem affects some 16 million people of whom 800,000 have the more serious form of skeletal fluorosis.

National geochemical maps of China are currently being prepared for puposes of mineral exploration. Such maps would have an obvious additional application to health studies in a country where the population largely live on locally produced foodstuffs and in which research into geochemistry and health receives increasing attention.

4.) Scandinavia

Two symposia have brought together Scandinavian interests in 'geomedicine', under the auspices of the Norwegian Academy of Science and Letters (LÅG 1980, 1984), at which geochemists, epidemiologists and several other disciplines have exchanged information on a wide range of projects in Norway, Finland, Sweden, Denmark and Iceland. The subject has also been reviewed by LÅG (1983). Extensive geochemical mapping, using mainly soil and drainage samples, has been completed in the Nordic countries, and comparisons are currently being made with the distribution of cancers (BØLVIKEN, OTTESEN & GLATTRE 1980), and several other diseases. Maps have also been compiled from the analysis of humus layers of Norwegian forest soils showing distribution patterns for several trace elements including selenium. The application of geochemical maps to medicine is an active research area in the Scandinavian countries, though the relationships between geochemistry and disease proposed are as yet empirical.

5.) India

There are at present element maps for the subcontinent of India, though the area of geochemistry and health is active both in relation to human health and to agriculture. Extensive surveys have been made of micronutrients in Indian agricultural soils and a similar scheme is planned for micronutrients in livestock (ARORA 1986). Both skeletal and dental fluorosis are widely distributed in India in communities located according to the distribution of fluoride bearing rocks, and it has been suggested that geochemical mapping would greatly help in identifying the extent of the human fluorosis problem (KRISHNA-MACHARI 1986).

Conclusions

Geochemical mapping and disease mapping programmes are currently active in many parts of the world, and multi-disciplinary studies into the use of trace element maps to delineate areas in which the chemical composition of foodstuffs and drinking water may influence human health and disease will continue to fluorish into the 21st century. Such studies are likely to be most useful in the developing countries where people live "close to the land" and in areas of low population where the effects of urbanization are small. In countries such as the U.K. and U.S.A., where the majority of the population lives in large cities and towns, geochemical maps are less likely to relate to human health problems, and there is a need to develop research in urban geochemistry and establish techniques for fingerprinting the multielement status of the urban environment.

Many of the correlations established between the geochemical environment and human health are empirical and the question has been raised as to when a correlation or association becomes a causation (UNDERWOOD 1979). It is also true that some such published relationships are based on very small numbers of deaths and can be misleading (SHAPER 1979). Causal relations between geochemistry and health are still very limited and probably only firmly established between iodine and goitre and fluorine and fluorosis worldwide, and selenium and selenium-responsive diseases in China. New techniques to study possible relationships between element levels and disease in man using image analysis and multi-variate statistics hold promise for the future, and provide the basis for utilizing the full multielement nature of geochemical data bases in the search for the cause and cure of disease.

References

ARORA, S.P. (1986): Livestock problems related to geochemistry in India including selenium toxicity and goitre.- In: [THORNTON, I.]: Proc. 1st Int.Symp.Geochemistry and Health. London, Science Reviews Ltd.

BARLTROP, D.; STREHLOW, C.D.; THORNTON, I. & WEBB, J.S. (1975): Adsorption of lead from dust and soil.- Postgrad.medic.J. **51**: 801-804.

BARLTROP, D. & STREHLOW, C.D. (1982): Cadmium and health in Shipam.- Lancet Dec. 18.

BERROW, M.L. & URE, A.M. (1976): Trace element mapping of Scottish soils.- In: [THORNTON, I.]: Proc. 1st Int.Symp.Geochemistry and Health. London, Science Reviews Ltd.

BØLVIKEN, B.; OTTESEN, R.T. & GLATTRE, E. (1980): Comparison of geochemical and epidemiological data from south-eastern Norway.- In: [HEMPHILL, D.D.]: Trace substances in environmental health - XIV, 19-26, Columbia, Univ. of Missouri.

[BOWIE, S.H.U. & THORNTON, I.] (1985): Environmental geochemistry and health.- Dordrecht, Reidel.

CANNON, H.L.; CONALLY, C.G.; EPSTEIN, J.B.; PARKER, J.G.; THORNTON, I. & WIXSON, B.G. (1988): Rock: the geological source of most trace elements.- In: Geochemistry and the environment III: distribution of trace elements related to the occurence of certain cancers, cardiovascular diseases and urolithiosis. 17-31, Washington, DC, National Academy of Sciences.

COMSTOCK, G.W. (1986): Water quality and cardiovascular disease: a review of recent studies in Canada and the United States.- In: [THORNTON, I.]: Proc. 1st Int.Symp.Geochemistry and Health. London, Science Reviews Ltd.

CROUNSE, R.G.; PORIES, W.J.; BRAY, J.T. & MAUGER, R.L. (1983): Geochemistry and man: health and disease.- In: [THORNTON, I.]: Applied environmental geochemistry, 267-333, London, Academic Press.

DAVIES, B.E. (1983): Heavy metal contamination from base metal mining and smelting: implications for man and his environment.- In: [THORNTON, I.]: Applied environmental geochemistry, 425-462, London, Academic Press.

DAVIES, B.E. (1986): Baseline survey of metals in Welsh soils.- In: [THORNTON, I.]: Proc. 1st Int.Symp.Geochemistry and Health. London, Science Reviews Ltd.

[EDITORIAL COMMITTEE FOR THE ATLAS OF CANCER MORTALITY IN THE PEOPLE'S REPUBLIC OF CHINA] (1979): Atlas of cancer mortality in the People's Republic of China.- China Map Press.

GARDNER, M.J.; WINTER, P.D.; TAYLOR, C.P. & ACHESON, E.D. (1978): Atlas of cancer mortality in England and Wales, 1968-1978.- Chichester, John Wiley.

GARDNER, M.J.; WINTER, P.D. & BARKER, D.J.P. (1984): Atlas of mortality from selected diseases in England and Wales, 1968-1978.- Chichester, John Wiley.

GARDNER, M.J. (1986): Mapping cause-specific mortality in England and Wales.- In: [THORNTON, I.]: Proc. 1st Int.Symp.Geochemistry and Health. London, Science Reviews Ltd.

[HEMPHILL, D.D.] (1966-1985): Trace substances in environmental health.- **1-19**, Colombia, Univ. of Missouri.

HOWARTH, R.J. & THORNTON, I. (1983): Regional geochemical mapping and its application to environmental studies.- In: [THORNTON, I.]: Applied environmental geochemistry, 41-73, London, Academic Press.

KOVALSKY, V.V. (1970): The geochemical ecology of organisms under conditions of varying contents of trace elements in the environment.- In: [MILLS, C.F.]: Trace metabolism in animals, 385-396, London, Livingstone.

KOVALSKY, V.V. (1979): Geochemical ecology and problems of health.- Phil.Trans.roy.-Soc.Lond. **B288**: 185-191, London.

KRISHNAMACHARI, K.A.V.R. (1986): Geographical distribution of human fluorosis in India.- In: [THORNTON, I.]: Proc. 1st Int.Symp.Geochemistry and Health. London, Science Reviews Ltd.

KUBOTA, J.; LAZAR, V.A.; LANGEN, L.N. & BEESON, K.C. (1961): The ralationship of soils to molybdenum toxicity in cattle in Nevada.- Soil Sci.Soc.Amer.Proc. **25**: 227-232.

LÅG, J. (1980): Geomedical aspects in present and future research.- Oslo, Universitets forlaget.

LÅG, J. (1983): Geomedicine in Scandinavia.- In: [THORNTON, I.]: Applied environmental geochemistry, 335-353, London, Academic Press.

LÅG, J. (1984): Geomedical research in relation to geochemical registration.- Oslo, Universitetsforlaget.

[LOS ALAMOS NATIONAL LABORATORY] (1983): The geochemical atlas of Alaska.- Los Alamos.

MASON, T.J.; MCKAY, F.W.; HOOVER, R.; BLOT, W.J. & FRAUMENI, J.F.Jr. (1975): Atlas of cancer mortality for U.S. Counties: 1950-1969.- DHEW Publ. (NIH) **75-780**, Washington, DC, U.S. Government Printing Office.

MASON, T.J.; MCKAY, F.W.; HOOVER, R.; BLOT, W.J. & FRAUMENI, J.F.Jr. (1975): Atlas of cancer mortality among U.S. non-whites: 1950-1969.- DHEW Publ. (NIH) **76-1204**, Washington DC, U.S. Government Printing Office.

McGRATH, S.P.; CUNCLIFFE, C.H. & POPE, A.J. (1986): Lead, zinc, cadmium, copper, nickel and chromium concentrations in the topsoils of England and Wales.- In: [THORNTON, I.]: Proc. 1st Int.Symp.Geochemistry and Health. London, Science Reviews Ltd.

McNEAL, J.M. (1986): Regional-scale geochemical mapping in the United States.- In: [THORNTON, I.]: Proc. 1st Int.Symp.Geochemistry and Health. London, Science Reviews Ltd.

MIESCH, A.T. (1976): Geochemical survey of Missouri - methods of sampling, laboratory analysis and statistical reduction of data.- U.S.geol.Surv.prof.Pap. **954-A**, Washington.

MILLS, C.F. (1979): Trace elements in animals.- Phil.Trans.roy. Soc.Lond. **B288**: 51-63, London.

MITCHELL, R.L. (1964): Trace elements in soils.- In: [BEAR, F.E.]: Chemistry of the soil, 2nd edn., 320-368, New York, Reinhold.

MITCHELL, R.L. (1974): Trace element problems in Scottish soils.- Neth.J.agric.Sci. **22**: 295-304.

MOYNAHAN, E.J. (1979): Trace elements in man. Phil.Trans.roy. Soc.Lond. **B288**: 65-79, London.

[NATIONAL ACADEMY OF SCIENCES] (1974, 1977, 1978): Geochemistry and the environment.- **1**, 1974; **2**, 1977; **3**, 1978. Washington, DC, National Academy of Sciences.

PLANT, J. & STEVENSON, A.G. (1985): Regional geochemistry and its role in epidemiological studies.- In: [MILLS, C.J.; BEMNER, I. & CHESTERS, J.K.]: Trace elements in man and animanls, 900-906, Farnham Royal, Commonwealth Agricultural Bureaux.

POCOCK, S.J.; SHAPER, A.G., COOK, D.G.; PACKHAM, R.F.; LACEY, R.F.; POWELL, P. & RUSSELL, P.F. (1980): British regional heart study: geographic variations in cardiovascular mortality and the role of water quality.- Brit.medic.J. **280**: 1243-1249.

[ROYAL SOCIETY] (1983): Environmental geochemistry and health. A report in summary.- London, Royal Society.

SHACKLETTE, H.T.; HAMILTON, J.C.; BOERNGEN, J.G. & BOWLES, J.M. (1971): Elemental composition of surficial materials in the conterminous United States.- U.S.geol.Surv.prof.Pap. **574-D**, Washington.

SHACKLETTE, H.T. & BOERNGEN, J.G. (1984): Element concentrations in soils and other surficial materials of the conterminous United States.- U.S.geol.Surv.prof.Pap. **1270**, Washington.

SHAPER, A.G. (1979): Epidemiology for geochemists.- Phil.Trans. roy.Soc.Lond. **B288**: 127-136, London.

SHERLOCK, J.C.;SMART, G.A.; WALTERS, B.; EVANS, W.H.; MCWEENY, D.J. & CASSIDY, W. (1983): Dietary surveys on a population at Shipam, Somerset, United Kingdom.- Sci.total Environ. **29**:121-142.

TAN, J.A. et al. (1982): The relation of Keshan Disease to the national environment and the background of selenium nutrition.- Acta Nutrimenta sin. **4**: 175.

THOMSON, I.; THORNTON, I. & WEBB, J.S. (1972): Molybdenum in black shales and the incidence of bovine hypocuprosis.- J.Sci. Fd.Agric. **23**: 871-891.

THORNTON, I. & WEBB, J.S. (1979): Geochemistry and health in the United Kingdom.- Phil.Trans.roy.Soc.Lond. **B288**: 151-168, London.

THORNTON, I. & PLANT, J. (1980): Regional geochemical mapping and health in the United Kingdom.- J.geol.Soc.Lond. **137**:575-586, London.

THORNTON, I.; JOHN, S.; MOORCROFT, S. & WATT, J. (1980): Cadmium at Shipam - a unique example of environmental geochemistry and health.- In: [HEMPHILL, D.D.]: Trace Substances in environmental health - XIV, 27-37, Columbia, University of Missouri.

THORNTON, I. (1983): Geochemistry applied to agriculture.- In: [THORNTON, I.]: Applied environmental geochemistry, 231-266, London, Academic Press.

THORNTON, I. (1985): Environmental geochemistry and health in the U.K.- In: [LÅG, J.]: Geomedical research in relation to geochemical registration, 125-136, Oslo, Universitetforlaget.

THORNTON, I.; CULBARD, E.; MOORCROFT, S.; WATT, J.; WHEATLEY, M. & THOMPSON, M. (1988): Metals in urban dusts and soils.- Environ. Technol.Lett. **6**: 137-144.

UNDERWOOD, E.J. (1977): Trace elements in human and animal nutrition.- 4th edn., New York, Academic Press.

UNDERWOOD, E.J. (1979): Trace elements and health: an overview.- Phil.Trans.roy.Soc.- Lond. **B288**: 5-14, London.

WEBB, J.S. (1964): Geochemistry and life.- New Scient. **23**: 504-507.

WEBB, J.S.; NICHOL, I. & THORNTON, I. (1968): The broadening scope of regional geochemical reconaissance.- Proc. 13 int. geol. Cong. **6**: 131-147.

WEBB, J.S.; THORNTON, I.; THOMPSON, M.; HOWARD, R.J. & LOWENSTEIN, P.L. (1978): The Wolfson Geochemical Atlas of England and Wales.- Oxford, Univers. Press.

XU, G. & JIANG, Y. (1986): Selenium and the prevalance of Keshan Disease and Kaschin-Beck Diseases in China.- In: [THORNTON, I.]: Proc. 1st Int.Symp.Geochemistry and Health. London, Science Reviews Ltd.

ZHU, Z.Y. et al. (1984): Researches on the geographical cause of Kaschin-Beck Disease and Keshan Disease.- Scient.geogr.sin. 4: 365.

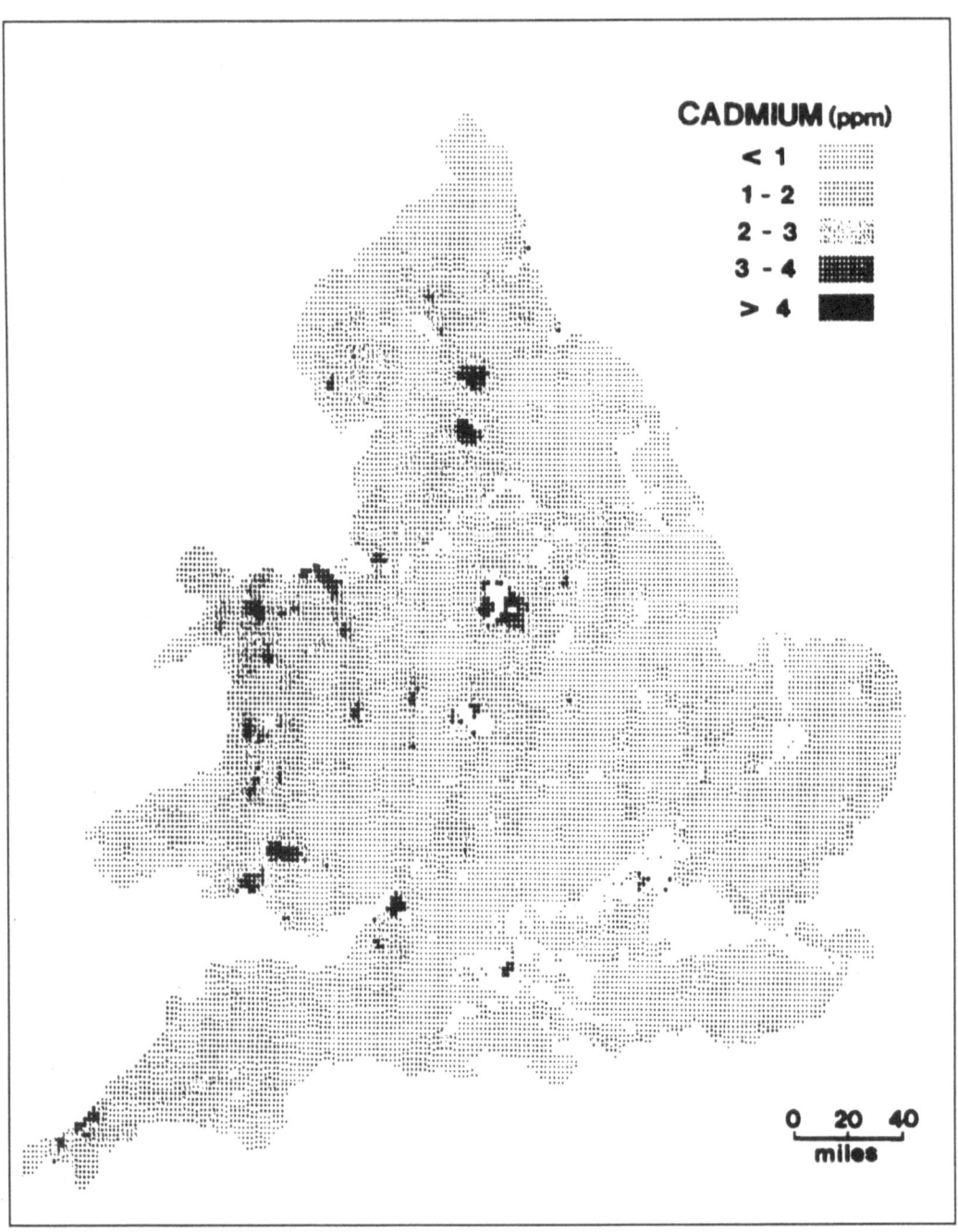

Figure 1: The distribution of cadmium in stream sediments in England and Wales

Elements	Ultramafic Igneous	Basaltic Igneous	Granitic Igneous	Shales and Clays	Black Shales	Limestones	Sandstones
Arsenic	0.3-16 / 3.0	0.2-10 / 2.0	0.2-13.8 / 2.0	-- / 10	--	0.1-8.1 / 1.7	0.6-9.7 / 2.0
Barium	0.2-40 / 1	20-400 / 300	300-1,800 / 700	460-1,700 / 700	70-1,000 / 300	-- / 10	-- / 20
Cadmium	0-0.2 / 0.05	0.006-0.6 / 0.2	0.003-0.18 / 0.15	0-11 / 1.4	<0.3-8.4 / 1.0	0.05	0.05
Chromium	1,000-3,400 / 1,800	40-600 / 220	2-90 / 20	30-590 / 120	26-1,000 / 100	-- / 10	-- / 35
Cobalt	90-270 / 150	24-90 / 50	1-15 / 5	5-25 / 20	7-100 / 10	-- / 0.1	-- / 0.3
Copper	2-100 / 15	30-160 / 90	4-30 / 15	18-120 / 50	20-200 / 70	-- / 4	-- / 2
Fluorine	--	20-1,060 / 360	20-2,700 / 870	10-7,600 / 800	--	0-1,200 / 220	10-880 / 180
Iron	94,000	86,500	14,000-30,000	47,200	20,000	3,800	9,800
Lead	-- / 1	2-18 / 6	6-30 / 18	16-50 / 20	7-150 / 30	-- / 9	1-31 / 12
Mercury	0.004-0.5 / 0.1	0.002-0.5 / 0.05	0.005-0.4 / 0.06	0.005-0.51 / 0.09	0.03-2.8 / 0.5	0.01-0.22 / 0.04	0.001-0.3 / 0.05
Molybde-num	0.3	0.9-7 / 1.5	1-6 / 1.4	2.5	1-300 / 10	0.4	0.2
Nickel	270-3,600 / 2,000	45-410 / 140	2-20 / 8	20-250 / 68	10-500 / 50	20	2
Selenium	0.05	0.05	0.05	0.6	--	0.08	0.05
Vanadium	17-300 / 40	50-360 / 250	9-90 / 60	30-200 / 130	50-1,000 / 150	-- / 20	-- / 20
Zinc	-- / 40	48-240 / 110	5-140 / 40	18-180 / 90	34-1,500 / 100	-- / 20	2-41 / 16

Table 1: Range and mean concentrations of some elements in igneous and sedimentary rocks (ppm) Adapted from table compiled by M. FLEISCHER and H.L. CANNON (CANNON et al. 1978).

	Published atlases	Elements
Applied Geoche-mistry Research Group, Imperial College	Northern Ireland	Al, As, Ba, Ca, Cr, Co, Cu, Ga, Fe, Pb, Mg, Mn, Mo, Ni, K, Sc, Si, Sr, V, Zn.
	England and Wales	Al, As, Ba, Cd, Ca, Cr, Co, Cu, Ga, Fe, Pb, Li, Mn, Mo, Ni, K, Sc, Sr, Sn, V, Zn.
	Shetland	Ba, Be, B, Cr, Co, Cu, Fe, Pb, Mn, Mo, Ni, U, V, Zn, Zr.
British Geological Survey	Orkney	Ba, Be, B, Cr, Co, Cu, Fe$_2$O$_3$, Pb, Mn, Mo, Ni, U, V, Zn, Zr.
	South Orkney and Caith-ness	Ba, Be, B, Cr, Co, Cu, Fe, Pb, Mn, Mo, Ni, Sr (partial data), Ti (partial data), U, V, Zn, Zr.
	Sutherland	Be, B, Cr, Co, Cu, Fe, Pb, Mn, Mo, Ni, U, V, Zn, Zr.
	Lewis/Little Minch	Ba, Be, Bi, B, Ca, Cr, Co, Cu, Fe, K, La, Pb, Li, Mg, Mn, Mo, Ni, Sr, Ti, U, V, Y, Zn, Zr.
	Provisional maps available for purchase through NGDB/GISA Great Glen	Ba, Be, B, Ca, Cr, Co, Cu, Fe, K, La, Pb, Li, MgO, Mn, Mo, Ni, Rb, Sr, Ti, U, V, Y, Zn, Zr.
	Argyll	
	Moray/Buchan Tay/Forth Lake District	

Table 2: Summary of chemical elements available in geochemical atlasses of the United Kingdom

Soil (ppm)	House dust (ppm)	Blood Child (g/100 ml)	Blood Mother (g/100 ml)	Hair Child (ppm)
130-900	190-2450	15-33	9-40	2-36
420 (24)	531 (20)	21 (29)	14 (25)	8 (29)
1050-9100	390-25000	13-45	11-44	2-62
3390 (40)	1564 (36)	24 (43)	19 (41)	11 (43)
10000 - 28000	1000-5100	21-43	13-19	10-39
13969 (8)	2582 (8)	29 (10)	15 (8)	29 (10)

Table 3:   Lead content of garden soil, housedust, blood and hair
Range and geometric mean; number of samples in brackets.

Industrial Minerals and Rocks: Present Trends in Exploration, Exploitation and Use

GERD W. LÜTTIG [*]

Abstract

Today the industrial minerals and rocks are the most important mineral resource group as far as quantity goes and after the energy carriers the most significant as far as value goes. Their value is rising constantly and in their use there are possibilities for projects with low investment costs and quick cash flow. This is important for the developing countries in particular. Since it is partly a matter of near surface bulk raw materials; their use involves local conflicts with other utilization claims. It is necessary for the geoscience to work out suggestions for solutions to these conflicts, to simultaneously mobilize research and training capacities and in face of the present desperate situation to improve them so that a better contribution than is presently being made can be made by this professional field for the public welfare.

[*] Author's address: Prof. Dr. G.W. LÜTTIG, Chair of Applied Geology, University of Erlangen-Nuremberg, Schloßgarten 5, D-8520 Erlangen, FRG.

1)   The role of industrial minerals and rocks in the global socioeconomic network

In the public discussion on the extent of the exhaustibility or limits to natural resources, which began with the slogan "limits to growth" (CLUB OF ROME, see MEADOWS et al. 1972, PESTEL 1975 and others), an opinion prevails at present according to which the warnings expressed there are supposed to have been exaggerated. And the shock that shook mankind with the so-called oil crisis has given way to a more balanced cool attitude in which decreased consumption, increased exploration for resources, improved economic regulation and better and more ecologically conscious planning play a role. We do not want to discuss which of the two standpoints at long last will be the right one, but the indisputable fact of exhaustibility should be taken into consideration so that economics, science and politics be found in an area of conflict (Fig. 1) in which they must prove themselves. Talking about it will not make the problem go away, namely that human socioeconomics will be driven by events in the constantly growing economies of the developing countries, marked by increasing consumption of natural goods, no matter whether the industrial nations can control their problems or not. The consumption figures shown in Table 1 will be true for the developing countries in the near future, too.

Within this problem area it has been made evident in recent years, as a result of the efforts of just a few resource experts (also of the author in 1979, 1983), that in this development as far as the mineral resource group energy carriers, metallic and non-metallic raw materials go, not the first two groups, which usually stand in the foreground in the crisis discussions, but the latter, better called industrial minerals and rocks, possesses particularly dynamic behavior which changes the picture constantly. First of all, a fact previously overlooked, the industrial minerals have a leading role in the consumption statistics of important industrial nations (Fig. 2). Then the statistics first presented by the author even showed for the consumption figures of the year 1977 good front places in the world statistics on the industrial minerals, not only with reference to quantity but also to the production value (so called Raw Materials Snake).

This tendency has become even clearer in recent years, as a study prepared for 1987 by a scholar of the author, J. LAWATSCHECK (1990) shows. According to this, the following industrial minerals rank among the first 10 minerals in the order of rank in quantity:

sand and gravel	1st,
hard and dimension stone	3rd,
limestone and dolomite	5th,
clays	9th and
peat	10th (Fig. 3).

Groundwater, which certainly has an important place, was not considered.

In the statistics on value, the industrial minerals occupy the following important places:

sand and gravel	4th,
limestone and dolomite	7th,
hard and dimension stone	8th, (Fig. 4).

Among the 50 most important minerals as to value,

27 are industrial minerals,
18 are metallic raw materials and
5 are energy carriers.

Compared to 1977, there is an increasing trend as to quantity for the industrial minerals sand and gravel, limestone and dolomite, peat (primarily used for amelioration, thus not counted as an energy carrier), gypsum, kaolin, bentonite, soda, talcum, titanium ores, and fluorspar. As fas as quantity goes, the importance of the industrial minerals found in the top group (Tables 2 and 3) increases, in comparison the increase in value (Table 4 and 5) is not as significant.

The following should be noted in these statistics: the fact that for both the metallic as well as the non-metallic raw materials a few dominate the statistics does not become sufficiently clear. For the metallic raw materials the following dominate the statistics as to quantity:

iron ore	with	85.9 %,
bauxite	with	8.7 %,
manganese ore	with	2,0 %
		96,6 %.

As to value it is not so significant:

iron ore	with	27.7 %,
gold	with	25.5 %,
bauxite	with	13.6 % and
copper ore	with	12.7 %
		79.0 %.

For the industrial minerals as to quantity,

sand and gravel	with	54.2 %
limestone and dolomite	with	21.0 % and
hard and dimension stone	with	16.3 %
		92.5 %

dominate absolutely, less significant as to value:

sand and gravel	with	32.4 %,
limestone and dolomite	with	16.9 % and
hard and dimension stone	with	15.3 %
		64.6 %.

For the energy carriers the important quantity and value percentages (Table 6) are in the hands of oil, coal and natural gas.

An important fact should be added. The statistical comparison of production and value of both mineral resource groups (metallic and non-metallic) shows that since 1950 the non-metallics have overtaken the metallics and are outstripping them.

The high percentage of the exploitation of non-metallic raw materials in overall mining has considerable consequences for society and the geosphere it lives in.

It is true that especially for industrial minerals one can usually speak of "universality throughout the landscape", i.e. of an only temporal impact on the structure of the landscape and the presence of numerous possibilities for reclaiming the landscape. Two thirds of the important industrial minerals types as to quantity and value, and if one counts the total production amount far more than 90 %, are exploited from near surface deposits. Even with respect to the surface are claimed here, this raw material exploitation is noteworthy for society. It is true that one should not exaggerate in this respect, for:

1)   The consumption of area is not as significant as assumed in some circles. In 1987 in the Federal Republic of Germany the near surface resource exploiting industry only claimed

0.28 % of the total area.

In contrast about

10 % of the total area

is taken up by building and business area and the traffic industry (Source: Statistisches Bundesamt, Wiesbaden).

2)   The largest open-cast mines are not created by the industrial mineral exploiting industry but by lignite, iron ore, bauxite, copper ore and oil shale mining.

Nevertheless, it is indisputable that this near surface resource exploitation causes impacts in the locations concerned, in fact in whole landscape sections of the pedosphere, hydrosphere, atmosphere and biosphere. Specifically this is done to

-- the pedosphere, by removal of + fertile soil, possibly making it unusable for re-use later, and by contamination with waste from mining or dressing,

-- the hydrosphere by groundwater lowering or contamination, disturbance to the run-off regime, etc.,

-- the atmosphere by emission of dust,

-- the litho- and hydrosphere by combustion residues,

-- the biosphere by destruction or expulsion of organisms from the biotops that have been built up, biological and medical damage even as far as destruction of species by impacts on the quality of life of the human environment itself. Here the acoustical pollution (blasting, noise from mining machines and transport) caused by near surface mining and quite unpleasant at times for human beings should not be forgotten (LÜT-TIG 1988).

Since minerals only occur at specific locations due to geological laws and the geological history of the lithosphere, and only at these places, it is in the nature of the matter that natural resources which cannot be displaced and are important for the economy and found in locations that at the same time could be of interest for other types of utilization. Of these utilization types, groundwater extraction, agricultural and forestry industry (with the special branches horticulture, fruit and wine-growing), leisure time and recreation, nature conservation and landscape protection are of special interest to society. In an individual case it might seem that compromises are possible among the utilization types, especially in such cases where the raw material exploitation only takes place for a short period of time and in which more or less good reclamation of the structure of the landscape is possible afterwards. Fig. 5 shows such a tolerance diagram for the Federal Republic of Germany. Without a doubt there are examples in which compromises are not possible.

This leads to considerable problems in land-use planning. How they can be solved will be discussed in Chapter 6.

Generally one imagines that here it is a question of problems which mainly play a role in the industrial nations. Experience, however, shows that they also occur in the developing countries, and in fact to an increasing extent. It is really a dream, which many a propagandist of the so-called North/South dialog dreams, that namely developing policy could originate from the free development of the natural environment's potential. The problem of development is simply seen by these naive philosophers and technicans as if the superior technology of the industrial countries must simply be applied in the developing countries to their exceptional geopotential, and then nature would open her treasure chest and general prosperity would be the unavoidable consequence.

The author has along with others (LÜTTIG 1978, ARCHER, LÜTTIG & SNEZHKO 1987) frequently referred to the fact that this attitude is a fateful error. Not only with respect to minerals but also water and soil potential, the situation is such that a whole number of developing countries has literally nothing. Apart from the large number of natural resource haven-nots, there are only a few developing countries that are relatively rich in metallic raw materials and energy carriers. Thus, a survey of the world resources of a few minerals (Table 7) shows that the developing countries are over-proportionally well equipped with only a few, e.g.

bauxite, cobalt, copper, nickel, phosphate, tin,

so that the statement of DAHRENDORF (1981): *"It is simply not true that we cannot live without the developing countries,"* may seem exaggerated but in no way unrealistic.

What is true is that especially in the area of industrial minerals development possibilities for a number of raw materials lie virtually unused on the street. This raw material group is not only wider spread there than generally assumed. It is also extremely important for the developing countries since corresponding projects are relatively easy to handle technically, no great investment costs are necessary and they lead to quick cash flow (LÜTTIG 1987).

2) Present trends in use

From a technological point of view, the group of the industrial minerals and rocks is extremely dynamic. Their types of application change rapidly. New ones are added daily. Thus, it is a matter of a raw material group with fast-moving applications. This fact is true for resource technology in general, it presents high demands for resource science and technology.

A large portion of the industrial rocks are bulk goods, the price of which is relatively low and in essence determined by transport costs. Thus the consumer quite likes having these raw materials really lie at his doorstep. In fact the impression prevails predominantly in public that they really do lie at one's own doorstep unlimited in quantity and quality, with easy technical use, and thus completely problem-free as far a supply goes. The fact that this attitude is completely wrong has been elaborated on numerous time, especially with regard to availability (ARNOULD 1984, BECKER-PLATEN 1986, HOFMEISTER 1976, LÜTTIG 1976, 1977 ff., PAULY 1980, STEIN & HOFMEISTER 1977).

In the meantime, however, in this interesting group of raw materials, in which one has become increasingly aware of their value and their decreasing availability, a tremendous change has occurred. The era of practically planless exploitation, which in part willfully destroyed nature, has ended. On-site selection, well-directed exploitation and improvement of quarry and pit, improvement of low-quality raw materials, better use of deposits, decrease in mining losses, careful separation of individual components into mineral of value and waste materials, better use of waste materials, and decrease in environmental impact during exploitation can be observed almost everywhere today. Only the following are to be mentioned as examples of the transition from bulk goods to sophisticated products:

-- In ordinary and fine ceramics low-quality raw materials are being used to an increasing extent. Here sophisticated dressing methods help (KROMER, MÖRTEL, OEL & SCHOBER 1987, DITZ & MÖRTEL 1987, KROMER & MÖRTEL 1988, MÖRTEL 1989).

-- In sand and gravel production there are increasing attempts to effect an improvement of quality during exploitation, sorting and dressing on the spot.

-- The on-site exploitation and dressing of bentonite, kaolin, quartzsand and feldspar have been so refined that high-quality final products do not originate at the customer's but directly at the pit.

This development has also been supported of course since new applications for minerals have arisen, which one would not have dreamt of before. With modern techniques cheap raw materials suddenly got types of uses which improved them by several price classes at the same time. There are many examples for this, but only a few are to be mentioned as representatives, like the use of

-- special clays for manufacturing the ceramic engine,
-- bentonite in catalytic converter technology, as a food conditioner, a feed additive,
-- zeolites in numerous partially unknown fields of application,
-- strontianite in the glass industry,

-- flint as a street brightener,

-- white lime as a paper coating material, etc.

It is especially these altered uses which have accelerated sales and production rapidly. One can see these changes quite well in e.g. the production statistics of the titanium minerals, which rose tremendously with the beginning of their use as pigment in the sixteens. In this area the development above all of plastics, ceramics and catalytic converter technology has had a tremendous influence on the development of the industrial minerals. One only has to mention the numerous plastic fillers, like asbestos, feldspar, mica, limestone, kaolin, diatomite, nepheline syenite, perlite, quartz powder, barytes, talcum, clays (fine ceramic quality), vermiculite, wollastonite, a.s.o..

A development that has influenced the yield factor favorably in the area of the industrial minerals and that can be led back to the restrictions based on increased environmental protection in the long run should not be overlooked. This is integrated and multipurpose mining. Often in the past and even today in the exploitation of **one** raw material valuable geological horizons are moved into the mining waste, and this since they can only be removed with difficulty, since they occur in the operation at the wrong place or the wrong time or because the company involved is not interested in them. This causes valuable geopotential to be wasted. Mining engineers usually counter geologists who point this out with economic or technical arguments that can only be eliminated with difficulty or successfully if public or governmental pressure is used. The fact that this can be done differently can be shown by numerous examples, of which only one is to be mentioned here: In the non-ferrous metal integrated plants of Norilsk and Ust-Kamenogorsk 15 - 20 chemical elements are extracted additionally from the ores (VARTANYAN 1989).

This shows that the efficiency of mineral use can be promoted by changes in methods of processing them. The composition of both basic products and wastes will then be changed so that the latter is suitable as a raw material for other plants. This may be illustrated by the design of the recently developed region in the central area of the Baikal-Amur railway. The following raw materials serve as source products for the planned plant there: the Yakut gas, apatite ore and carbonate overburden rocks of the Seligar field, and also synnerite deposits -- ultra-potassium and aluminium-bearing rocks -- in the adjacent China region, pyrite concentrates of the Transbaikal non-ferrous metallurgy works, limestones. The plant will produce the following products: different kinds of fertilizers: nitrogen, phosphate, potassium, combined and complex (not acidizing soils); limestone (to deacidate soils), alumina, silica-gel, cement, bellite sludge concrete products and other construction material.

In the technical procedure of the plant the following raw materials are used: pyrite concentrates for producing sulphuric acid used in the combined scheme of synnyrite treatment; potash (as one of the initial components in sintering) regenerated from silicon-potash solution formed in the combined scheme of synnerite processing: apatite concentrate of the Seligdar deposit, used as an initial product in the modified sintering scheme of synnerite processing.

The combination of products manufactured from synnerites of Seligdar apatites and Yakut gas leads to combined and complex mineral fertilizers and ferroconcrete products (using overburden carbonate rocks of the Seligdar deposit; cement obtained out of wastes of treated synnyrites and local sands).

As a result of such combination the annual volume of the treated rock mass may be decreased by about 40 million m^3 -- this constitutes half of the production.

A further problem should be mentioned at the end of this chapter. It currently influences the sales of industrial minerals to a considerable extent: the true or assumed health hazards caused by a few industrial minerals. Here the problem of industrial dusts can be skipped since it is known and proven, whereby of course the fact of the presence of numerous natural form of dust must also be referred to (Table 8). But even for the well-known asbestos problem, the matter begins to become questionable. It is certain that needle shaped particles of about a 5 micron fiber length can be considered responsible for the physical cause of asbestosis and pleuropulmonary cancer when certain living conditions are present, but it is also certain that not all asbestos varieties are the cause here. You cannot dump the baby out with the bath water and the global condemnation of minerals, as done by various health agencies in the USA, e.g. of diatomite, silica gel (what is that?), talcum, attapulgite, and bentonite is an irresponsible procedure of no scientific basis.

3) Research and exploration

The presumed health problems that are supposed to arise when working with some industrial minerals have only been really illuminated by the activities of geoscientists. A number of medical doctors have namely carried out the corresponding tests with material not tested by them. The partly inexact and even incorrect mineral description printed on the package of the samples by the supplier involved was sufficient for the medical doctors. Thus, they maintained, e.g. that attapulgite was carcinogenic although the mineral involved was proven to be sepiolite. The same false conclusion occurred for talcum (GRANGE 1978, 1989).

The same thing happened with diatomite, in which particle sizes < 5 u do not occur at all. A medical doctor, however, had read that diatomite consisted of amorphous silicic acid, thus causing a circular argument with silica gel and the association of ideas with (natural and of course not very healthy when very fine) quartz dust.

These examples show that in the area of industrial research on industrial minerals representatives of applied geology and mineralogy as well as materials sciences should be approached more. This is necessary in particular due to improved exploration in economic geology and geotechnology, i.e. the process technology connected with geoscientific methods or based on geoscientific findings. It is in fact indisputable that the best knowledge on the engineering properties of minerals is available in this field. So when in a practical situation a certain material with well-defined properties is needed or when a traditionally used one is missing for reasons of natural depletion or a change in enginee-ring needs, it will be best for the practician to enlist the help of geologists, mineralogists and materials scientists. He generally consults the latter two professional groups, but the discussion with the geologists is sometimes lacking. This is in part because they don't offer themselves or they often have proven to be not sufficiently knowledgeable in the past. Here deficiencies in training are the cause, as the author has pointed out numerous times (LÜTTIG 1985, 1987 a, b, 1989).

In the field of the prospecting and exploration for nonmetallic resources, the search for an improved dialog between geology and practice, to be sought by both parties, proves to be very important in practical situations. It also simply should not happen that an owner of a gravel pit undertakes the prospecting for a new deposit on his own and only with the help of a backhoe and good common sense -- which of course should not be scoffed at. Here too, deficiencies and lacks in the training of geologists, but also in research, are noticeable. Thus it is e.g. a real shortcoming to not have been successful in achieving problem-free

-- determination of geochemical prospecting methods for finding certain industrial mine-rals (e.g. talcum, soapstone, graphite),

-- application of promising radar prospecting methods even for multilayer problems or

-- progression in geoelectrics from the differentiation of clasticity or conductivity groups to mineral differentiation (e.g. clay - marl - brown coal).

A particular lack lies in the neglect of applied Quaternary geology in some countries. The author has been trying to eliminate this for years since the practical problems of applied geology are usually of Quaternary nature. A large portion of the industrial mine-rals and rocks also occurs mainly in the Quaternary (Fig. 6).

4) Exploitation and Dressing

Mining and transportation costs influence the sales price of a mineral resource in a decisive way. The mining costs are all the higher, the higher the wages are. With the increase of mechanical activities in general in modern economy exploitation by hand has for this reason given way to mining with appropriate equipment. Modern mining has been shifted in many fields in which underground mining was the classical type of exploitation to open cast mining. The power companies prefer brown coal exploitable in open cast pits even if it is inferior in quality to the hard coal mined underground even with respect to its calorific value. The inferiority is compensated for by possibilities of conditioning, standardization, modern boiler technology and finally the low production price. In copper mining also similar developments can be seen in the use of disseminated copper ores; in iron mining in the use of itabirite minerals. Everywhere the tendency towards open cast mining is evident even for industrial minerals.

The fact that the largest open cast pits created are not those in which industrial minerals are exploited has already been mentioned.

With the development of open cast technology the limits to minability, given in the seam/overburden (S/O) ratio, has changed in favor of the overburden. Thus, for a number of industrial minerals there are considerably different S/O values today than in the past. Of course, the figures vary greatly from industrial area to industrial area (and economic system).

It has already been stressed that the possibility to use the correspondingly cheaper industrial minerals usually available in larger amounts depends to an exceptional degree on the improved dressing methods.

In the area of re-use and recycling, which can extend the lifetime of the corresponding resource decisively for a few metals (e.g. iron and steel, copper, lead), not all that could be conceivable has been done for the industrial minerals. Nevertheless, the recycling of glass contributes decisively to the saving of quartz sand additions since the recycled glass reduces the energy costs considerably. In this country one reckons with an average of 30%. In the Federal Republic of Germany there have also been great efforts to recycle rubble. This has reduced the consumption of aggregates. In the Federal Republic of Germany there is an annual amount of

      15.2 million t of rubble
      13.2 million t of waste road surface
      <u>105.0</u> million t of excavated material
      132.0 million t.

The recycling capacity amounts to only 10 million tonnes at present and is thus limited to the first two groups listed. In addition, there are also 20 million tonnes of hard coal heaps, 10 million tonnes of blast furnace slag, 9 million tonnes of used asphalt and 4 million tonnes of steelworks slag which can be used in road construction like rubble -- not everywhere for the roadway paving but usually with the excavated material for the verge.

One of the most interesting branches is the use of gypsum, as it occurs in flue gas desulfurization and phosphoric acid production. Currently the large amounts occurring (2 million tonnes of flue gas gypsum in the old Federal Republic of Germany) cannot be used yet for reasons of quality but in numerous countries, such as Japan, the Federal Republic of Germany, Finland, etc., research projects are being carried out which promise a solution soon.

For other industrial minerals there are also numerous possibilities in which geoscientific expertise is called for, also with respect to by-products, i.e. in the treatment of bentonite, phosphate and feldspar.

5) Environmental Handling

It should be emphasized again and again (LÜTTIG 1984, 1989) that especially mining belongs to those branches of industry which very early -- much much earlier than the origin of the terms environment and environmental protection -- were aware that they had an impact on the structure of nature and the landscape. Mining is certainly not the antagonist, which some circles of nature protection portray, of course with a forced over-- emphasis that is not meant seriously. The oldest environmental protection law goes back to EDWARD I of England and was enacted in 1306 (DOWN & STOCKS 1977), since QUEEN ELEANOR couldn't bear the pollution caused by the smoke of the forge fires (produced with the help of hard coal) during a visit to Nottingham in the year 1257. GEORGIUS AGRICOLA, the founder of mining science, geology and mineralogy from Saxony, described the first processes for treating mine water in 1544 and 1566 (FRAU- STADT & PRESCHER 1974), a matter of concern which the father of all European geologists ABRAHAM GOTTLOB WERNER (1749 - 1817) later took up at the famous Freiberger mining academy. The Somerset Colliery Lease of 1791 also gave exact regulations on what was to be done with an abandoned hard coal pit: It was supposed to be filled and the mining area sewn with rye grass. The oldest peatbog research station in Bremen, which was founded in 1877, goes back to the oldest citizen action committee in the sense of environmental protection the "association against the burning of the bogs". At that time in the area of Bremen the home fires were supplied with the raised bog peat from the peat deposits available in plenty in the surroundings. At times the air in Bremen was supposed to have been so full of smoke that it was impossible to breathe freely (OVERBECK 1975).

Here one of the environmental impacts should be referred to which has made mining and even more so the processing of mining products the target of public criticism: emissions. They are of course significant in particular for the exploitation and combustion of energy carriers, such as hard coal, brown coal and peat, less so hydrocarbons and not at all the nuclear fuels (apart from steam emissions). Mind you, there is also the emission and immission problem with industrial minerals and rocks. Just think of the cement and limestone works in this context!

However, the acoustic pollution caused by blasting and mining machinery, above all for the hard and dimension stone but also the cement and limestone industry, can be rated higher than this. Although the impairment from mining and dressing machinery and transportation vehicles are in the long run certainly less pleasant permanently than blasting, due to the temporal distribution and the noise level reached (up to 110 db, e.g. for compressed air rock drills, compressed air fans and compressed air hammers; this is of course not more than the noise of a pop music concert !), humans consider these especially disturbing. This is connected with the manner the human ear responds (DOWN & STOCKS 1977).

Very strict and rather unified standards with respect to goals have been enacted in the whole world against this acoustic pollution. They are of great significance for deposit exploitation and mining planning in so far as they cause safety zones between the working and the adjacent settlements (and transportation routes) which locally go far beyond the usual safety measures in mining and thus reduce the lifetime of the pits.

Beside the impact on the environment by mining caused by acoustic and dust, smoke and gas emission, there is optical pollution which can be traced to the not favorable or wise installation of open cast pits, quarries and above all treatment plants which in part appear damaging on purpose. How else should one assess the case of a brown coal power plant being placed on a hill with a stack 100 m high to be seen from a long way off, spreading smoke over the countryside? If the industrial architect then still proudly announces he wanted to place an index finger in the landscape in order to tell human society, "Here are those who are producing your electricity!" then the author can only turn away with contempt at so much clumsiness and arrogance. We also know of other examples which show how wisely and inconspicuously a quarry can be placed in the landscape, how waste piles can disappear into cavities so that they cannot be found at all later, how mining can go through the landscape without being seen. HERRMANN & RAUEN (1976) have described some exemplary solutions for the exploitation of gypsum: The geologist who later mapped the area involved was able to find for example only a portion of the former open cast pit; it had been filled so well that it could no longer be identified morphologically or pedologically.

The exploitation of surface-near resources can certainly be carried out in such a way that changes in the lithosphere can be reduced to a minimum. Naturally the pedosphere is always affected, but one can strip away the soil lying on top of a deposit and place it to the side momentarily so that successful use can be made of it later if one is of good will and values recultivated locations with high agricultural and forestry yield. In individual cases one can even produce soil locations with increased yield, as HEIDE (1973), HEIDE & SCHALICH (1977) and HEIDE & WERNER (1982) have shown for the case of recultivated brown coal open cast pits in the area of brown coal of the Lower Rhine.

More severe than the impact on the pedosphere are those on the hydrosphere. UNESCO has devoted a research program of its own "The impact of mining on the environment" to this problem (VARTANYAN, 1989), which can be referred to.

In addition, there are the impacts on the biosphere. They are so evident and lamentable that it should be the need for the geosciences, sailing of course between Scylla and Charybdis (LÜTTIG 1975) as they mediate between mining and nature protection, to refer to possibilities of avoiding impacts, yes, of living with each other, and to remain in search of further solutions.

This also concerns their role in establishing regions in which land-use priority is to be given for the exploitation of mineral resources, especially surface-near ones, instead of other claims. Although the geosciences are a nature-loving branch of science and have been able to provide an abundance of useful maps for designating nature reserves and environmental protection areas due to their very long cartographic geognostic tradition, -- in contrast to biology, which has carried out biotope mapping only for a few years -- they do find the exaggerations in the nature conservation claims quite regulating. Here two maps of nature reserves and natural parks from the old Federal Republic of Germany are quite informative (Fig. 7 and 8). The problem involved is shown particularly well in Fig. 8. If one transferred to the map the large extraction regions of hard and dimension stone, as well as limestone and cement marls and other mineral resources extracted near the surface, one would note that they correspond to a large portion with the areas reserved and protected for ecological reasons. One arrives at a similar conclusion when transferring the regions with high groundwater potential. This means that in these regions numerous claims on the natural environment's potential collide with one another.

The consequence is that if one treats the nature conservation claim as the priority, the accessibility to mineral resources must be restricted severely, a problem which has been referred to numerous times (ARNOULD 1984, BECKER-PLATEN & PAULY 1984, LÜTTIG 1975, 1977, V. STEIN 1982). The result is an anthropogenic, i.e. artificial, shortage of resources.

The fact that this administrative miscalculation must and can be met is shown e.g. by the successful resource conservation maps in Lower Saxony (STEIN & HOFMEISTER 1977, STEIN et al. 1981).

On the other side, i.e. that supporting nature conservation, geology can make a contribution by adding to the number of biologically interesting objects those which are to be protected from a geoscientific point of view (BECKER-PLATEN 1982, 1983, BECKER-PLATEN et al. 1979 a, b, LÜTTIG 1976, 1990 among others).

It cannot be concealed, however, that resistance from some interest groups to extraction has in the meantime taken on even hysterical forms for a number of resources, especially peat. The search for sensible compromises is especially urgent in the case of this resource in particular, which e.g. is needed for producing important growth substrates for horticulture and for gaining peat important for medical and therapeutic purposes (GROSSE--BRAUCKMANN 1989, J. GÜNTHER 1982, 1988, LÜTTIG 1979, 1989, SCHNEE-KLOTH 1980).

Geologists and the scientists related to them are faced with a very important task in this respect, which they must and can solve as scientifically honest and economically experienced experts, if one listens to them.

6) Future scientific approach

Attentive listeners can be found if one is particularly talented at articulating oneself. In this respect there is a need for a great deal of coaching for the honest but well-behaved conservative and somewhat untalented geologists. The traditional song of the German geologist says

> *"We geologists are*
> *terribly well-behaved*
> *(plain, blunt) people...,"*

and this is how we express ourselves in general. The consequence for those active in teaching and training must be to not only train the guild for polished speeches and understandable statements but also to convince it that in order to "sell" scientific findings -- and that involves the spoken and written word -- one must restrict oneself to the essentials for the outsider and the things to be explained, leaving out the academic terms, the specific scientific details that aren't of interest to the "customer", and avoid tech-talk (at least in this case) (BAIRD 1968, HORST 1963, KUTSCHER 1961, LÜTTIG 1987, 1989, 1990, MARTINSSON A. 1969 and others).

Another area concerning the field of the industrial minerals and rocks in particular is the necessity to help the underrepresented scientific fields of research blossom. These include Quaternary geology, in fact all shades of geology of unconsolidated rocks including subsoil and hydrogeology, likewise limnogeology, marine geology, and materials sciences with reference to geology.

An important field of activity is geotechnology, i.e. the process technology based on geoscientific methods. It can help

-- find out minerals and rocks needed for certain industrial uses,
-- find substitutes for conventionally used minerals and rocks which are running short,
-- find new industrial uses for resources used in certain well defined areas of application,
-- recycle materials occurring in the processing of mineral resources
-- or provide non-toxic waste disposal.

There are numerous examples of innovations for the benefits of activities in this area, from the "Frankfurter Pfanne" (a special type of roof-tiles), expanding clay, the use of fine-grained industrial minerals as fillers, whitening, and conditioners all the way to the processes developed by geologists of radioactive waste disposal in clay rocks, granitic bodies and salt domes.

Since the geosciences have turned to technology in this way, a traditional and characteristic field of collaboration with other sciences can also be taken on again, one which arose at the beginning of the activities of geology, mineralogy and mining sciences with GEORGIUS AGRICOLA, but which in the meantime has partly been lost, namely mining geology. Although it is indisputable that geologists play an important role in many companies and enterprises, there are numerous mining companies on the side which get along (or believe they can) without the help of geologists. And the exploitation of surface-near resources is carried out in many cases, as has already been said, "without geologists"; the expert is only called in when "the waggon is stuck in the mud", e.g. the "dirt which is covering geology" (LÜTTIG 1985).

The fact that the resource industry recognizes this lack is also connected with the increase in points of contact in mining with environmental and nature conservation problems, whereby we are again at the focal point of this observation. And actually, geoscientists ought to be happy about this fact because of the reasons described.

One of the necessary consequences is the intensification of the dialog between the geosciences, environmental science and mining. This is a current topic which the UN-ESCO, in addition to others, is so highly involved in with the already mentioned project "Geoscience and the Environment" initiated by the Soviet Minister for Geology, Prof. E. A. KOZLOVSKY.

The vehicle with which the necessary collaboration can be driven forward is without a doubt planning geology and we will dwell on it for a moment at the end.

The improvement of the dialog between the geosciences, the resource extracting and processing industry and environmental protection and nature conservation is best possible on the basis of map works. Since the map, the plan, is the basic tool in land-use and regional planning, the latter also forms the braces. And those parts of geology providing the information, called planning geology by the author, have the best possibility to articulate themselves via the geoscientific map.

Meanwhile in most countries of the world and in fact under the auspices of the IUGS Subcommission on Maps of Environmental Geology (SC-MEG), the work of the Geoscientific Map of the Natural Environment's Potential has best proven itself for this for some time (BECKER-PLATEN, LÜTTIG & MEINE 1979, LÜTTIG 1972, 1975, 1976, LÜTTIG & PFEIFFER 1974, VON DANIELS & LÜTTIG 1982 among others).

To summarize briefly, this is a continuously revolving (every 10 - 15 years) work built-up on 5 levels
-- scientific base maps,
-- maps of elements relevant to planning,
-- land-use claim maps,
-- land-use conflict maps and
-- land-use priority maps

at scales of 1 : 100 000 to 1 : 500 000, primarily of the scale of 1 : 200 000, like the first region-wide geological map of the natural environment's potential of Lower Saxony and Bremen (BECKER-PLATEN 1983, LÜTTIG 1990). It comprises the maps listed in Table 9. It is clear that due to varying conditions in individual states, no strict legend standard can be valid. Thus, in the neighboring country of the Netherlands a somewhat different system has been used (Tab. 10, see also DE MULDER 1988).

The fact that these maps and their manner of representation are very dependent on the scale is self-evident but the individual planning levels also vary greatly. This is shown by Fig. 9.

The attempt to raise this map to a level of the same type which can be used internationally is being carried out in a type of pilot strip by the already mentioned SC-MEG. The legend added in Fig. 10 can form a usable draft for this, but it is also true here that for the pilot strip planned at first for the E - W direction over the FRG - the Netherlands - Belgium - the UK - and Ireland the local conditions relevant to planning vary greatly and the legend must be adapted. Therefore, Fig. 10 should only be considered a preliminary draft.

Experience shows that these maps have been accepted by land-use planning in general but there is a long way to go before this has been accomplished completely. Obstacles on this path have proven to be the inadequate and uncoordinated training of especially land-use planners but also the unwillingness of some geoscientific circles who remain in their ivory tower of so-called pure doctrine.
The great socioeconomic problems of human society call for the geologist in the role of mediator in the area of conflict between resource extraction and environmental protection and this means that we should not remain offside.

References

AGRICOLA, GEORGIUS: (1544): De ortu et causis subterraneum. Librum II. Basel 1544. - Übersetzt von FRAUSTADT, GEORG: GEORGIUS AGRICOLA: Schriften zur Geologie und Mineralogie. Berlin 1956.

--- (1556): De re metallica libri XII. Basel 1556. - Übersetzt von FRAUSTADT, G. & PRESCHER, H.: GEORGIUS AGRICOLA, Bergbau und Hüttenkunde, 12 Bücher, 931 S., 300 Abb., Berlin 1974.

[ARCHER, A. A., LÜTTIG, G. W. & SNEZHKO, I. I.]: (1987): Man's Dependence on the Earth. The Role of the Geosciences in the Environment. - 216 p., 83 fig., 15 pl., Stuttgart (Schweizerbart) + Nairobi + Paris 1987.

ARNOULD, MARCEL: (1984): Bilan des inventaires de sables et graviers en France de 1971 à 1984. - Bull. intern. Ass. engin. Geol. **29**: 5 - 9, Paris 1984.

BAIRD, D. M.: (1968): Geology in the public eye. - Roy. Soc. Canada spec. publ. **11**: 222 - 230, Toronto 1968.

BECKER-PLATEN, JENS D.: (1982): Zur Kartierung schutzwürdiger geowissenschaftli cher Objekte in Niedersachsen. - Laufener Seminarbeitr. **7/82**: 44 - 57, Laufen/Salzach 1982.

--- (1983): Geowissenschaftliche Karten des Naturraumpotentials -- Forsch. dt. Landesk de. **220**: 119 - 164, Trier 1983.

--- (1986): Protection and categorization of near-surface mineral resources in the Federal Republic of Germany, especially in Lower Saxony. - In: [BENDER, FRIEDRICH]: Georesources and Environment: 125 - 144, Stuttgart (Schweizerbart) 1986.

BECKER-PLATEN, J. D. et al.: (1979 a): Geowissenschaftliche Karte des Naturraumpo tentials von Niedersachsen und Bremen 1 : 200 000. - Schutzwürdige geowissenschaftli che Objekte. Blatt CC 3926 Braunschweig, Hannover 1979.

--- (1979 b): Geowissenschaftliche Karte des Naturraumpotentials von Niedersachsen und Bremen 1 : 200 000. - Schutzwürdige geowissenschaftliche Objekte. - Blatt CC 3110 Bremerhaven, Hannover 1979.

BECKER-PLATEN, J. D., LÜTTIG, G. & MEINE, K.-H.: (1979): Geoscientific Maps for Planning. Natural Resources Forum **3**: 167 -177, New York 1979.

BUNDESMINISTERIUM FÜR WIRTSCHAFT: (1979): Einheimische Rohstoffe -- Steine, Erden und Industrieminerale. - 48 S., Bonn 1979.

DAHRENDORF, R. (1981): Es ist schlicht nicht wahr, daß wir ohne die Entwicklungsländer nicht leben können. - Erz **22, 4:** 4 - 5, Bonn 1981.

DANIELS, C. H. von & LÜTTIG, G.: (1982): Geowissenschaftliche Karten des Naturraumpotentials als Unterlagen für Raumordnung und Landesplanung. - In: [Energierohstoffe im Alpen-Adria-Raum: 151 - 168, 2 Taf., Graz (Amt der Steiermärkischen Landesregierung) 1982 (1980).

DITZ, H. & MÖRTEL, H.: (1987): Veränderung von Rohstoffeigenschaften durch bakteriellen und mykotischen Angriff. - Fortschrber. dt. Keram. Ges., CFI-Beih. 2, 1: 221 - 220, Coburg 1987.

DOWN, C. G. & STOCKS, J.: (1977): Environmental impact of mining. - 371 p., London (Appl. Sci. Publ.) 1977.

GRANGE, JEAN-PIERRE: (1978): Talc and Health. - Preprint 3rd industr. Min. intern. Congr., 3 pp., Paris 1978.

--- (1989): Health regulations. Let realism replace prejudice. - Industr. Min. 256: 41 - 42, London 1989.

GROSSE-BRAUCKMANN, G. (1989): Zur Einführung. - Telma, Beih. **2:** 7 - 11, Hannover 1989.

GÜNTHER, J. (1982): Torf. Wissenswertes zu einem aktuellen Thema. - 14 S., Bad Zwischenahn (Torfforschung GmbH) 1982.

--- (1988 b): Pressekampagne des Bundes für Umwelt und Naturschutz Deutschland (BUND) gegen Torfverwendung -- ein Spiel mit falschen Karten? - Gartenb. u. Gartenw. **18/88:** 786 -787, Hamburg 1988.

HEIDE, GÜNTHER: (1973): Pedological investigations in the Rhine brown-coal areas. - In: [HUTNIK, R. J. & DAVIS, G.]: Ecology and Reclamation of Devastated Land 2: 295 - 313, 3 fig., 1 tab., New York 1973.

HEIDE, G. & SCHALICH, J.: (1977): Boden. - In: [DÜRO, F. et al.]: Tagebau Hambach und Umwelt: 65 - 84, Krefeld 1977.

HEIDE, G. & WERNER, H.: (1981): Verfahren zur schadlosen Deponierung von sulfathaltigen Kraftwerksaschen im rheinischen Braunkohlenrevier. - Braunk. **33,** 2: 7 - 11, Düsseldorf 1981.

HOFMEISTER, ERICH: (1976): Der Schutz oberflächennaher Lagerstätten, eine wichtige Aufgabe der Daseinsvorsorge. - Bergbau 1976, **7:** 275, Essen 1976.

HORST, ULRICH: (1963): Bemerkungen zu unserer Fachsprache. - N. Jb. Geol. Paläont. Mh. 1963, **10**: 570 - 587, Stuttgart 1963.

KROMER, H. & MÖRTEL, H: (1988): Refinement and enrichment of clay and kaolin raw materials by dressing methods. - Elsevier appl. Sci. 1988: 39 - 48, Amsterdam 1988.

KROMER, H., MÖRTEL, H., OEL, H. J. & SCHOBER, G.: (1987): Deferrizing clay by high-gradient magnetic separation, as exemplified for Oberpfalz raw materials. - C. F. I., Ber. dt. Keram Ges. **64**, 12: 15 - 18, Coburg 1987.

KUTSCHER, F. (1961): Anweisungen für die Verfasser beim Druck wissenschaftlicher Arbeiten in den Veröffentlichungen des Hessischen Landesamtes für Bodenforschung. - Notizbl. hess. L.- Amt Bodenforsch. **89**: 489 - 502, Wiesbaden 21.7.1961.

LAWATSCHECK, J. (1990): Mengen- und Wertbetrachtung der 50 wichtigsten minerali-schen Rohstoffe der Welt. - Inaug. Diss., 767 S., 69 Abb., 269 Tab., Erlangen 1990.

LÜTTIG, G. (1972): Naturräumliches Potential I, II und III. - In: Niedersachsen, Industrie-land mit Zukunft: 9 - 10, 3 Karten, Hannover (Nds. Min. Wirtsch. öff. Arb.) 1972.

--- (1975): Geoscience and the potential of the natural environment. - Geoscientific Studies and the Potential of the Natural Environment: 29 - 40, Köln (Deutsche UN-ESCO-Komm.) 1975.

--- (1975): The geologist's role in planning for the future. - Natural Resources and Development **1**: 23 - 30, Tübingen 1975.

--- (1976): The International Map of the Natural Environment's Potential and its Im-portance for the Economic Classification of Peatlands. - Proc. 5th intern. Peat. Congr. 1: 65 - 75, Poznan 1976.

--- (1976 a): Geoscience and the potential of the Natural Environment. - Natural Re-sources and Development 3: 93 - 107, Tübingen 1976.

--- (1976): Der Naturschutz in Niedersachsen aus geowissenschaftlicher Sicht in 30 Jahren Naturschutz und Landschaftspflege in Niedersachsen: 78 - 92, Hannover (Nds. M. ELF) 1976.

--- (1977 c): Zur Lebensdauer, Verfügbarkeit und Vorratssicherung der feuerfesten und keramischen Rohstoffe in der Bundesrepublik Deutschland. - Sprechsaal, Ceramics, Glass, Cement, **110**, 3: 126 - 128, Coburg 1977.

--- (1978): Die Entwicklungsländer mit geringem Geopotential -- aus der Sicht des Geo-wissenschaftlers. - 174 S., 1 Taf., Hannover (Nds. Landeszentrale polit. Bildg.) 1978.

--- (1979): Zum Problem der Rohstoffsicherung bei Torf-und Peloid-Rohstoffen. - Z. angew. Bäder- u. Klimaheilkunde **26**, 1: 77 - 84, Stuttgart 1979.

LÜTTIG, G. (1983): Industrie-Minerale und Gesteine -- Bedeutung für die Volkswirtschaft und Aufgabe für die Forschung. - Weltenburger Akad., Erwin-Rutte-Festschr.: 151 - 155, 1 Abb., Kelheim & Weltenburg 1983.

--- (1984): Influence of mining activities on the environment. - Abstr. 27. intern. geol. Congr. **9, 1**: 324 - 325, Moskwa 1984.

--- (1985): Über den "Dreck, der die Geologie verhüllt" (STILLE). - Nachr. dt. geol. Ges. **32**: 90 - 105, Hannover 1985.

--- (1987): Approach to the problems of mineral resources' extraction, environmental protection, and land-use planning in the industrial and developing countries. - In: [ARNDT, P. & LÜTTIG, G. W.]: Mineral resources' extraction, environmental protection and land-use planning in the industrialized and developing countries: 7 - 13, 2 fig., Stuttgart (Schweizerbart) 1987.

--- (1987): Conclusions: Geology versus mineral, groundwater and soil resources' management. -- Aproach to the public -- Education and training questions -- Types and acceptance of geopotential maps. - In: [ARNDT, P. & LÜTTIG, G. W.]: Mineral resources extraction, environmental protection and land-use planning in the industrialized and developing countries: 319 - 331, Stuttgart (Schweizerbart) 1987.

--- (1988): The influence of mining activities on the environment. - 27th intern. geol. Congr. gen. Proc.: 167 - 176, Moscow 1988 (1984).

--- (1989): Quaternary deposits: Suppliers of mineral raw materials and prerequisites for human development. - In: [MULDER, E. F. J. DE & HAGEMAN, B. P.]: Applied Quaternary Research: 83 - 104, Rotterdam + Brookfield (Balkema) 1989.

--- (1989 a): Historical development of the mining/environment dialog. - Proc. intern. Worksh. "The Impact of Mining on the Environment" USSR, Tallinn-Leningrad, June 18 - 25, 1986, 1: 262 - 279, Moscow (GKNT) 1989.

--- (1989 b): History of the utilization of peatlands in Europe with special reference to Germany. - Suo **40, 5**: 169 -175, Helsinki 1989.

[---]: (1990): Das Naturraumpotential in kartographischer Darstellung (Bericht des Arbeitskreises "Karten des Naturraumpotentials" der ARL) -- Textliche Beschreibung und Generallegende. - ARL Arbeitsmat., Einzelveröff. **168**: 1 - 247, 48 Abb., Hannover 1990.

LÜTTIG, G. (1990): Quaternary research in view of modern requirements of applied geology. - Striae **29**: 15 - 29, 4 Fig., 2 Tab., Uppsala 1990.

MARTINSSON, A. (1969): Den vetenskaplige publiceringers dilemma. - Nordisk Forum 4: 232 - 245, Stockholm 1969.

MEADOWS, D. et al. (1972): Die Grenzen des Wachstums. - 180 S., Stuttgart (Deutsche Verlags-Anstalt) 1972.

MÖRTEL, H. (1989): Veredelung von Roh- und Werkstoffen. - In: [DECHEMA-AR-BEITSAUSSCHUSS]: Mikrobiologische Materialzerstörung und Materialschutz, Studien zur Forschung und Entwicklung: 134 - 140, Frankfurt a. M. 1989.

MULDER, E.F.J. DE: (1988): Engineering geological maps: A cost-benefit analysis. - In: [MARINOS, P. G. & KOVKIS, G. C.]: The engineering geology of ancient works, monuments and historical sites: 1347 - 1357, Rotterdam + Brookfield (Balkema) 1988.

OVERBECK, F. (1975): Botanisch-geologische Moorkunde. - 719 S., Neumünster (Wach-holtz) 1975.

PAULY, E. (1980): Bewertung und Sicherung mineralischer Rohstoffe aus oberflächen-nahen Lagerstätten in Hessen, Methoden und Stand. - Erzmetall 33, 4: 222 - 225, Wein-heim 1980.

PESTEL, E. (1975): Weltkrise und organisches Wachstum -- Perspektiven der gegen-wärtigen Situation. - Universitas 30, G: 561 - 570, Stuttgart 1975.

ROBINSON, E. & ROBBIN, S. R. C.: (1971): Emission, concentration and fate of parti-culate atmospheric pollutans. - Final Rep. Stanford Res. Inst. Proj. SCC-8507, Menlo Park, Calif. (S.R.I.) 1971.

SCHNEEKLOTH, H. (1980 c): Konflikte zwischen Naturschutz und Torfabbau sind ver-meidbar! -- Ergebnisse der neuen Moorbewertung in Niedersachsen. - Telma 10: 149 - 157, 1 Tab., Hannover 1980.

STEIN, V. et al. (1981 a): Begründung für die Ausweisung von Gebieten besonderer Bedeutung und von Vorranggebieten für die Rohstoffgewinnung in Niedersachsen. - Mskrpt., Archiv Nr. 88450, 52, + 3 S., Hannover (NLfB) 1981.

STEIN, V. & HOFMEISTER, E.: (1977 a): Die Darstellung oberflächennaher Rohstoff-vorkommen in Rohstoffsicherungskarten. - Geol. Jb. (D) 27: 121 - 132, 3 Abb., Han-nover 1977.

[VARTANYAN, G. S.]: (1989): Mining and the Geoenvironment. - In: [KOZLOVSKY, E. A.]: Geology and the Environment, an International Manual in Three Volumes. 2: 1 - 201, Paris + Nairobi (UNESCO + UNEP) 1989.

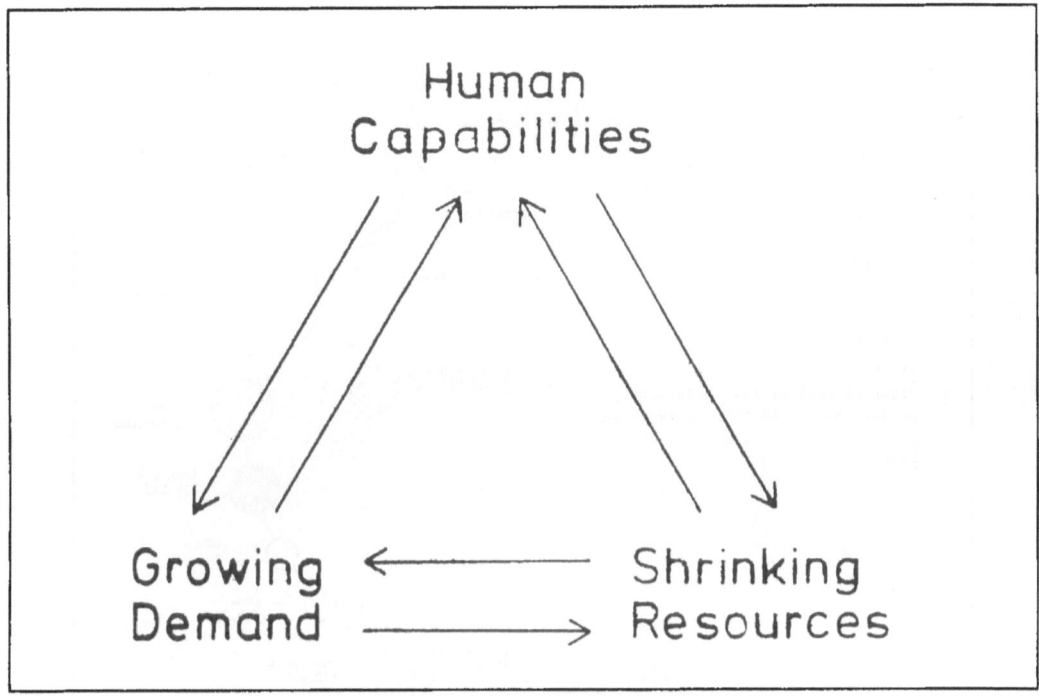

Figure 1:   Diagram of current socioeconomic interdependence

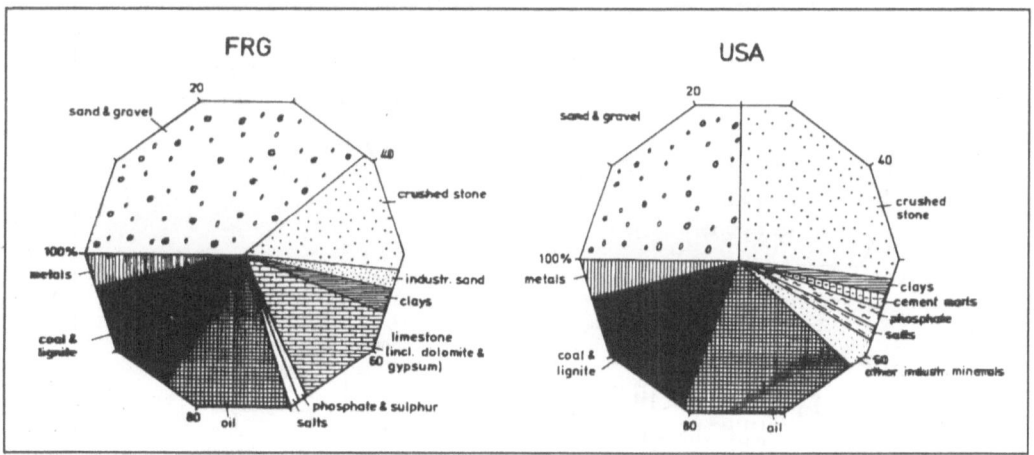

Figure 2:   Percentage consumption of important mineral and energy resources in the
Federal Republic of Germany and the USA, 1984

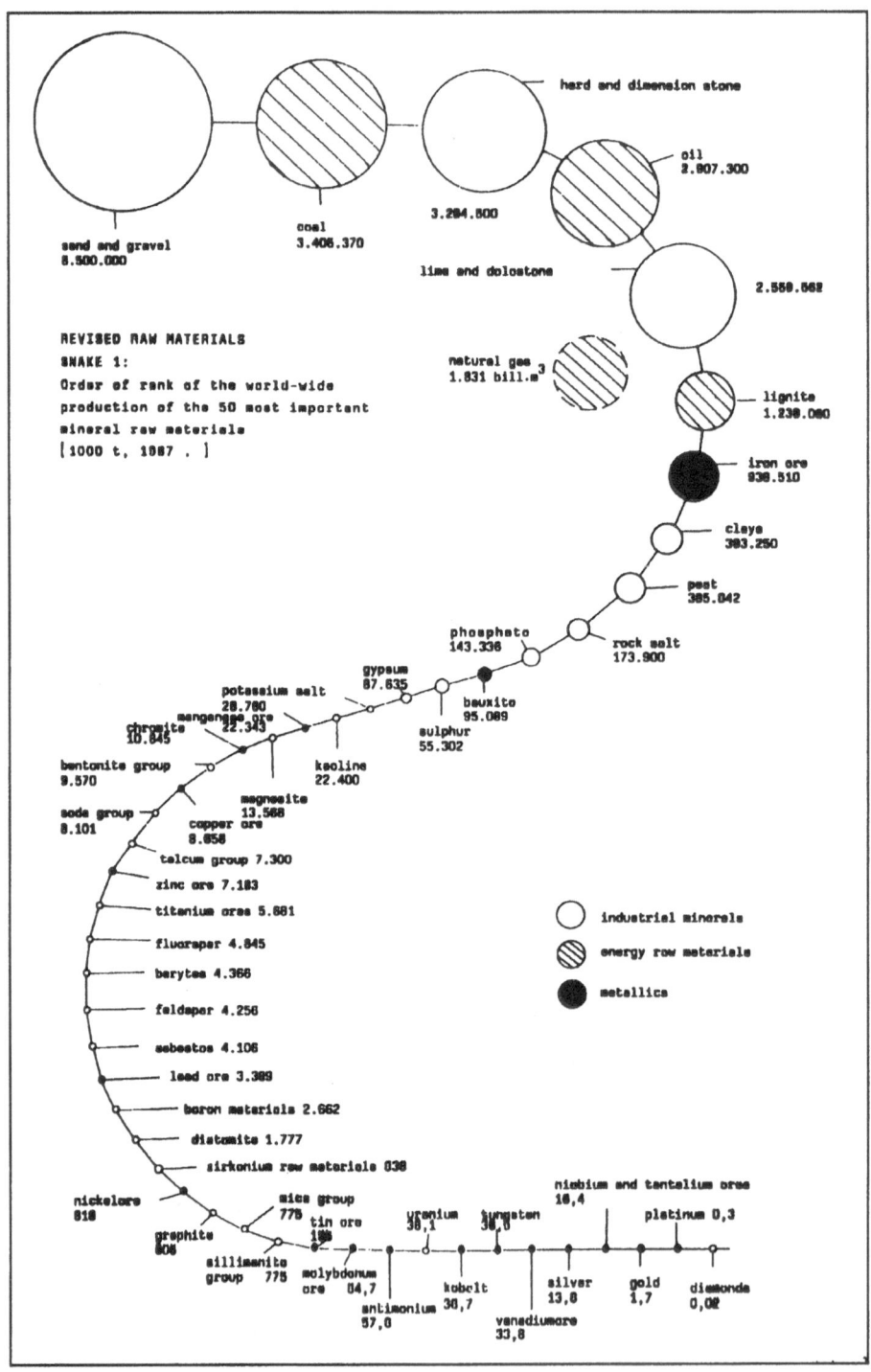

Figure 3: Order or rank in quantity of the 50 most important minerals in 1,000 t, 1987 (after LAWATSCHECK (1990)

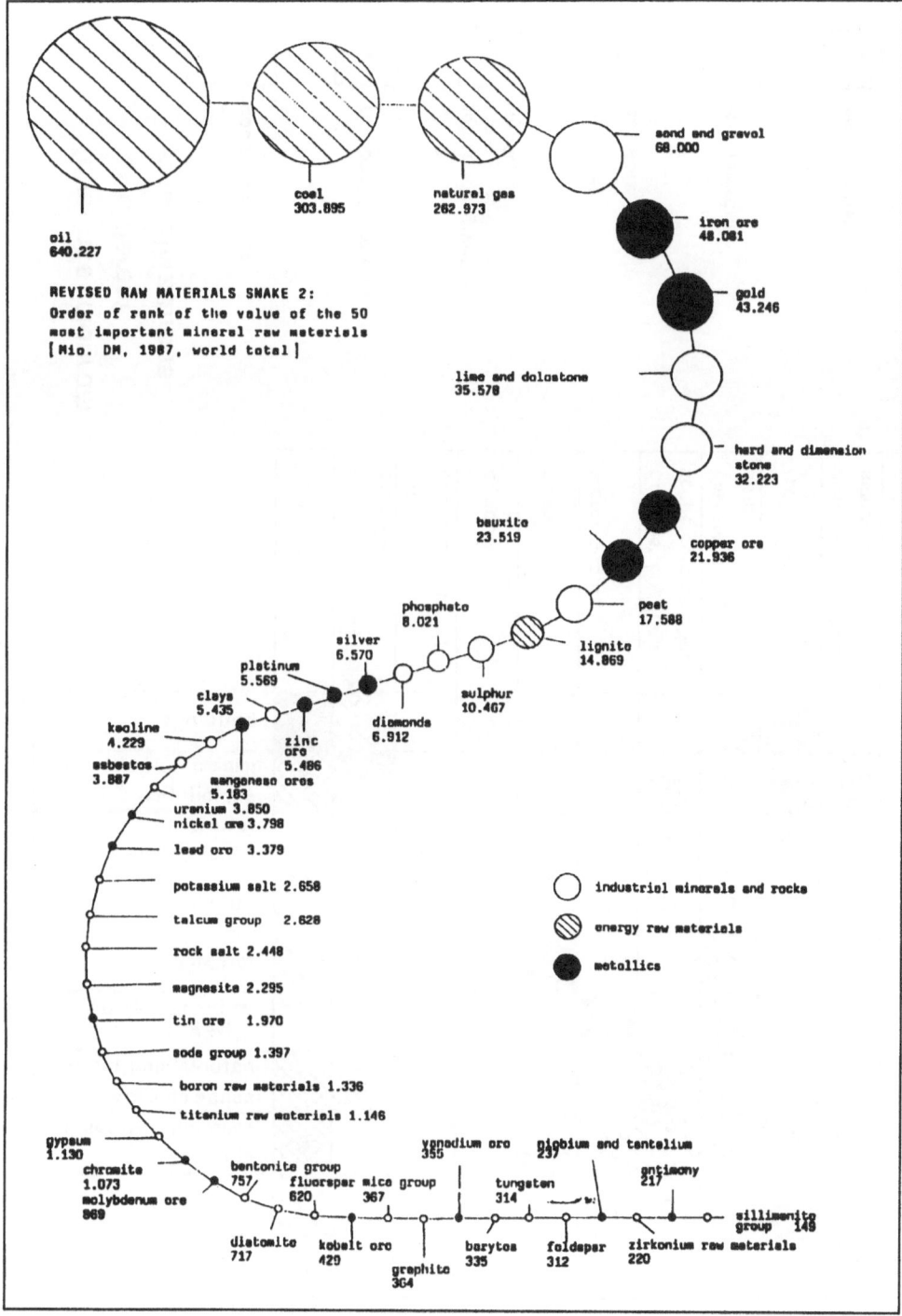

Figure 4: Order of rank in value of the 50 most important minerals in million German marks [DM], 1987, after LAWATSCHECK (1990)

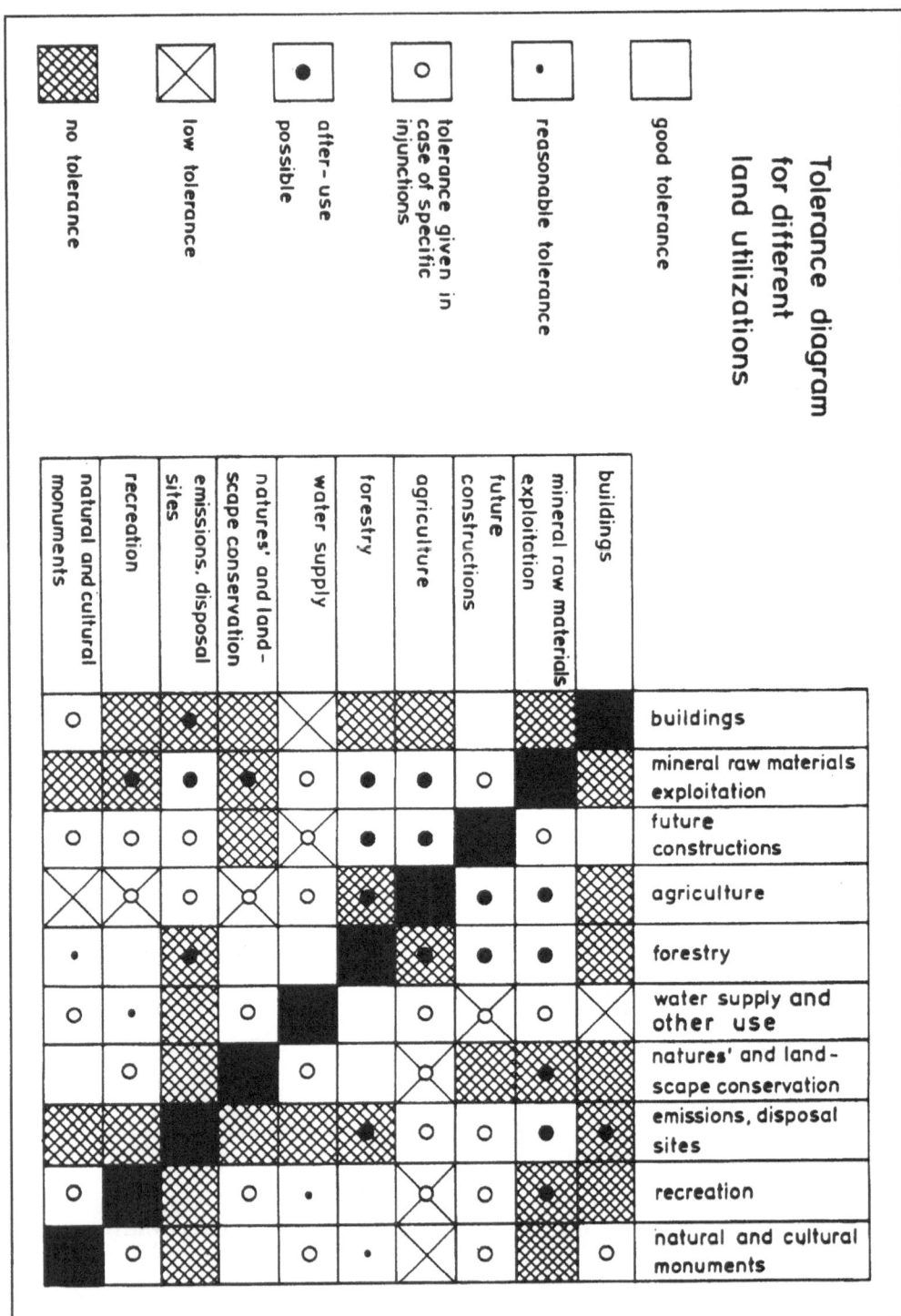

Figure 5: Explanation with the Figure

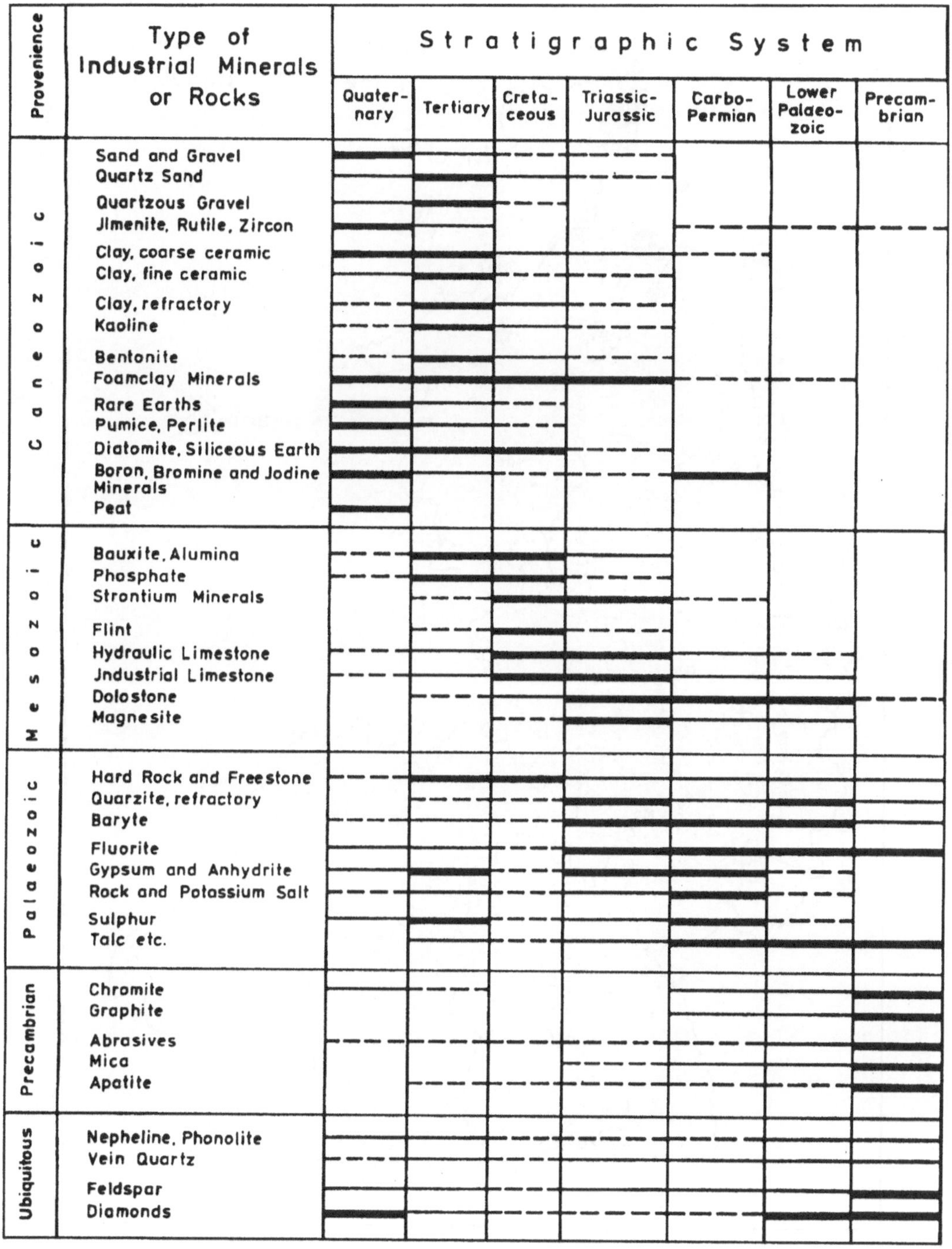

315

Provenience	Type of Industrial Minerals or Rocks	Stratigraphic System						
		Quaternary	Tertiary	Cretaceous	Triassic-Jurassic	Carbo-Permian	Lower Palaeozoic	Precambrian
Cenozoic	Sand and Gravel / Quartz Sand / Quartzous Gravel / Jlmenite, Rutile, Zircon / Clay, coarse ceramic / Clay, fine ceramic / Clay, refractory / Kaoline / Bentonite / Foamclay Minerals / Rare Earths / Pumice, Perlite / Diatomite, Siliceous Earth / Boron, Bromine and Jodine Minerals / Peat							
Mesozoic	Bauxite, Alumina / Phosphate / Strontium Minerals / Flint / Hydraulic Limestone / Jndustrial Limestone / Dolostone / Magnesite							
Palaeozoic	Hard Rock and Freestone / Quarzite, refractory / Baryte / Fluorite / Gypsum and Anhydrite / Rock and Potassium Salt / Sulphur / Talc etc.							
Precambrian	Chromite / Graphite / Abrasives / Mica / Apatite							
Ubiquitous	Nepheline, Phonolite / Vein Quartz / Feldspar / Diamonds							

Figure 6: Main areas of stratigraphical provenience of the most important industrial minerals and rocks. Thick line = center of frequency, thin line = frequent, dashed line = occasional

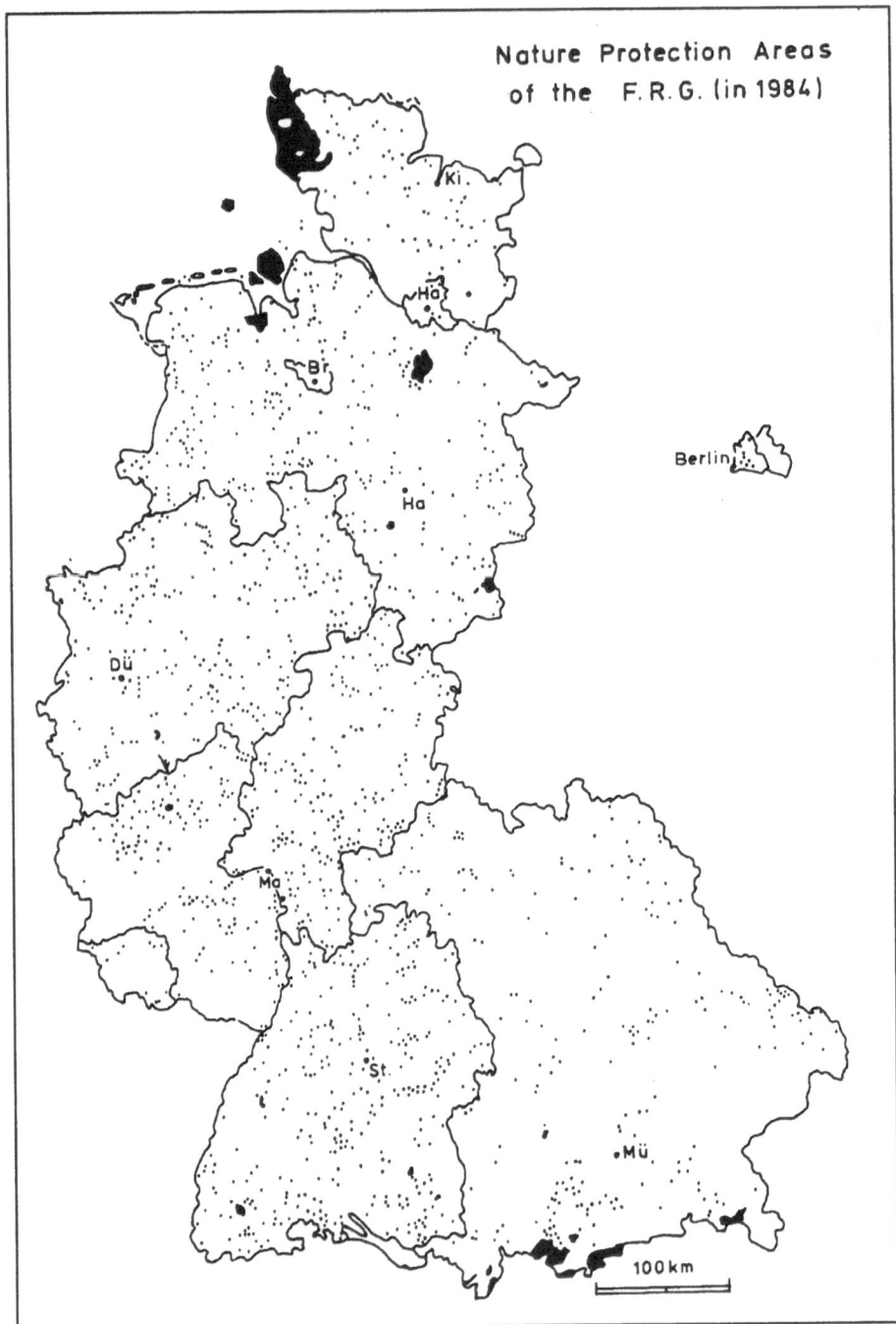

Figure 7: Nature protection areas in the (old) FRG, state 01.01.1984, including natural parks (cross hatching). Source: Bundesforschungsanstalt für Naturschutz und Landschaftsökologie, Bonn-Bad Godesberg

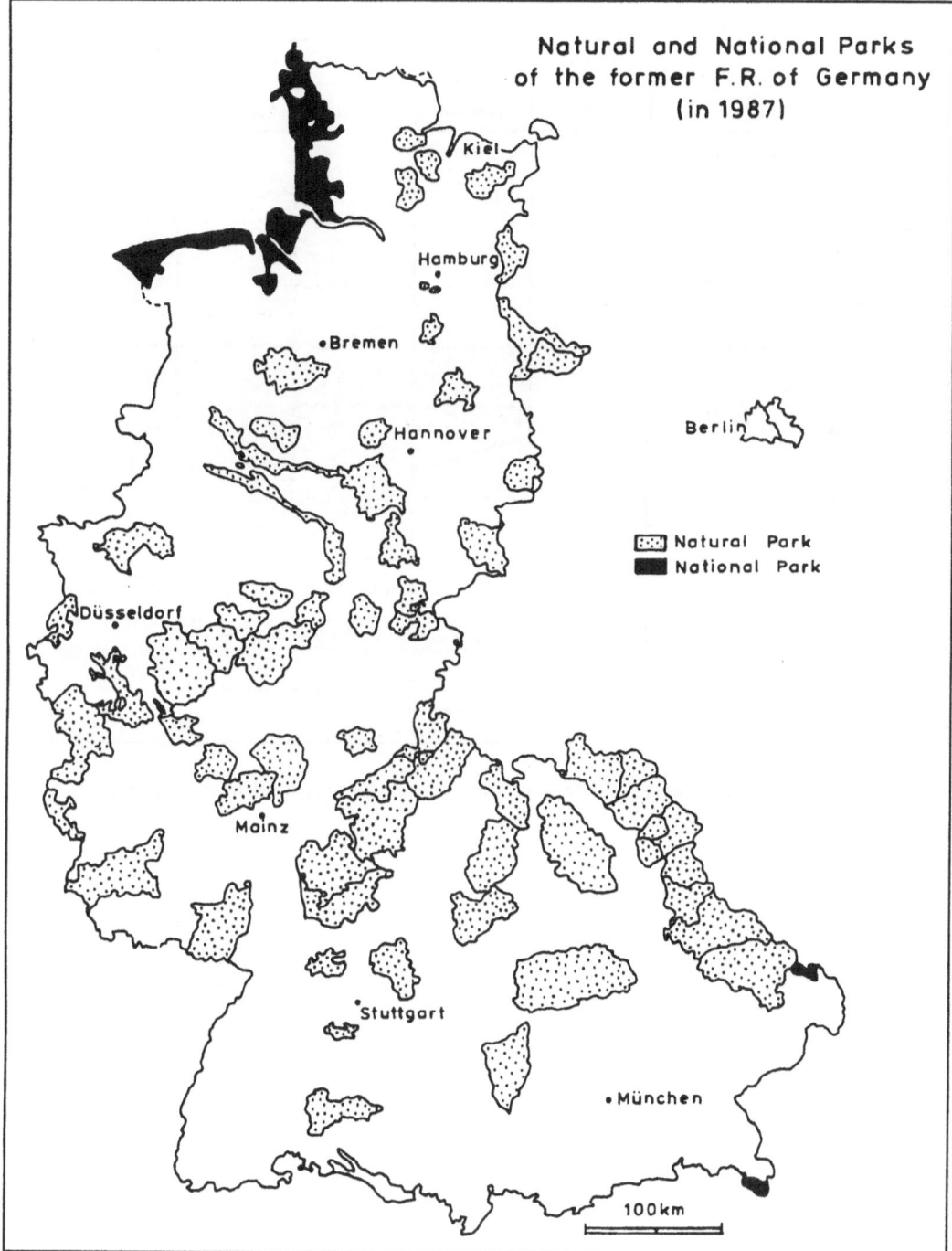

Figure 8: Natural and national parks in the (old) FRG, state 01.01.1987. Source: Bundesforschungsanstalt für Naturschutz und Landschaftsökologie, Bonn-Bad Godesberg

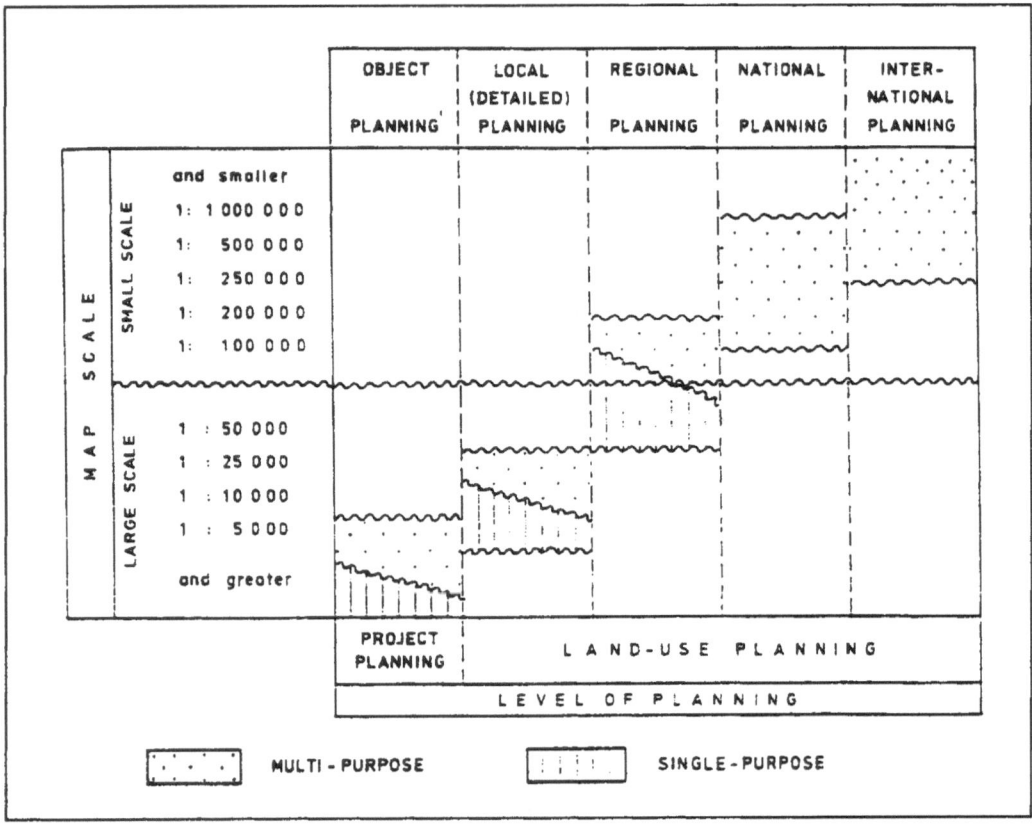

Figure 9:   Levels and scales of maps of environmental and land-use planning

319

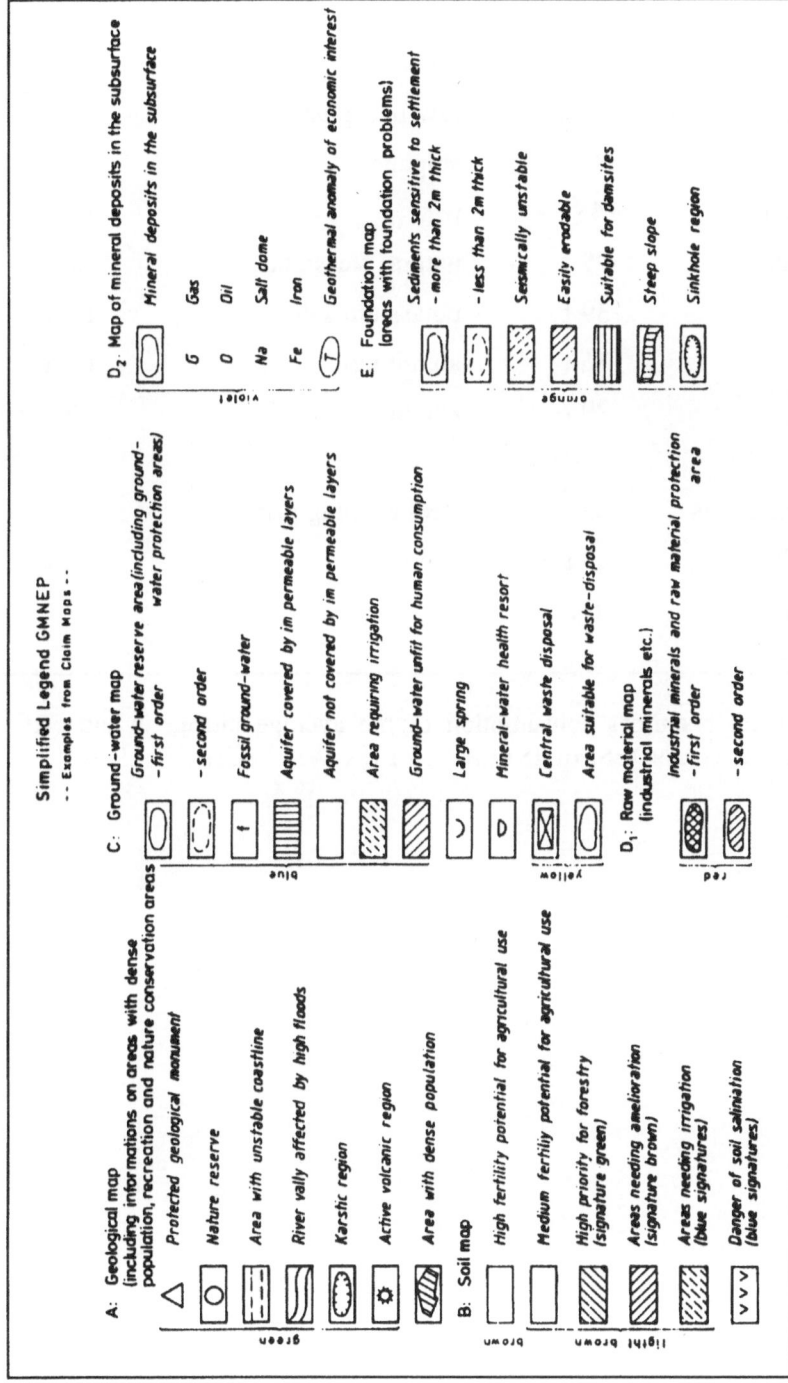

Figure 10: Simplified general legend for a Geoscientific
            Map of the Natural Environment's Potential

sand and gravel	460 t	dolomite	3.5 t
petroleum	166 t	raw phosphates	3.4 t
hard rock	146 t	sulphur	1.9 t
brown coal	145 t	peat	1.8 t
limestone	99 t	natural freestone	1.8 t
steel	39 t	potassium salts	1.6 t
cement	36 t	aluminium	1.4 t
clays	29 t	kaolin	1.2 t
industrial sands	23 t	steel refining elements	1.0 t
rock salt	13 t	copper	1.0 t
gypsum	6 t		

Table 1:   Mineral resources consumption of the average citizen in the FRG in his lifetime (BUNDESMINISTERIUM FÜR WIRTSCHAFT 1979)

Type	Production in 1980	Production in 1987
Sand and gravel	7,550,000,000	8,500,000,000
Coal	2,799,640,000	3,405,370,000
Hard and dimension stone	2,956,821,000	3,284,500,000
Oil	3,089,100,000	2,907,300,000
Limestone and dolomite	2,406,145,000	2,559,562,000
Lignite	1,007,120,000	1,239,080,000
Iron ore	917,000,000	939,510,000
Clays	394,725,000	393,250,000
Peat	307,122,000	385,842,000
Rock salt	168,881,000	173,900,000
Phosphate	136,900,000	143,336,000
Bauxite	92,744,000	95,089,000
Gypsum	71,027,000	87,635,000
Sulphur	54,983,000	55,302,000
Potassium salts	27,627,000	28,760,000
Kaolin	18,200,000	22,400,000
Manganese ore	26,391,000	22,343,000
Magnesite	13,106,000	13,568,000
Chromite	9,902,000	10,845,000
Bentonite group	9,700,000	9,570,000
Copper ore	7,737,400	8,657,500
Soda group	6,666,000	8,101,000
Talcum group	7,205,000	7,300,000
Zink ore	6,191,900	7,182,700
Titanium ores	5,352,700	5,681,000
Fluorspar	5,023,500	4,844,600
Barytes	7,406,000	4,366,000
Feldspar	3,202,000	4,256,000
Asbestos	4,699,306	4,105,520
Lead ore	3,589,200	3,389,300
Boron materials	2,609,800	2,662,000
Diatomite	1,724,200	1,777,400
Zirconium materials	739,864	837,961
Nickel ore	738,400	818,200
Mica group	752,895	775,152
Graphite	602,258	607,824
Sillimanite group	443,325	395,595
Tin ore	235,600	184,900
Molybdenum ore	110,670	84,650
Antimony group	63,287	57,584
Uranium ores	44,149	38,058
Cobalt ores	31,333	36,677
Tungsten ores	53,485	36,481
Vanadium ore	38,645	33,825
Niobium and Tantalum ores	17,561	16,356
Silver	11,101	13,824
Gold	1,237	1,676
Platinum ores	212	263
Diamonds	9	19
Total	22,122,425,637	24,343,424,065
Natural gas (in m³)	1,514.9	1,861.1

Table 2: Comparison of production (in t) of the 50 most important minerals resources in 1980 and in 1987

Type	Rank	Production	Share [%]
Sand and gravel	1	8,500,000,000	34,9
Coal	2	3,405,370,000	14,0
Hard and dimension stone	3	3,284,500,000	13,5
Oil	4	2,907,300,000	11,9
Limestone and dolomite	5	2,559,562,000	10,5
Lignite	6	1,239,080,000	5,1
Iron ore	7	939,510,000	3,9
Clays	8	393,250,000	1,6
Peat	9	385,842,000	1,6
Rock salt	10	173,900,000	0,7
Phosphate	11	143,336,000	0,6
Bauxite	12	95,089,000	
Gypsum	13	87,635,000	
Sulphur	14	55,302,000	
Potassium salts	15	28,760,000	
Kaolin	16	22,400,000	
Manganese ore	17	22,343,000	
Magnesite	18	13,568,000	
Chromite	19	10,845,000	
Bentonite group	20	9,570,000	
Copper ore	21	8,657,500	
Soda group	22	8,101,000	
Talcum group	23	7,300,000	
Zink ore	24	7,182,700	
Titanium ores	25	5,681,000	
Fluorspar	26	4,844,600	
Barytes	27	4,366,000	
Feldspar	28	4,256,000	1,7
Asbestos	29	4,105,520	
Lead ore	30	3,389,300	
Boron materials	31	2,662,000	
Diatomite	32	1,777,400	
Zirconium materials	33	837,961	
Nickel ore	34	818,200	
Mica group	35	775,152	
Graphite	36	607,824	
Sillimanite group	37	395,595	
Tin ore	38	184,900	
Molybdenum ore	39	84,650	
Antimony group	40	57,584	
Uranium ores	41	38,058	
Cobalt ores	42	36,677	
Tungsten ores	43	36,481	
Vanadium ore	44	33,825	
Niobium and Tantalumores	45	16,356	
Silver	46	13,824	
Gold	47	1,676	
Platinum ores	48	263	
Diamonds	49	19	
Total (in t)		24,343,424,065	100,0
Natural gas (in Mrd. m^3)		1,861.0	

Table 3: Order of rank in quantity of world production of the 50 most important minerals in 1987 [t]

Type	Rank	Value 1987 in DM	Share in %
Oil	1	640,226,500,000	39,8
Coal	2	303,895,210,000	18,9
Natural gas	3	262,973,430,000	16,3
Sand and gravel	4	68,000,000,000	4,2
Iron ore	5	48,081,087,000	3,0
Gold	6	43,245,697,000	2,7
Limestone and dolomite	7	35,577,911,000	2,2
Hard and dimension stone	8	32,223,132,300	2,0
Bauxite	9	23,519,132,000	1,5
Copper ore	10	21,935,565,000	1,4
Peat	11	17,588,365,000	1,1
Lignite	12	14,868,960,000	0,9
Sulphur	13	10,466,486,000	0,7 ·
Phosphate	14	8,021,080,200	0,5
Diamonds	15	6,912,205,600	
Silver	16	6,569,710,000	
Platinum ores	17	5,568,664,800	
Zink ore	18	5,486,146,000	
Clays	19	5,435,388,700	
Manganese ore	20	5,182,856,500	
Kaolin	21	4,229,393,200	
Asbestos	22	3,887,298,800	
Uranium ores	23	3,849,815,100	
Nickel ore	24	3,798,337,600	
Lead ore	25	3,378,454,240	
Potassium salt	26	2,658,214,300	
Talcum group	27	2,628,000,000	
Rock salt	28	2,447,446,200	
Magnesite	29	2,294,450,500	
Tin ore	30	1,969,898,800	
Soda group	31	1,396,996,200	4,8
Boron materials	32	1,335,519,500	
Titanium ores	33	1,145,755,760	
Gypsum	34	1,129,886,200	
Chromite	35	1,072,581,300	
Molybdenum ore	36	869,162,660	
Bentonite group	37	757,186,050	
Diatomite	38	716,499,110	
Fluorspar	39	620,263,050	
Cobalt ores	40	429,153,440	
Mica group	41	366,562,580	
Graphite	42	364,329,050	
Vanadium ore	43	355,212,830	
Barytes	44	334,450,090	
Tungsten ores	45	313,568,670	
Feldspar	46	312,248,070	
Niobium and Tantalum ores	47	237,415,060	
Zirconium materials	48	220,283,140	
Antimony group	49	217,242,750	
Sillimanite group	50	148,755,220	
Total		1,609,261,906,570	100,0

Table 4: Order of rank in value of the most important minerals in 1987

Industrial minerals and rocks	production	rank in value
Sand and gravel	1	4
Limestone and dolomite	5	7
Hard and dimension stone	3	8
Peat	9	11
Sulphur	14	13
Phosphate	11	14
Diamonds	49	15
Clays	8	19
Kaolin	16	21
Asbestos	29	22
Potassium salts	15	26
Talcum group	23	27
Rock salt	10	28
Magnesite	18	29
Soda group	22	31
Boron materials	31	32
Titanium ores	25	33
Gypsum	13	34
Bentonite group	20	37
Diatomite	32	38
Fluorspar	26	39
Mica group	35	41
Graphite	36	42
Barytes	27	44
Feldspar	28	46
Zirconium materials	33	48
Sillimanite group	37	50

Table 5

Name	Production quantity 1987 in t or m³	SKE units (average)	Production quantity 1987 in SKE	Share SKE in %
Oil	2,907,300,000	1.44	4,186,512,000	35.8
Coal	3,405,370,000	1.0	3,405,370,000	29.1
Gas	1,861.1	1.33	2,475,263,000	21.2
Uranium	38,058	28,000	1,065,624,000	9.1
Lignite	1,239,080,000	0.45	557,586,000	4.8
Total without gas	7,551,788,058		11,690,355,000	100.0
Resource	31% (excl. gas)			

Table 6a: Comparison of energy resources (production)

Resource	Value 1987 in DM	Share in %
Oil	640,226,500,000	52.2
Coal	303,895,521,000	24.8
Gas	262,973,430,000	21.5
Lignite	14,868,960,000	1.2
Uranium	3,849,815,100	0.3
Total value in DM	1,225,813,915,100	100%
% share of total value of all resources	76%	

Table 6 b: Comparison of energy resources (value)

	WEC	PEC	DC
		Percentage share	
Asbestos	54	35	11
Barytes	47	18	35
Bauxite	25	3	72
Chromite	65	6	29
Cobalt	10	27	63
Copper	29	13	58
Fluorspar	58	10	32
Gold	62	24	14
Ilmenite	68	15	17
Iron	35	34	31
Lead	67	18	15
Manganese	53	38	9
Molybdenum	52	10	38
Nickel	22	30	48
Phosphate	17	13	70
Potash	41	54	5
Platinum group metals	83	17	0.1
Rutile	48	10	42
Silver	50	25	25
Tin	8	26	66
Tungsten	23	64	13
Vanadium	52	46	2
Zinc	73	15	12

WEC = Industrialized market economy countries
PEC = Industrialized centrally planned economy countries
DC = Developing countries
(Source: USBM and BGR)

Table 7: Share of World resources of some minerals in 1981 (ARCHER, LÜTTIG & SNEZHKO 1987)

MAN-MADE	Millions of Tonnes/Year
Particles	92
Gas-particle conversion: sulphur dioxide	147
nitrogen oxides	30
Photochemical, from hydrocarbons	<u>27</u>
	296 11 %
NATURAL	
Soil dust	200
Gas particle conversion: hydrogen sulphide	204
nitrogen oxides	432
ammonia	269
Photochemical, from terpenes, etc.	200
Volcanoes	4
Forest fires	3
Sea salt	<u>1000</u>
	2312 89 %

Table 8: The particles in the atmosphere (after ROBINSON & ROBBINS, 1971)

-- geological overview,

-- soil locations,

-- drought danger,

-- agricultural yield potential,

-- subsoil,

-- groundwater -- basis map,

-- groundwater -- use,

-- surface-near raw materials -- deposits and occurrences, -- ditto -- resource conservation areas,

-- deep-lying resources -- ores, coal, industrial minerals,

-- ditto -- salt,

-- ditto -- mineral oil, natural gas,

-- geoscientific objects worthy of protection,

-- priority uses from a geoscientific point of view.

Table 9: The individual maps belonging to the Geoscientific Map of the Natural Environment's Potential of Lower Saxony and Bremen at a scale of 1 : 200 000

Natural resources
    8 maps on national level
    19 maps on regional level

Agricultural use
    2 maps on national level
    3 maps on regional level

Groundwater
    7 maps on national level
    16 maps on regional level

Vulnerability for pollution
    4 maps on national level
    4 maps on regional level

Foundation stability
    2 maps on national level
    3 maps on regional level
    3 maps on local level

Vulnerability for settlement
    1 map on national level
    3 maps on regional level
    6 maps on local level

Natural hazards
    3 maps on national level
    4 maps on regional level

Geological monuments
    3 maps on national level
    12 maps on regional level

In total: about 100 types of maps/map-sets on a national, regional or local level

Tab. 10: Inventory of Environmental Geological Maps of the Netherlands -- Items

Mining Potential of the Inner Continental Shelf

ROGER H. CHARLIER *

The Inner Continental Shelf

The topic which has been proposed requires that a certain number of conventions be adopted, or definitions be made. First what is to be understood as mineral resources? Frequently, where the ocean is concerned, authors have included all "non-living" resources; these range from energy conversion, to recreational and touristic use, to ocean space utilization, and include mining and desalination. We will limit ourselves to minerals **sensu stricto**, but whatever their origin.

Next we ought to delimit the inner continental shelf. The shelf itself has been defined by jurists and geologists; the latter's limits will prevail. This gently-sloping shallow-water platform extents from the coast out to a "break" which occurs on the average at 130 m (450 feet) although there is considerable variation in depth from a few tens to several hundreds of meters. There are several types of continental shelves which together cover 25.9 g 10^6 km . They are a continuation of the adjacent land areas but no means a featureless plain.

From here on definitions are more fluid: terms like inner (ICS) and outer (OCS) continental shelf appear frequently in the literature and are of common usage when leases are granted but what they actually mean remains unclear: what is the boundary between

---

* Author's address: Prof. Dr. R.H. CHARLIER, Avenue du Congo, 2, B-1050 Brussels.

ICS and OCS? A survey of four encyclopedias and over 25 texts by recognized authorities yielded no answer, nor did four glossaries. Under the circumstances we have adopted as an arbitraty limit of 25 km (13.7 naut. mi) from there, which appears the distance commonly used in surveys of the ICS for mineral exploitation of unconsolidated deposits (particularly sands and gravel).

Distance, however, may not be the only factor as in some cases great depths are rapidly reached where the shelf is narrow. As we also accept depth as a limit, we will not discuss polymetallic nodules, found in some cases close to shore and well within the E.E.Z. (Hawaii, Clarion Island, Clipperton Island, South Africa). The near-shore environment, including beaches, deltas, inlets and channels, barrier islands, river mouths and further to sea the inner continental shelves are high-energy areas and consequently privileged areas for mineral concentration.

Renewed interest in marine mining

After years of wrangling the U.N.O. passed a law of the sea, ratified by an overwhelming majority of members. But it did not unleash a non-hydrocarbon exploration feverish activity. Nor did deep-sea mining bring funds into the accounts of the Seabed Authority. Years went by and finally a renewed interest is shown for marine minerals.

An International Symposium on Coastal Ocean Space Utilization, in New York in May, proposed to establish a central para-governmental organization, funded by the U.S. Congress at its inception, which should initiate, coordinate and foster a partnership of interests to promote new enterprise projects of significant value to the nation. The organization would allay private sector financial reticence through a financial commitment to specific projects; it would, thus, incube these through the initial high-risk stage and bridge the gap between the basic technology and the acceptance of the risk of establishing a derivative business enterprise by the private sector, such as mining and energy production companies (CHARLIER 1983 : 120).

While great rewards have been touted for decades, the risks scale has acted as the major inhibiting factor. The symposium recognized the vast promise held by heavy minerals mining, ocean energy conversion, offshore waste treatment plans and giant ocean platforms construction; this trove of riches has been detailed for decades but, except for hydrocarbons, hardly tapped (CHARLIER 1978, CHARLIER & GORDON 1978, CHARLIER 1982, 1987, HOAGLAND & BROADUS 1987, DEHAIS & WALLACE 1988, CHARLIER 1989). Today, overall, barely 1 % of the "natural resources" that Americans consume come from the sea; the share of "non living" resources is even smaller. Greater aggressivity has been shown by the Japanese faced, it must be said, by a shortage of space and diwindling land resources. An understanding of the world marine minerals markets is needed (HOAGLAND & BROADUS 1987). This reviewed interest came on the heels of a growing concern of potential shortages. New York State sponsored a conference on marine sand and gravel mining (WISE & DUANE 1988) and an international Marine Mining Society had been organized (SIAPNO 1987). At this state already 25 % of total natural aggregates and 10 % of all aggregates in Japan are extracted from the sea; so do 25 % of the fine aggregates, used in concrete. Mineral resources assessments have been conducted somewhat everywhere: iron, titanium, chromium and nickel deposits have been identified in Greek waters while ocean leaching of discharges from an iron-nickel smelting plant created enriched, high grade oves, deposits of iron, chrome, nickel, cobalt and manganese deposits (VARNAVAS & PAPTHEODOROU 1987, VOUTSINAU-TALIADOURI & VARNAVAS 1987). Deposits of heavy minerals are estimated at 2,200,000 tons dry weight, worth about $ 700 million. Large chrome deposits have been identified in Mexico (CASTELLANOS TRUJILLO 1986). Important deposits of phosphates, glauconite and titanium exist along South Africa's coasts (BARTLETT 1987).

The U.K. Department of Trade and Industry, according to Minister ERIC FORTH

(June, 1989) has launched a four-year program to develop ocean-linked markets. A sum of £ 9 million (U.S. $ 15 million) has been ear-marked for the purpose of surveying, exploring, developing and managing, besides food and space, ocean-energies and minerals. Efforts of academe, research institutions and private companies are to be supported and coordinated, and British organizations are to be helped to take full advantage of E.C. oceanographic programs, e.g. MAST and EUREKA. The Marine Technology Centre of the University of Strathlcyde estimated that marine mining would represent, in 20 years, a $ 260 million (£ 168 million) industry. The United Kingdom aims at capturing 10 % of that market, hence it suggested encouraging private companies' interest (MARINE TECHNO-LOGY CENTRE 1988).

In the U.S. The Department of the Interior finally included marine mining interests and offshore support industry representatives in its OCS advisory board. Here to help and advice is sought on exploration, development, evaluation, environmental impact and protection of resources. The Minerals Management Service opened its Environmental Studies Program to proposals dealing with development of minerals other than petroleum or gas.

It should thus not be overlooked that there is a simultaneous rekindling of interest in deep-sea mining, particularly of polymetallic nodules and manganese crusts. The U.S.S.R. has taken up contacts with U.S. and Finnish interests. Norwegian shipping interests have created a marine mining consortium. A Japanese test operation is scheduled for 1994 and unproved dredges (line-bucket) have been tested off the Market Islands. An European marine minerals mining consortium came into being early in 1989. India is planning to map its sites and to launch, in 1990, a pilot plant for extractive metallurgy. And yet if heavy mineral places are found on many beaches and extent for the Konkan coast of Maharashtra State (GUJAIR et al. 1988). Black sand deposits of Kerala Beach have been described by MALLIK et al. (1987); black sands of Guatemala were already studied by BOSS (1940) (Table 3).

The products

There are several ways to classify products that can be extracted from the ocean, p. ex. according to their nature or after their geographical occurence (Table 4). Of interest to the economist and the exploration geologist are metallogenic provinces; these include stannous provinces, black beach sand areas, magnetic and titanomagnetic zones, chrome provinces, auriferous provinces, diamond provinces and hydrocarbon provinces (Table 5).

Inner shelf mineral resources include sand, gravel, shells, tin, heavy minerals, diamonds, gold and, of course, hydrocarbons (PEPPER 1958, MERO 1968, ESTRUP 1978, HOAGLAND & BROADUS 1987) (Table 3).

Extraction technologies include dredging, water processing and conventional mining (Tables 1 and 2). Land technologies to tap oil and gas wells are naturally adapted to the ocean environment. No less than six consortia, some grouping as many as twenty-six companies, are involved in ocean mining activities not including oil and gas operations; exclusive of deep sea operations, close to fifty individual companies are active in marine mining (FORD et al. 1987, CRUICKSHANK 1988). Obviously the interest in marine mining is very much alive and technologies exist, yet little extraction of minerals from the sea has been carried out except for salt. The oldest modern marine mining activity is retrieval of cassiterite and of sand and gravel. Bromine and magnesium are extracted from water and, sporadically, diamonds have been mined. However, as land reserves become depleted attention for marine mining will be rekindled.

Mining

Methods range from an elementary approach to quite sophisticated technologies. Evidently the type of method used plays a major role in the economics of the undertaking

(ATTANASI 1987). Choices are governed by water depth, distance from shore, weather conditions. Nuclear techniques have been recently suggested (RANASINGHE et al. 1988). Visual methods have been employed, sampling by dredger or corer and so have free fall samplers; geophysical methods include precision depth recording, use of the side scan sonar and a seismic approach. Finally exploration can be carried out using electrical methods (with natural or man-made electrical current), magnetic methods, gravity and heat flow. Geochemical methods encompass in situ analyses (natural radioactivity; Californium - 252) and analysis of recovered material; in the latter for metalliferous sediments there will be determination of dispersion in sea water and in sediments.

Exploitation depends on ore grade and quality, distribution and economic value. Mining of aggregates is by suction dredge, of placers also, but it may be done with buckets or grab dredges, using also mobile dredging platforms. This is of course shallow water operation. In the deep sea -- and in deeper waters of the shelf -- the continous line bucket system may be used, which is a dredge bucket along a cable loop. In the Red Sea, where metalliferous muds are considered for exploitation, suction dredges are used.

Four undersea mining techniques - dredges - are commonly employed: bucket-ladder, surface-pump hydraulic, wire-line (a.k.a. drags and grabs), and airlift-hydraulic. The first two are quite adequate for inner continental shelf depths.

Mineral Resources

The ocean waters contain exploitable, dissolved quantities of magnesium, sodium, calcium, bromine, potassium, sulfur, and uranium. Also present are copper, lead, zinc and silver. The continental shelf has deposits of sands, gravel, barite, aragonite, phosphorite and numerous heavy metals, plus gold, diamonds, platinum, and native copper. The continental slope has phosphorite, iron-containing minerals and mud containing zinc, copper, lead and silver, while the deep sea bottom is the resting place of nodules and various oozes. Consolidated materials include coal, limestones, sulfur, tin, gold and hydrocarbons (SHIGLEY 1951, MERO 1968, CHARLIER 1978, 1983, 1987 a, b).

Little extraction of minerals from the sea has been carried out with exception of salt (Table 6). The oldest modern mining activities are the retrieval of cassiterite, a tin ore, and of construction materials such as sand and gravel. Sporadically bromine and magnesium, even diamonds have been mined. But we may well soon be compelled to turn to the sea for many more minerals as land reserves beecome depleted.

Dissolved Substances

An interesting statistic is provided by JOHN MERC (1966) who calculated that with an extraction efficiency of 25 % in the production of fresh water from sea water, 416 km^3 of ocean water would contain 6.4 billion tons of sodium chloride, 240 million tons of magnesium, 160 million tons of sulfur, 800,000 tons of boron, 2,000 tons of aluminium, 400 tons of manganese, 56 tons of copper, 560 tons of uranium, 2,000 tons of molybdenum, 40 tons of silver and one ton of gold. With a recovery of only 10 % the needs of 100 million people would be satisfied in water, molybdenum, boron and bromine and they would have too much of the other products; there would also be enough thorium and uranium torun a nuclear conversion plant to furnish the necessary power. Practically speaking however, concentrations of less than 0.01 are not economical which, for the present, would limit exploitation to sodium chloride, magnesium, sulfur, potassium, bromine and boron. Gypsum has also been considered for extraction (PETTERSON 1928).

Problems associated with chemical extraction of minerals from sea water are low concentration and technical difficulty of selective extraction. If we consider chlorine as having a concentration factor of 1 (166,000 lbs/min gal.= 1981600 kg/mln hl) then gold has a factor 57 (0.000041 kg/mln hl or 0.0047 g) and tin 26 (0.00035 kg/10^6 hl or 0.35 g). The stake is not small: a conservative estimate puts a $ 200 million yearly tag on marine extracted chemicals. A dissolved minerals recovery plant, if operated in series with a saline water conversion plant, could prove economically profitable; other advantages could accrue as well: single energy source, consolidation of services and utilities, relief for waste discharge.

Salt

Since earliest times man has extracted common salt from sea water and used it for both domestic and industrial purposes. It is currently extracted in many parts of the world, for instance in France, Italy, Bulgaria, the United States. Of the minerals dissolved in sea water, salt, which has been extracted at least for 30 centuries, is the most abundant. While today still an important part of the salt consumed throughout the world, even in the United States, the mining of sea salt is of little economic importance beyond the coastal area. It is an open ocean, international water mineral resource, but of little significance here.

Magnesium: bromine

Magnesium salts have been located along the coast of Brazil and in the Congo Basin. Bromine and magnesium compounds are extracted simultaneously: 70 % of the total production of bromine and 60 % of the magnesium used are of marine origin; about 75 % of the United States' needs in bromine come from sea water and so does most of the magnesium (PETTERSON 1928, SEATON 1931).

The discovery of bromine is creditted to the Frenchman A.J. BALARD who identified it in the bitterns of the "marais salants" of southern France some 150 years ago. Over 99 % of the lithosperic bromine is dissolved in ocean waters. Land deposits exist in Germany and the United States. It owes its commercial importance to the demand for "no-knock" gasoline.

Bromine has been extracted "at sea" by direct precipitation from unconcentrated water. The water borne plant, working with an efficiency of 70 %, could produce about 29,000 kg per month. There is only a 0.1 % concentration of bromine in sea water; its extraction necessitates the use of highly corrosive reagents and it appears that a coastal land-based operation resolves the many problems encountered with a ship-based operation.

There is a slightly higher concentration of magnesium in sea water than of bromine (1.3 %), but it is minimal as compared to that of land ores. Nevertheless the largest part of United States primary magnesium metal supply is of marine origin. Its extraction is less costly than using land deposits of magnesite or dolomite, or well-brines. While extraction of magnesium from sea water requires for processing only one twentieth as much water as extracting bromine, near-by abundant supplies of lime, fuel and power are necessary.

Iodine

Iodine has been extracted from seaweed (which concentrate it by a factor of 30,000), and from the Chilean nitrates. Seaweed can also be used as a source of potassium and sodium but land deposits provide cheaper sources; it is still harvested to provide sodium alginate used in the feed industry. Patents have also been taken out to extract iodine from seawater (VIENNE 1949).

Potassium

Potash deposits are large and reserves of potassium run into billions of tons. Large potassium oxide deposits exist in the North Sea. Gypsum is extracted by solar evaporation from San Francisco Bay water and is a by-product of petroleum, and other well brines.

Potash beds of little economic value exist in the Gulf of Mexico in very deformed evaporite beds. Potassium may be precipitated by selective chelation. Glauconite is discussed farther down.

Brines: muds

The estimated value of the minerals present in the Red Sea hot brines exceed several billion dollars. These hot brines can be located by seismic reflection due to their difference in density compared with water. Since 1976 problems of extraction, pollution and treatment of wastes have been systematically investigated. Mud deposits lie at depths of 2,200 m on a basalt bed covered by a 200 m thick zone of concentrated brine. A proposed retrieval scheme includes a plunger pump and a vibrating suction sieve, and a conveyor belt. Testing was done on a river bed and apparently the system works.

Metalliferous hot brines have been identified in the Gulf of California. They are rich in copper and zinc. In the Red Sea some elements are in concentrations from 1,000 to 50,000 times greater than elsewhere, an observation which led to wonder whether metal enriched waters may not also be present in rift zones along plate boundaries. The better known deposits are those currently accumulating along the Red Sea axial deeps; they are underlain by thick salt deposits. At least fourteen sites have been identified there.

Red Sea muds extraction methods were successfully tested in Scottish lochs by the German concern PREUSSAG. The company believes it could now process 10,000 tons per day of mud at 2,200 m depth. Indeed, muds are currently processed on board in the Red Sea. Some 96,000 tons of mud and water from hot brines sites are liquefied, in such operation, to 200,000 tons, and this mass grows, during floatation, by addition of seawater to reach 367,000 tons. The concentrate that is finally transported to the land site amounts to 1,650 tons.

Vanadium

Ocean waters have been mentioned as a source of vanadium. It is concentrated by tunicates in their tissues. Similarly some marine organism. Sunstantially concentrate copper, tin, zinc or lead (fishes), also iron, titanium, manganese, or chromium (algae). Even through the vanadium is present in tunicates at a 280,000 higher concentration than in the sea waters, it is still not commercially attractive (GOLDBERG 1951).

Deuterium and Hydrogen

The deuterium fuel necessary for nuclear fission can be extracted from the ocean. This could be an added incentive to locate nuclear powre plants off shore as was contemplated by the Netherlands.

Hydrogen, naturally, is available in huge quantities in the ocean, but its extraction requires another source of energy, whether conventional or alternative; perhaps little would be gained to extract hydrogen, an easily stored and transported fuel, unless the fuel necessary is not fossil. Hydrogen could be produced by electrolysis.

Minerals as suspended particles and in living matter

Filtering of sea water to recover such sispended particulate matter as gold, lead and iron is not economically rentable, even as part of another operation due to foreseable filter fouling by marine organisms.

Concentrations of other metals in sea water are small, but these could possibly be recovered as a by-product of desalination plants.

Consolidated Deposits

We will limit ourselves to discussion of "consolidated" deposits such as coal, limestone, oil (petroleum) and (natural) gas.

Even before the Christian era, coal mines were exploited by the Greeks beneath the sea at Laurium. Coal, limestone, iron ore, tin copper and nickel have been extracted in Great Britain, Greece, Turkey, Spain, Finland (Baltic Sea), Florida (Atlantic coast), Canada, Chile, Japan and China. Gold, mercury, barium and diamonds have been retrieved from the sea.

Coal

About 1530 a shaft was sunk in an artificial island off the Scottish coast to reach coal deposits. Some mines are as deep as 2400 m, located below 120 m of water, and some at more than 8 km from shore. Such mines are still being worked in Great Britain, Newfoundland, Nova Scotia, Chile, Japan, Turkey and China.

At least 100 sea mines are currently in operation, but coal resources are neither well known nor well quantified and may be large. Excellent prospects exist on the Siberian and Alaskan continental shelves, as well as near Southeastern Australia. Some 57 undersea coal mines located in off-shore Japan, Taiwan, Nova Scotia, Chile, England, Scotland, and Turkey produce over 33,5 million tons. In Japan such mining is done since 1860 off the West coast of Kyushu and new mines are being operated off East Hokkaido. Coal mines could be worked off Alaska, Canada, Brazil, Argentina, Australia, Greece, Norway, the Arctic Soviet Union, Spain and Israel. Artificial islands had to be constructed to establish ventilating shafts for sub-sea coal mining in Japan.

Calcium

Calcium carbonate has been dredged under the form fo coral sand off-shore from Hawaii and Fiji. It is recovered near Arkanes (Iceland) and the United States (Gulf Coast and near San Francisco) under the form of lime shells. It is also recovered elsewhere as *Lithothamnium* and aragonite mud. Because it is bulky it may prove very economical to exploit local reefs rather than to import the product; it is used in the manufacture of cement and fertilizers, pulp and paper and in the extraction process of magnesium from sea water.

Natural gas

Natural gas is possibly present in organic-rich bottom sediments of Baltic, Black, Aegean and Adriatic seas, and may offer an important source of methane; neither preliminary studies nor technology are very advanced. According to Soviet reports, natural gas and petroleum have been located east of Crimea, along the Romanian Coast and on the submarine plateau from Varna (Bulgaria) to the Bosporus. Direct observations by this writer during 1979 revealed very little petroleum and gas between the Soviet and Turkish borders on the eastern Black Sea coast. He has, however, observed some modest evidence of this stretching from the Danube mouth to the Turkish frontier.

In the past decade 70 nations conducted drilling operations in the sea and over twenty are producers. Indonesia is the leading producer in Asia, although production has not lived up to expectations. A current search for offshore oil id being directed towards Southeast Asia, where potential for discovery exists and because it may provide an alternative source to Middle East oil, and the oil contains less sulfur.

All geologists do not share the gloomy view of end-of-century exhaustion: KENNETH EMERY believes huge reserves exist at greater depths and that technology will eventually permit tapping of these resources. It will be expensive, but a sky-rocketing cost of oil may make such fields economically exploitable. P.W.J. WOOD, in an April 1979 paper, shared this idea, although at the same time he advised the development of alternative sources.

Pipeline transport is possible for fields in the North Sea, Gulf of Mexico and near Northern Canada's coast. Gas was found in the Cook Inlet and Bristol Bay, near Trinidad, and in the Gulf of Guayaquil. Rich reserves occur in the Persian Gulf, the Insulindian coast, the Australian North-west Shelf, the Gulf of Tonkin, the Gulf of Thailand, and the Cook Straits. The United Kingdom's reserves are estimated at just below 1 billion m^3 and Canada's at 1'5 trillion.

Petroleum

Oil shales are thought to occur along the coasts of Southern California, Brazil, South Africa, and in the Mediterranean and Red seas; tar sands have been located near Southern California, Venezuela, Trinidad and Tobago. Estimates of one trillion barrels contained in shales and 200 billion barrels in tars have been made. Exploitation depends on technological developments on land-based deposits of which only Canada has started extraction. Present technology can only recover about 50 of the 967 billion barrels estimated to be present in the Athabasca River oil sands. The sand is separated from the bitumen in tumblers filled with a mixture of hot air, stream and water. Carbon, sulfur and nitrogen are then chemically removed from the bitumen. To produce a single barrel of this synthetic crude oil, some 2'5 tons of raw material must be processed.

Estimated remaining reserves of offshore petroleum, calculated in millions of barrels are: Africa 8,704; Oceania 2,070; South America 90; North America 12,110; the Caribbean 6,048; and Europe 11,569. The worldwide total exceeds 140,000 millions barrels.

Estimated recoverable gas reserves in billions of cubic meters are: Africa 540; Asia 530; North America 118; the Caribbean 10.5; Oceania 72.9; and Europe 1905. The total for offshore reserves is about 5054 billion m^3.

Raw Materials

For the United States alone in a few years the value of offshore mineral production may climb. For instance, from $ 539.4 million in 1960 it went to $ 1,343.9 in 1966, more than double. Excluding petroleum, gas, and sulfur, the value went from $ 115.8 (1960) to $ 166.2 (1966), nearly a 50 % increase. Aside from exploration activities, current marine mining operations include dredging of tin (Indonesia), of iron sands (Japan), diamonds (South Africa), shell sands (Iceland), aragonite (Bahamas), gold (USA), shells (USA), sand and gravel (France), sand (USA), sulfur (USA).

Aggregates are principally quartz, thus sands and gravel, though other minerals fall also in this category. Black sands, such as those found on the shores of the island of Hawaii are aggregates. French efforts have been made to assess a variety of inner continental shelf aggregates for close to fifteen years; they include sands and gravels off the coasts of the Atlantic Ocean and the Channel, calcareous sands and cassiterite bearing sands, both off the coast of Britanny, and iron sands near the eastern shores of the island of Elba (Italy). Overseas they examined the heavy minerals containing sands of (French) Guyana, the titanium bearing sands off Senegal, and the phosphatic sediments near the southern coast of Gabon. If the German enterprise PREUSSAG A.G. is the main contractor for the recovery of the Red Sea "muds", the French B.R.G.M. acted as technical consultant (BAUGSSE et al. 1977).

Sands and Gravel

It has been estimated that by the beginning of the 21st century most industrialized nations will have exhausted their land-base reserves of construction materials. Construction materials include sand, gravels, refractory muds, shells and coral (DEHAIS & WALLACE 1988). Thanks to adequate technology Canada and Japan are mining sand. Others are currently hauling sand and gravel from North Sea deposits (France, Great Britain, The Netherlands), and Belgium is using them for artificial beach rebuilding. In the Mediterranean, Lebanon and Israel attempted the same, but they mined too close to shore thereby accentuating beach retreat (UREN 1988).

Geographical distribution of on-going operations covers principally the United States, the Republic of South Africa, Japan, Denmark, Sweden, The Netherlands and Great Britain; in this last location the recuperation of sand and gravel is the largest and most advanced: 13.5 million tons then worth $ 32 million were extracted in 1971. Most promising US areas are off New York and New Jersey. Occasional operations have also been conducted in Thailand and near Hong Kong. Already in value and in quantity dredging of sand and gravel surpass all other seafloor mining operations (WILLIAMS 1986, BARTLETT 1987). The U.S.A. have large deposits of sand near urban areas where beach nourishment is needed.

A shortage of construction materials has developed over the last two decades in Britanny. The demand for offshore sand exceeded already fifteen years ago 4 million ton and some 18 million ton were extracted, of which 14 million from the sea. This led to a thorough examination of such new sources as offshore sand bars, offshore aggregates and tailings from mining sites. Already then a serious concern was voiced about the possible environmental impact of such recovery and guidelines were spelled out limiting exploitation to water 25 to 30 m deep, sites away from human occupied areas, fishing zones and maritime routes, and at appropriate distance from biological breeeding places. To abide by these specifications, sand mining can occur only at about 5 km (3 sta. mi.) from the coast (BLANC 1975, LE LANNM 1975, DE BYSER 1976).

In the United Kingdom some 18×10^6 t of sand are retrieved and exported, and marine sands represent 14 % of the national consumption (UREN 1988). Calcium carbonate comes from aragonite, coral, shells and *Lithothamnia*, an alga. In the USA operations involving shells recovery have been reported in alabama, Louisiana, Texas, Maryland (Baltimore) and California (San Francisco); some of these operations date back over a hundred years. *Lithothamnium* is exploited in Britanny.

Artificial islands have been built to recover aragonite in Bahamian waters and exploitation has been underway for some time at Ocean Cay.

Depth is an important element in gravel development and deposits under more than

35 m of water, or under thick layers of sand, may not be economically profitable at present. Sand is dredged in Iceland, and in New England as well. Sands could be retrieved as by-products of other mining operations. Offshore sands may need to be cleaned of salts and clays. **Coral** and **coral sand** can be used and are mined for instance in Bali, the Bahamas, Hawaii and Taiwan. Being coarse gained they require thus a grinding process. They contain on the average 90 % CO_3Ca (BROWN & DUNNE 1988).

At depths between 50 and 80 m, carbonate of calcium deposits of 85 % concentration have been identified on the coasts of Britanny and an estimate of 1 billion m^3 of economically interesting was made.

Calcareous shells have been mined near California, and along the Gulf of Mexico. Oyster shells have been exploited near San Francisco and on the Gulf of Mexico. Off the Isle of Man shell banks are exploited, so is *Lithothamnium* off Scotland, and shells in Faxa Bay (Iceland) and the Waddensea (The Netherlands).

Zeolithes (phillipsite, clinoptilotite, harmotome, analcine and some others) may be the result of chemical reactions within sediments, or have a volcanic origin either submarine or terrestrial.

Placers are metallic minerals; marine placers are virtually confined to the near shore area through sea-level changes may cause them to be found way out on the continental shelf, p.ex. in drowned valleys and former beaches. They originate generally from igneous rocks that have been weathered and whose material has been transported, then concentrated in traps (CRONAN 1980). Concentrations on beaches are due to the action of waves and currents, occasionally to submarine erosion but then rarely and poorly.

Placers have been found, and are exploited, on several coasts: Alaska (Platinum), Florida and South Carolina (light, heavy minerals), West and East Africa (ilmenite), Republic of South Africa (diamonds), Queensland and New South Wales (rutile, zircon, monazite), West and eastern Australia (ilmenite), southeast Australia and Tasmania (gold, tin), New Zealand (titanomagnetite, ilmenite, gold, magnetite, rutile, monazite, zircon, iron-bearing sands, garnet, cassiterite), Kerala (monazite, ilmenite), Sri Lanka (rutile, zircon) and Cornwall (cassiterite, stream tin) (KULM 1988, PETERSON & BIRMEY 1988). Rutile placers have been found on beach deposits in Australia and Africa; ilmenite near Mozambique and India. Chrome, zircon and titanium occur in the coasts of Tabasco (Mexico) (CASTELLANOS TRUJILLO 1986).

Barite is found in the form of nodules, faecal pellets, crystals, aggregates, etc. It is of volcanic origin, or organogenic, but occurs also within marine sediments after burial. Incidentally it is also found near rises. Deposits have been located off Indonesia, Sri Lanka and Southern California.

Placers may vary in composition; in the Pacific Northwest (USA) deposits should be addressed for their economic and strategic value. Deposits with over 50 % of heavy minerals may represent from 10^4 to 10^6 m^3 (PETERSEN & BIRMEY 1988). Some sands contain gold, others platinum. These have been exploited near Alaska, particularly in Goodnews Bay. Gold has been identified off the Oregon coast (KULM 1988). Platinum represents a $ 1 million operation. Gold and platinum placers occur preferentially in a gravel matrix, yet silt-size deposits have been found in bays and on shelves.

Little difference exists, between the concentration of gold in the open ocean and in coastal waters. Although there might be as much as eight trillion tons gold dissolved in the oceans, the usual concentration is barely over 0.001 mg/ton and only in the southern Atlantic does it climb to a bit over 0.044 mg/ton.

Most "economic" deposits of gold and platinum are alluvial in origin and found close to their original terrestrial source. Gravel and cobble deposits rarely hold marine placers except where submarine erosion of a lode may have occured or a placer-type concentration of fine fractions may have accumulated close-by.

Diamond market conditions brought to a halt diamond dredging near Walvis Bay off the coast of Namibia. Storm damage contributed to the decision (COLLINS & KEEBLE 1962). However, during the last decade, a flurry of artisanal activity of gathering them, has led to a resumption of limited operations since 1979 (BARTLETT 1987). A ton of gravel yields on the average one diamond. Besides Namibia, diamonds could be recuperated along the coasts of the Cape, Eastern South Africa, South America and off several Atlantic African countries (Ghana, Sierra Leone, Gabon, Congo, Zaire and Angola). Metalliferous sands and heavy minerals but also gems are recovered off China, Indonesia, Malaysia, the Philippines, Singapore, Thailand and Vietnam. Amber has been mined in the Baltic Sea (BERERSDORF 1972).

Muds can result from the sorting of material by tidal currents which deposit them when they loose their force; this is the origin of the deposits off the Cumberland coast. The metalliferous muds of the Red Sea represent deposits that formed during the early stages of ocean evolution at a time when the sea floor has spreading started again following a long period of quiescence. Of principal economic importance is the sulfide group of sediments which includes two facies: mixed sulfides of iron-copper-zinc, and pyrite. Other facies are the silicates group (smectite, chamosite and amorphous sulfates), the oxydes group (iron, manganese, goethite, hematite, limonite, etc.), sulfates (gypsum, anhydrite), and the carbonate group.

Iron Ore

The Japanese have retrieved some 30,000 tons of iron ore a month since 1963, from Ariake Bay. The Soviets announced plans to mine magnetite and titanomagnetite placers in the Baltic, Black and Azow seas.

Drowned valleys could be the site of valuable cassiterite bearing sands in Brittany, but little further exploration has been done since the economic profitability has been put in doubt.

Prospecting for several metals in coastal sands is currently carried on: Zirconium and monazite are extracted simultaneously in Florida, Sri Lanka and Australia, where it takes 750,000 t to get 4 per thous. Zircon; chromite, rutile, ilmenite, have been located in several countries. Barite is extracted near Alaskan shores, magnetite near the Kyushu Islands, ilmenite in Australia and South Africa (CARLSON 1914, COETZEE 1957).

If ilmenite has been mined along the Gambia coast, prospects for marine mining in Senegal have not been as promising. However, mineralized lenses, several meters thick, in relatively shallow waters (20-30 m) have yielded some titanium; ilmenite and zircon content ranges from 3 to 5 %, and chromium is present in a 0.18 % concentration (HORN et al. 1974). Chromite, ilmenite and zircon occur on the shelf off Oregon (KULM 1988). Here, and in California, the black sands are poorly suited for processing precisely because of the presence of ilmenite: it is hard to remove and this complicates processing of iron ore.

Titaniferous sands are mined off Kyushu (Japan), Luzon (Philippines), Java and New Zealand. South Africa has titanium-bearing placers (BARTLETT 1987). Magnetite is mined from a submarine vein 80 km south of Helsinki near Jussaro Island, the operation should be considered a consolidated deposit. It is also mined below the sea floor in Japan and on Bell Island (Newfoundland) (PEPPER 1958). Marine ores are available in numerous locations: British Columbia, Newfoundland, Hudson Bay, Alaska, Florida, Georgia, North Carolina, the west coast of the U.S., Japan, Fiji, Chile, England and Elba. Rutile and ilmenite have been identified on the US West Coast, in Florida, the former British West Indies, Mozambique and South Africa, barite off Castle Island (Alaska).

Some presence of limonite and siderite has been detected in the mineral sands off French Guyana but do not justify exploitation.

Tin

For close to 75 years, cassiterite, a tin ore, has been exploited in Southeast Asia. Recovery of tin-containing sands is reported in St. Ives Bay (Cornwall) and important deposits have been located off Britanny. Soviet reports claim numerous exploitable deposits exist in its northern waters (e.g. Laptev Sea) and in the Sea of Japan.

Most marine tin comes from cassiterite-containing placers. Such deposits are being dredged along the coasts of northern Burma, Thailand, Malaysia and Indonesia (e.g. Bangka, Belitung, Phuket, Singkep).

Several specially equipped ore processing ships have been built, apparently the largest by the Japanese.

Phosphorite

Phosphorite nodules have been located off the coasts of the United States, Mexico, South America, Asia, Japan, Spain and Northwest Australia. They are not currently competitive with the land deposits, but price rises may soon place them in a more favorable position, even though world supplies appear to be considerable (MARVASTI & RIGAS 1987). The best area for the U.S.A. would be the "Frying Pan" of North Carolina (id.).

If, as foreseen, offshore phosphorite deposits are mined, then uranium recovery may be undertaken as a by-product; uranium is also found in consolidated and non-consolidated deposits.

Phosphorites are found on the continental shelf, on off shore banks and on marginal plateaux. Usually very impure, their deposition is enhanced by upwellings. Although characteristically a continental shelf deposit, phosphorites are principally found at the margin of marine ocean basins (CRONAN 1980). These are the only regions where they are presently forming. It is believed that they result from direct inorganic precipitation of phosphorus, or by replacement of the carbonate by PO_4 in the carbonate to form a replacement deposit. The question has been raised of deposits on oceanic seamounts; here the origin has been attributed to guano deposits with phosphatization of the carbonate rock. Important deposits have been reported for Peru, Chile and between the 18th and 24th parallels along the coast of Southwest Africa. On the continental shelf they occur on top or along sides of banks, escarpments and submarine canyons. Deposits have been located along coasts of both oceans in North and South America, and off Japan and South Africa. Grab dredges could "harvest" California deposits.

Most phosphate deposits are located along the western coasts of Africa and America. Exploration has been carried out on Chatham Rise near New Zealand.

Besides on the deeper regions of the continental shelf and upper reaches of the continental slopes, phosphorite nodules are found on ridges and submarine banks, at depths of 30 to 300 m. Often mentioned deposits lay off-shore from Peru and Chile, Northest Mexico and southern California, the southeastern United States and the Republic of South Africa. The Californian reserves have been estimated at some 1.5 billion tons. Exploitation has been interrupted because the area was a dump for WWI ammunition.

Deposits of phosphatic sediments found along the coasts of Gabon and the Republic of Congo are of marginal interest only. Those along the Angolan enclave of Cabinda have not been fully assessed.

Ocean phosphorite also occurs in shallow waters as phosphatic mudbanks, sand, pellets and oolithes (Malabar, Baja California, Florida, Massachusetts, Southern California).

The greater concentration of phosphorite nodules on the shelf, and particularly near the coast, has been ascribed to the frequency of sudden temperature changes there, causing the death of organisms. Concentrations of phosphatic materials also occur where waters of different salinities mix, e.g. near upwellings and the mouths of great rivers. One rather commonly accepted hypothesis about the formation of these nodules is that the phosphate is dissolved, migrates off shore, is oxidized, and eventually precipitates in colloidal form. Finally, the colloids agglomerate to form nodules.

The amount of phosphorus dissolved in the sea depends upon such factors as deep sea water upwelling rate and the fraction of particles surviving oxidation as they fall to and reach the deep sea. It depends also on the source: what is the rate of sand run-off and what is the phosphorus content of the river water. The phosphate content of surface water is 30 times lower as that of upwelling deep water.

It is currently assumed that these nodules accumulate only as a single layer. Along the coast of California such nodules have been dredged at depths ranging from a shallow 72 m to as deep as 3,300 m, but the **Challenger** Expedition brought some up off the continental slope base of the Cape of Good Hope, from 6,300 m. This raises the possibility of slumps or turbidity current transport.

Glauconite is an authigenic hydrated potassium, iron, aluminium mineral. It is encountered on the continental shelf but has been reported only at depths exceeding 300 m. However it is found in near shore globigerina oozes along coasts with low sedimentation rates and free of large rivers. At depths below 30 m glauconite exceeds rarely 1 % of the total sediment and is not economically mineable.

Sulfur

Prospects for exploitation of marine sulfur are good. Some deposits have been mined off the Louisiana coast for over ten years; the 80,000 long tons yearly production covers 10 % of U.S. needs. More sulfur is recoverable from undersea gas and oil operations. Production is climbing towards two million long tons a year.

Cap rocks of salt domes contain sulfur. It can be recovered at reasonable cost by relatively simple technology: superheated water is piped down from the surface, the sulfur is thus melted and is then forced up with compressed air. However only one in twenty salt domes contains enough sulfur to warrant exploitation; two decades ago two deposits were being tapped on the continental shelf off Louisiana with a yearly production of 2 millions tons, worth $ 37 million (1973 $). Yet, the increasing demand for and decreasing supplies of sulfur make marine deposits more attractive. The discovery, some ten years ago, of sulfur-bearing domes in the deepest parts of the Gulf of Mexico kindled an active interest.

Offshore salt domes in the Gulf of Mexico were already exploited in 1965 when sulfur exploitation rights were acquired for $ 34 million on 29,140 ha at 30-60 m depth. Onshore technology was easily adapted to ocean mining. Commercial sulfur is being mined from offshore salt domes along the Gulf of Mexico coast of Louisiana, Mississippi, Texas and in the gulf itself. Roughly 10 % of the gulf domes are considered productive. Superheated water is forced down specially designed pipes and the sulfur slurry up to the recovery platform.

Environmental impact

Significant disturbance of the sediment and sessile benthic organisms result from dredging. Distribution and resedimentation of particles falling back is influenced by their density and any near-bottom current. An enrichment in certain compounds may take place in the near-bottom water and affect organisms living near the bottom; the strongest effect may well be for slow reproducing sessile fauna. Man-made sedimentation encompasses transportation of spores with subsequent colonization of new areas.

Mining activities on the continental shelf can disuput the sediment budget and interfere with sediment dispersal patterns, resulting in coastal erosion and the formation of navigational hazards (OWEN 1977). Also influenced will be biogeochemical processes. Alterations may destroy organisms and habitats, cause oxygen depletion and release sediment-held toxic substances. Dredging is a major factor; oil pollution is overwhelmingly due to accidents with carriers; mining from land shafts, underneath the seabed and chemical extraction pose negligible threats to the environment.

From a geological viewpoint, the near-shore energy regime altered by mining which reshapes bottom contours, may lead to erosion or accretion; this development is significant in water less than 20 m deep. It is believed that consequences will be less dangerous along straight coastlines than in embayments which constitute natural sediment traps. Large mining operations cause turbidity, thereby reducing light penetration and leading to lesser productivity and provoke fin-rot for species unable to escape the area or to follow atavistic migration routes.

The effects of marine mining are thus three-fold: those of exploration, of extraction and of the processing facilities. Coral mining caused the loss of 100 m of beach on Bali and sand mining 260 m in three years on Ciluicing Beach (Djakarta).

Activities will affect the water surface and column, the sea floor, beaches and harbors. Human health runs dangers due to foodstock chemical contamination and viral or bacterial infections. The ecosystem will be altered, degraded, stocks depleted and foodstocks lost; shore amenities will be reduced and degraded, marine cultural heritage possibly lost.

Impacts can be classified into two categories: those of the first order, vig. turbidity, seabed smathering, contamination and toxicity; those of the second order which include reduction of biological products, contaminant biomagnification, seafood pathologies and stock mortalities (ELLIS 1981, CHARLIER 1987a, b, WALDICHUK 1987, BROWN & DUNNE 1988). A recent study of HUME and PULLEN (1988) showed that mining of sand has no long term effects on the marine benthos if the surface sediments are similar to the original surface material when using it for beach nourishment. The final long term profile will have similar contours as the original ones and relative depths will remain about identical.

Cassiterite mining e.g. is environmentally harmful because of the large amount of tailings rejected. Benthic comunities are destroyed, but tailings are dumped on the spot and ores are washed. Sediments are put in the water column indifferently. The most serious damage has occured on the Thai coast. Consequences include such physical modifications as changes in temperatures, current patterns, presence of suspended matter, reduced light penetration and man-made benthic sedimentation. Among chemical changes are the introduction of nutrients, trace elements and toxic elements. Food supplies can be reduced by deposition of a layer of particles, bottom divellers can choke, detoxification processes hampered and shellfish contaminated.

Some steps can be taken to remedy or reduce pollution. They include gathering of geographical and oceanographic data, survey variables, establish a control area design site ecosystem model, and assess aestethic effect of operations.

Conclusion

There is apparently no doubt that the mineral wealth of submerged continental areas is considerable. It is most probable that these resources may eventually be called upon to make an important contribution to the world economy. Already dissolved substances constitute a valuable supplement to land reserves. Necessity has forced several countries to turn to the sea to supply construction materials; according to Emery this industry represents three times the value of the tin retrieved from the ocean. However, exploitation has in several instances been detrimental to the environment (Tables 9 and 10).

No longer is ocean mining, particularly on the continental shelf, considered, except for hydrocarbons, a marginal activity. As the economic picture changes, so will and do attitudes towards such mining. Current interest in studying deposits, exploration and exploitation methods and the development of new technologies bear witness to the upsurge of interest in ocean mining.

References

ARCHER, A. (1970): Sub-sea mineral and the environment.- New Scientist **48**, 278: 372-373, London.

ATTANASI, E.G. & DEYOUNG, J.H.Jr. (1987): Economics and the search for offshore heavy minerals deposits.- Mar. Min. **6**, 4: 323-327, London.

BARTLETT, P.M. (1987): Republic of South Africa coastal and marine minerals potential.- Mar. Min. **6**, 4: 359-383, London.

BEIERSDORF, H. (1972): Mining for amber on the sea-bed in Kursiumario.- Meerestechn. **3**: 100-101, Düsseldorf.

BLANC, J.J. (1975): Recherches sur les gites de sables et agrégats sur la marge continentale de la Provence.- Bull. Soc. geol. France **7**, 4: 521-528, Paris.

BLISSENBACH, E. & NAUWAB, Z. (1982): Metalliferous sediments of the seabed: The Atlantis II-Deep deposits of the Red Sea.- In: [BORGESE, E.M. & GINSBURG, N.]: Ocean Yearbook 3: 77-104, Chicago (University of Chicago Press).

BOSS, M.E. (1940): The black beach sands of Guatemala.- Bull. geol. Soc. Amer. **51**: 1915-1921, New York.

BOUYSEE, Ph. et al. (1977): Reconnaissance sédimentologique du plateau continental de la Guyane Française.- B.R.G.M. rapp. 77 SGN 078 MAR 2/77: 1-48, Paris.

BROWN, B.E. & DUNNE, R.P. (1988): The environmental impact of coral mining on coral reefs in the Maldives.- Environ. Conserv. 15, 2: 159-165, Brisbane, London.

BROWN, G.A. (1967): Australian offshore mineral exploration.- Proc. Offsh. Explor. Conf. Long Beach, CA 1967: 67-80, Long Beach.

CARLSON, O.J. (1944): Exploitation of minerals in beach sands on the south coast of Queensland.- Queensl. Govt. min. J. 49: 223-245, Brisbane.

CASTELLANOS, T.L. (1986): La costa de cromo en los Estados de Tabasco y Campeche, Mexico.- An. inst. Cienc. mar. Limnol. Univ. nac. auton. Mex. 13, 1: 79-90, Mexico.

CHARLIER, R.H. (1978): Other ocean resources.- In: [BORGESE, E.M. & GINSBURG, N.]: Other ocean resources; Ocean Yearbook 1: 160-210, Chicago (University of Chicago Press).

CHARLIER, R.H. (1983): Water, energy and non-living ocean resources.- In: [BORGESE, E.M. & GINSBURG, N.]: Ocean Yearbook 4: 75-120, Chicago (University of Chicago Press).

CHARLIER, R.H. (1987a): Planning for coastal areas.- Norg. geol. Unders. spec. Publ. 2: 12-24, Trondheim.

CHARLIER, R.H. (1987b): Marine mineral resources extraction in coastal areas and its impact on the environment, and consequences for land-use.- In: [ARNDT, P. & LÜTTIG, G.W.]: Mineral resources extraction, environmental protection and land-use planning in the industrial and developing countries: 53-70, Stuttgart (Schweizerbart).

CHARLIER, R.H. & GORDON, B.L. (1978): Ocean resources: an introduction to economic oceanography: 89-123, Washington (University Press of America).

COETZEE, C.B. (1957): Ilmenite-bearing sands along the west coast in the Vanrhynsdorp District.- Un. south Afr., Dept. Min. geol. Surv. Div. Bull. 25: 1-17, Pretoria.

COLLINS, J.V. & KEEBLE, P. (1962): Diamonds from the sea bed.- In: [STRYKOWSKI, J.G.]: Underwater Yearbook 1962: 12-14, Chicago ((Underwater Soc.).

CRESSARD, A.P. (1974): Consequential effects of industrial exploitation of sand and gravel on the marine environment and the economic activities of the marine field.- Paris (Centre National pour l'Exploitation des Océans).

CRESSARD, A.P. & DE BYSER, J. (1974): Influence de l'exploitation des sables et graviers sur le milieu marin. Présentation du programme français.- Rapport GA 14-9-74: 1-23, Brest (CNEXO-COB).

CRONAN, D.S. (1980): Underwater minerals.- London (Academic Press).

CRUIKSHANK, M.H. & HESS , H.D. (1975): Marine sand and gravel mining.- Oceanus 19, 1: 11-17, Woods Hole, MA. (Woods Hole Ocean. Inst.)

CRUIKSHANK, M.H. (1988): Marine sand and gravel mining and processing technolo gies.- Mar. Min. 7, 1-2:149-162, London.

CURRAY, R.J. (1966): Continental terrace.- Encycl. Oceanogr.: 207-213, New York (Van Nostrand - Reinhold).

DE BYSER et al. (1976): Les granulats marins silicieux et calcaires du littoral breton.- Note techn. 52, 7/76: 1-16, Paris (CNEXO).

DEHAIS, J.A. & WALLACE, W.A. (1988): Economic aspects of offshore sand and gravel mining.- In: [DUANE, B.B.]: Marine sand and gravel workshop proceedings.- Mar. Min. 7, 1-2: 35-48, London.

ELLIS, D.V. (1987): A decade of environmental impact assessment at marine and coastal mines.- Mar. Min. 6, 4: 385-417, London.

EMERY, K.O. & NBAKES, L.C. (1968): Economic placer deposits on the continental shelf.- CCOP techn. Bull. (UNECAFE) 1: 95, New York.

ESTRUP, C. (1978): Minerals from the sea.- In: [WILMOT, P.D. & SLIGERLAND, A.]: Technology assessment and the oceans, Guilford, Surrey, GB (IPC Science & Technology Press).

FORD, G. et al. (1987): The future for ocean technology.- Wolfeboro, NH (Future Sci. Techn. Services).

GEORGHIOU, L. & FORD, G. (1981): Arab silver from the Red Sea mud.- New Scientist 89: 470-472, London.

GOLDBERG, E.D. et al. (1951): The uptake of vanadium by tunicates.- Biol. Bull. 101: 84, New York.

GUJAR, A.L. et al. (1988): Marine minerals: the Indian perspective.- Mar. Min. 7, 4: 317-350, London.

HESS, H.D. (1971): Marine sand and gravel mining in the U.K.- Tiburon, CA (NOAA Marine Minerals Techn. Ctr.).

HOAGLAND, P. & BROADUS, J.M. (1987): Seabed material commodity and resources summaries.- Woods Hole, MA. (Woods Hole Ocean. Inst.).

HURME, A.K. & PULLEN, E.J. (1988): Biological effects of marine sand mining and fill placement for beach replenishment.- Mar. Min. 7, 1-2: 123-136, London.

KULM, M.D. (1988): Potential heavy mineral and metal placers on the Oregon continental shelf.- Mar. Min. 7, 4: 361-395, London.

LE LANN, F. (1975): Recherche de granulats marins au sud de la Bretagne.- Rapp. SGN 20 (MAR) 75, Paris (BRGM).

MALLIK, T.K. et al. (1987): The black sand placer deposits of Kerala Beach, Southwest India.- Mar. Geol. 77, 1-2: 29-150, New York.

MARINE TECHNICAL CENTER UNIVERSITY STRATHCLYDE (1988): Oceans of opportunity.- Engineering 228, 10: 547-549, Strathclyde, GB.

MARVASTI, A. & RIGGS, D. (1987): Potential for marine mining of phosphate within the U.S. E.E.Z.- Mar. Min. 6, 3: 291-300, London.

MERO, J.L. (1965): The mineral resources of the sea.- 102 pp., Amsterdam (Elsevier).

OWEN, R. M. (1977): An assessment of the environmental impact of mining on the conti nental shelf.- Mar. Min. 1, 1-2: 85-102, London.

PEPPER, J.F. (1958): Potential mineral resources of the continental shelves of the western hemisphere.- In: An introduction to the geology and mineral resources of the continental shelves of the Americas, U.S. geol. Surv. Bull. 1067: 43-65, Reston, VA.

PETERSON, D.O. (1928): Production of gypsum and magnesium from seawater.- Swedish Pat. 65: 434, Stockholm.

PETERSON, D.O. & BINNEY, S.E. (1988): Compositional variations of coastal placers in the Pacific Northwest, USA.- Mar. Min. 7, 4: 397-416, London.

PHELPS, W.B. (1940): Heavy minerals in the beach sands of Florida.- Proc. Florida Ac. Sci. 5: 168-171, Tallahassee, FL.

RANASINGHE, V.V.C. et al. (1988): Application of nuclear techniques to marine exploration.- Mar. Min. 7, 4: 291-316, London.

SEATON, M.Y. (1931): Bromine and magnesium compounds drawn from western bays and hills.- Chem. met. Eng. 38: 638-641, New York.

SHEPARD, F.P. (1986): Continental shelf; classification.- Encycl. Oceanogr.: 202-204, New York (Van Nostrand-Reinhold).

SHIGLEY, C.M. (1951): Minerals from the sea.- J. Metals **3**: 25-29, New York.

SIAPNE, W.D. (1987): The International Marine Mining Society.- Mar. Min. **6**, 3: 245-252, London.

TSURUSAKI, K. et al. (1988): Seabed sand mining in Japan.- Mar. Min. **7**, 1-2: 49-68, London.

[UNITED NATIONS ORGANIZATION] (1970): Mineral resources of the sea.- 49 pp., New York (U.N.O., Dep. of Economic & Social Affairs).

UREN, J.M. (1988): The marine sand and gravel dredging industry in the United Kingdom.- Mar. Min. **7**, 1-2: 69-85, London.

VARNAVAS, S.P. & PAPATHEODOROU, G. (1987): Marine mineral resources in the eastern Mediterranean Sea (1): iron, titanium, chromium, and nickel deposits in the Gulf of Corinth, Greece.- Mar. Min. **6**, 1-2: 37-70, London.

VIENNE, G. (1959): Extracting iodine from seawater.- French Pat. 945, 347, Paris.

VOUTSINOU-TALIADOURI, F. & VARNAVAS, S.P. (1981): Marine mineral resources in the eastern Mediterranean Sea (2): an iron, chromium and nickel deposit in the northern Embolkos Bay, Greece.- Mar. Min. **6**, 3: 259-290, London.

WALDICHUK, M. (1987): Mineral extraction from the sea and potential environmental effects.- Mar. Poll. Bull. **18**, 7: 378-379, New York.

WILLIAMS, S.J. (1986): Sand and gravel deposits within the U.S. E.E.Z.: Resources assessment and uses.- Ann. Offsh. Technol. Conf. Proc. **18**: 377-386, Durham, NC.

WISE, W.M. & DUANE, B.B. (1988): An introduction to the sand and gravel workshop.- In: [DUANE, B.B.]: Conf. Proc.; Marine sand and gravel mining.- Mar. Min. **7**, 1-2: 1-6, London.

WOOD, P.W.I. (1979): New slant on potential petroleum resources. - Ocean Industry **14** (4): 59-72.

1. Dredging

    (a) clamshell (open-mouth box): rather shallow waters

    (b) dragline bucket (max. 1600 m)

    (c) ladder bucket: shallow water

    (d) airlift dredge

    (e) wire-line dredge

    (f) trailing suction hopper dredge

2. Water processing

    (a) lime treatment: magnesium extraction

    (b) sulfurization: bromine extraction

    (c) bacterial leaching: gold

    (d) chemical process: uranium

    (e) evaporation: salt

3. Conventional mining

    (a) land access, or island access: coal

    (b) artificial island: oil, also: coal, aragonite

    (c) non-explosive rock breaking

Table 1:   Technologies

Technique / Nature of deposit	Unconsolidated	Consolidated
Trailing suction hopper dredge	sand, gravel, shells	
Anchored suction dredge	heavy minerals, phosphorite	
Cutterhead pipeline dredge	sand, gravel, shells	
Bucket ladder dredge	heavy minerals	
Continous line bucket	phosphorite	
Drag dredge	phosphorite	
Sub-bottom mining		various (e.g. coal)
"solution" mining		various (e.g. sulfur)

Table 2:   Mining techniques on the continental shelf according to geological nature of resource

(a)   ores: e.g. cassiterite

(b)   placers: e.g. gold

(c)   dissolved substances: e.g. bromine, magnesium

(d)   sub-bottom deposits: e.g. coal, tin

Table 3:   Extractive marine products

(a) Unconsolidated	(α) continental shelf
↓	(β) continental slope
surficial, authigenic	(γ) deep ocean
in situ	

(b)   consolidated: surficial, authigenic, in situ

(c)   dissolved: ocean-wide

Table 4:   Geographical occurrence

1. Stannous provinces

    (a)   usually associated with granitic plutons, or with volcanic rocks or subvolcanics in Bolivia.
    (b)   never are alluvions at more than 5 km from primary deposits on land.
    (c)   coastal areas include SE Asia (10 mi.+), Cornwall (2 mi.), Tasmania & Queensland (300,000), Galicia & No. Portugal (20,000), Britanny (10,000), Alaska.
    (d)   it is necessary to prospect at sea, near known primary deposits. However, a lithology different from that on land, may cause other deposits to occur on the continental shelf.

2. Black beach-sands minerals

    (a)   ilmenite mined is found for about 50 % on beaches.
    (b)   rutile and natural zircon are nearly all of littoral origin.
    (c)   ilmenite, rutile, zircon, monazite do not occur in well individualized deposits. They are alteration products.

3. Magnetite and titanomagnetite

    (a)   they are part of detrital iron deposits, occasionally made up of hermatite or iron hydroxyde.
    (b)   such deposits are usually linked to post mesozoic volcanic layers.
    (c)   they are also black beach-sands, and like them found on the storm beach, dune, back-beach river deposits and on submarine accumulation strings; less known original concentrations are found on beaches and sand banks.

4. Chrome provinces

    (a)   found in basic rocks in S. Africa, Turkey, Philippines, Greece, Iran, Cuba, Western U.S.A., New Zealand and Papua-New Guinea.
    (b)   also beaches in Oregon, Sakhalin, New Caledonia.

5. Auriferous provinces

    (a)   Western U.S.A., E. Canada, Africa, Mediterranean, Far East Australia, Pacific (Fiji, Solomon), S. New Zealand.
    (b)   submarine sites are thalwegs, drowned beaches, eluvions.

6. Diamond provinces

    (a)   Africa, also W. India coast, Siberia, Crimea.
    (b)   Maurie Diamond Corp. quit operations in 1972.

7. Hydrocarbon provinces

Table 5:   Metallogenic provinces

Occurence	Mineral	Water depth (m) [up to]
Dissolved	Magnesium	any
	Salt	any
	Calcium	any
	Bromine	any
	Potassium	any
	Sulfur	any
Unconsolidated	Sand & Gravel	35 to 50
	Shell	35
	Aragonite	50
	Phosphorite	35 to 200
	Barite	35 to 200
	Glauconite	35 to 200
	Magnetite	10 to 200
	Rutile	< 35
	Zircon	< 35
	Cassiterite	< 130
Consolidated	Diamonds	< 70
	Platinum	< 170
	Gold	< 170
	Silver	< 170
	Coal	< 130
	Monazite	0 to 70
	Sulfur	< 70

Note: Oil and gas not included

Table 6: Recovery depth of exploitable principal marine minerals on continental shelf

Dissolved	Unconsolidated	Consolidated
Metals and Salts:	Non-metals:	Varous types of occu-rence (1):
Magnesium (x)	Sand, gravel, shells (x)	Coal (x)
Sodium (Salt) (x)	Semiprecious stones (x)	Limestone
Calcium (x)	Aragonite (x)	Irons (x)
Bromine (x)	Phosphorite	Sulfur (x)
Potassium (x)	Barite (x)	Tin (x)
Sulfur (x)	Glauconite	Metallic: sulfides
	Heavy metals:	salts
Boron	Magnetite (x)	
Uranium		
	Ilmenite (x)	Gold
Brines:	Rutile (x)	
		Hydrocarbons (x)
Zinc	Monazite (x)	
Copper	Chromite	
Lead	Zircon (x)	
Silver	Cassiterite (x)	
(2)	Rare & precious metals	
	Diamonds (x)	
	Platinum (x)	
	Gold (x)	
	Native Copper	

(1)   disseminated, vein, tabular, stratified, massive
(2)   some others but in very small quantities
(x)   ongoing production, some experimental

Table 7:   Mineral resources of the continental shelf

Resource/Year	1930	1960	1990
Iron (and steel)	289	407	537
Copper	7	11	14
Manganese	4.3	6	8
Lead	4.3	5.4	6.1
Zinc	3	5	8.1
Chromium	1	2.6	4.3
Tin	0.6	0.5	0.5
Nickel	0.2	0.6	1.2
Aluminium	0.7	9	32
Petroleum (bbl)	(7.5)	(17.7)	(25)

Table 8: Past and projected U.S. per capita consumption (per annum) of mineral resources found in the ocean (in kg or barrels)

Resource	U.S.A.	Rest of the World
Iron	1.8	1.9
Copper	1.6	3.5
Tin	1.5	1.5
Nickel	3.0	2.3
Aluminium	7.2	7.8
Phosphores	2.2	6.0
Petroleum	2.3	4.3

Table 9:   Mean ratio (US and World) of projected to current (1970-90) demand for some mineral resources

Earth Science Conservation in Europe
Present Activities and Recommended Procedures

GERARD P. GONGGRIJP [*]

Abstract

The geological as well as the biological landscapes have been affected by the impact of Man's activities, but the attention paid on the conservation of the abiotic nature in contrast with the biotic has been neglected for a long time. Only in the last few decades a change took place. In several countries inventories were started and an earth-science conservation policy was developed on national level. Although geology does not end at borders, the international contacts were few, excepted between specialists. In 1988 and initiative was taken in organizing an international meeting, which turned out to be a great success. On this meeting twelve scientists from seven coutries decided to co-operate in the European Working Group on Earth Science Conservation. Besides, exchange of information, mutual support, promotion of Earth-science conservation, organization of meetings and the production of a newsletter, they started the execution of common projects.

Three projects: a promotion paper, a manual and a European earth science site list will be commenced very soon.

[*] Author's address: Dr. G.P. GONGGRIJP, European Group of Earth-Science Conservation, Research Institute for Nature Management, POB 46, 3956 ZR Leersum, The Netherlands.

Recognition of the importance of Earth-science conservation and an active policy of supranational bodies like the Council of Europe, EEC, UNESCO and IUCN is one of the main purposes of the working group.

Introduction

Our planet Earth was born about 4,600 million years ago and the composition and surface form of its outer shell are the result of the prolonged operation of mighty internal or endogenic forces such as volcanism and tectonism which produced its large scale features. Superimposed on these, are smaller and more detailed features which are the results of superficial or exogenic agencies, principally climatic in character (wind, water, glacial action), and, to a lesser extent, of extraterrestrial agents such as meteorites. Together, all these agencies have built up an endless number of geological landscapes, which differ from place to place and change with time.

Since the arrival of Man, nature has been manipulated. Initially, Man's impact was small, but after the introduction of farming, his effect on geological processes increased and led to the erosion and denudation of slopes and to the aeolian erosion on overgrazed areas of sand. Later, active exploitation of natural resources, for example the winning of peat, began, as did the first attempts to control geological processes, e.g. by the construction of dykes.

In many places on Earth the impact of Man has radically changed - and is still changing - local geology, geomorphology and pedology (figure 1). The first attempts to preserve Earth science sites from these changes were made at the end of the nineteenth century, but the enthusiasm behind such attempts was much less than that enjoyed by their biological equivalents. This, however, changed in the second half of this century during a general revival of interest in nature conservation. After a necessary period of national consolidation, the time has now come for Earth science conservation to develop as a strong international movement - geological landscape do not stop at national borders. The establishment of the European Working Group on Earth Science Conservation has been the first step in this direction for this working group has set itself the task of improving earth science conservation policy at both a national and a international level.

Threats to the Geological Landscape

Man's activities have proved highly potent in changing most of the natural landscape. Over vast areas, in a few thousands of years, Man has replaced most of the natural vegetation by cultivated plants. In some particularly vulnerable areas, this shift has favoured the process of denudation and erosion, causing new geological landscapes such as the badlands of the Mediterranean countries. But Man has also left his mark upon the geological landscape by changing the structure and composition of its component strata and altering their forms on the surface. A wide range of activities is responsible for this change and degradation (GONGGRIJP 1981).

Intensive farming methods often require modifications to be made in the local geomorphology; the commonly practiced levelling of small landforms provides a good example. Tillage activities, such as deep ploughing and the lowering of groundwater levels by drainage, lead to the destruction of soil-profiles developed over hundreds, or thousands, of years.

Measures taken to **regulate rivers** interfere with the natural stream systems (figure 2). In the Netherlands almost all river and brook systems have been embanked and regulated to a greater or a lesser extent. Artificial canals can be accepted as beautiful only because there are also natural meandering rivers!

Along much of the coast, the **construction of schemes** designed to prevent marine erosion has had a marked effect through obscuring exposures of earlier strata and preventing the operation of natural processes of erosion and deposition.

Artificial fixation of coastal and inland dune areas by plantation is another impact on landscapes where geological processes are still active.

Intensive recreational use of vulnerable landscapes like alpine skiing areas and dune systems causes degradation of the vegetation, followed by denudation and erosion (figure 3).

Still increasing **urbanization** denies access to many geological sequences and mask, or even destroy landforms.

The **extraction of raw materials** such as clay, sand, gravel, peat and chalk leads to the creation of many pits and quarries. Until recently, except in areas of peat working, most of these excavations were small and did not greatly influence the morphology. But today these activities often are concentrated in, and have a pronounced influence on, relatively large areas.

Such human impacts sometimes lead to the partial or total destruction of complete geological sections and geomorphological features and soils. Notorious examples are to be found in large parts of the Salpausselkä moraines in Finland, a Saalian esker system and the building of a city quarter upon a unique 'fossile' braided river system in The Netherlands, the areas of peat across Northwestern Europe, the areas of brown-coal working in Germany and several of the original type-sections, such as that for the Tiglien.

JAHN (1974) warned of the possibility that the so-far almost untouched areas of the Arctic and sub-Arctic, with their vulnerable vegetation and soils, could be destroyed for, in these regions, Man's activities cause rapid erosion, slope-wash and deflation.

In most cases, the creation of exposures showing the internal structure and composition of landforms through mineral extraction is only small compensation for the eventual destruction of the landform itself. Furthermore, such exposures, even where they display geological and pedological phenomena, are only temporary for, like all exposures, whether natural or artificial, they are likely to become obscured by natural processes of denudation or, in case of disused and neglected pits and quarries, by 'restoration' involving infilling and re-vegetation (figure 4).

In short, the rapid growth of population and of technical development has led to considerable increases in the human impact on geological landscape and has caused irreversible changes to the landscape. Because many geological features are, in effect, 'fossil' rather than still actively developing like the Ice age phenomena, a significant proportion of geological landscapes, once destroyed or damaged, cannot be replaced or repaired.

Aims for Conservation

Over the past few decades, the continuing degradation of the geological landscape, through which more and more traditional and new study areas have been affected by all kinds of human activities, has increasingly alarmed Earth scientists. One by one, geomorphological features have disappeared partly, or even completely, so that the landscape has changed little by little. Sections used for reference, or as examples, vanished through being buried, and through the filling-up and grassing over of pits, quarries and cuttings (figure 5). All these activities are hostile to the needs of research and education.

JØSANG (1979) summarized the aims of conservation as follows:

1. The documentation of geological history throughout all parts of the geological column.

2. The protection and conservation of geological and geomorphological sites and areas of special interest where these are rare, where these are specially threatened, where these are of value to research, where these are of value to education and where these possess some distinctive quality and beauty (figure 6).

3. The protection of a sufficiently comprehensive and durable suite of localities to provide for the needs of research, education and recreation.

In addition to these motives for the preservation of important sites, there are also others. For example, untouched or nearly untouched areas which retain their original soils and vegetation are very important for studies in ecological development and, even if the vegetation is not original, but has been replaced, the area still can be of potential value to ecology. There are also ethical considerations. Should we, for instance, transform the natural geological landscape, formed in the course of thousands, or even millions, of years, into an artificial one? Such arguments supplement not only the purely Earth science arguments for conservation, but also those based on ecology and the perception of landscape as an amenity. It is only natural for Earth scientists to think first of geological motives for conservation, but they, like the biologists and the experts in landscape amenity, should not forget that there are other reasons available to justify conservation; even if these are of particular interest to other scientists, their existence should be bourne in mind.

Research on landscape perception in The Netherlands indicated the importance of variation in landscapes, particularly variation in the (natural) relief.

Site Selection

Our natural environment in general performs several roles, of which the provision of information regarding geological processes and their results, is only one. Because the other roles played by the environment have to be respected as well, compromises have to be made, especially where the different roles, like the extraction of raw materials and the conservation of geomorphology are clearly incompatible. In other cases, different roles can be compatible, such as mineral extraction and the conservation of many geological sections.

For this reason it is necessary to select those earth-scientific features which merit conservation; such a selection is based on a series of criteria which have been devised to accord with nature conservation policy.

More or less the same criteria for the selection of important Earth science sites are used in most countries and these include rarity, present condition, representativity, diversity, and scientific and educational importance. In addition, other aspects such as size, clarity, accessibility and the vulnerability of threatened areas are seen as important. Finally, it must always be remembered that most Earth science sites are irreplaceable for example those formed in the past under quite different climatic conditions like in the Ice age; once destroyed, they are destroyed for ever (figure 7, 8)! After sites have been assessed through the application of the different criteria, they often have to be categorised in some way, for example into those of international, national, regional, and local significance (BJÖRK-LAND 1987).

Development of Earth Science Conservation

At the end of the nineteenth century, the increasing impact of Man on the landscape led to the rise of nature conservation movements all over the world; in this movement, biologists took the leading role. Earth scientists many of whom were involved in the exploitation of natural resources, were, in general, not fired with the same enthusiasm. Moreover, at that time mineral exploitation was not extensive by present-day standards, was littele regulated, and the damaging 'restoration' of disused workings was not commonly practised. Nevertheless, individual Earth scientists and members of nature conservation societies gradually became more and more involved in Earth science conservation.

One of the earliest examples of earth science conservation was the protection in 1840 of the famous 'Agassiz Rock' on Blackford Hill in Edinburgh, a striated rock face where Agassiz had recognised evidence for the former existence of glaciers in Scotland (GORDON 1987). Elsewhere, erratic blocks had already impressed and challenged Earth scientists for many years and it is no wonder that such blocks were among the earliest features to receive protection. In 1887 the Geological Commission of the Swiss Nature Research Society (Geologische Kommission der Schweizerischen Naturforschenden Gesellschaft) proposed to protect erratic blocks and, shortly after, the Swiss State bought the most important (BRANCA 1915).

The influence of societies on the conservation policy of government often had been very great. The first Dutch nature conservation society was established in 1905, and recognised the importance of earth science conservation; not long afterwards, in 1907, the State established its first nature reserves, among which was an inland dune area showing active aeolian processes.

A meeting of the Norwegian Geographical Society in 1909 recommended that the protection of scientifically and historically important geological and mineralogical sites should be included in the first Norwegian Nature Conservation Act which became law in 1910. At a meeting in 1905 of the Geological Society of Stockholm, DE GEER drew attention to the need for the conservation of natural monuments. This led to the preparation of the first official specific inventory of Earth-science sites in Sweden by MUNTHE (1920). The cases listed above provide examples of the initiatives taken early in the development the (Earth-science) conservation movement in Europe. For a time all seemed well, but in practice it took many years before Earth-science conservation could obtain a standing equal to that of other conservation subjects like natural history and archaeology.

Meanwhile, it was mostly individual action which led to the protection of sites and, although in some countries small scale site inventories were prepared, such initiatives did not result in the adoption of active national Earth-science conservation policies. More than a century separates the protection of the Agassiz Rock from the first official recognition, in 1949, of Earth science conservation as an equal partner in the conservation policy of Great Britain, the first such recognition in Europe. The Nature Conservancy (now: Nature Conservancy Council) established in 1949 as an official conservation body introduced shortly afterwards a Geological and Physiogeographical Section (now: Earth-Science Division) responsible for Earth-science sites; a unique event in Earth-science conservation in Europe. Many inventories and other projects have been carried out under the Section's responsibility (NATURE CONSERVATION COUNCIL 1976). But even then, to be recognised as an equal did not mean to be treated as an equal. In most European countries, the adoption of a more active earth science conservation policy dates from the end of the sixties and early seventies, the timing being influenced by a general revival of the nature conservation movement.

In Sweden, at a wide-ranging discussion on the protection of eskers, the opinion emerged that the Earth scientific importance of (parts of) eskers deserved to be taken into consideration. This discussion led to the acknowledgement of Earth science conservation in the Nature Conservation Act of 1952 and in new initiatives (SOYEZ 1973). Several studies focussed on conservation were set up in the sixties and seventies and are still going on, for example, the Geomorphological Inventory of North Sweden (SOYEZ 1971, STATENS NATURVÅRDSVERK 1973).

In connection with the establishment of the Norwegian Ministry of Environment, the preparation of plans on a county-by-county basis to show areas and sites of conservation value, including those of value to the Earth science, was commenced (ERICSTAD 1984). Conservation studies and the preparation of inventories are still in progress (MILJØVERN-DEPARTEMENTET 1983, JANSEN 1983).

Since 1972, the eskers of Finland have been the subject of nation-wide inventories which are prepared in order to supply a factual basis for decision in physical planning, such as allocation for gravel extraction, recreational use and conservation (KONTTURI & LYYTIKÄINEN 1985). A national inventory on earth science sites in general is now in preparation.

The preparation of a National Heritage Inventory in Ireland in the early seventies led to the results being published in 1981. This publication includes Earth science sites, but the Wildlife Act (Nature Conservation Act) does not cover geological sites. Such sites are left to depend for their protection on planning control by the local authorities, a system which, in practice, does not function satisfactorily (DALY 1988).

The Durch working group Gea commenced the preparation of the first com-prehensive inventory on a national scale in 1969 (GONGGRIJP & BOEKSCHOTEN 1981) and completed this task in 1988. As a result of these activities, Earth science conservation was acknowledged in 1982 in a Structural Scheme for Nature and Landscape Conservation. Although this scheme was not directly translated into a concrete conservation policy, it had implications for the future. In the Nature Policy Plan, published in 1989, Earth science conservation is one of the four conservation objectives and proposals for further projects in Earth-science conservation have been formulated (GONGGRIJP 1989).

In 1984, the Ministry of the Environment in Danmark published a list of 197 sites of national Earth scientific interest, as a base for conservation (FREDNINGSSTYRELSEN 1984, LARSEN 1988). But also on county level inventories have been finished (NIELSEN 1981, 1988).

The Spanish Geological Survey published several well-produced guides with Earth-science sites as a base for conservation (INSTITUTO GEOLOGICO Y MINERO DE ESPAÑA 1985).

In the following federal countries the initiatives have been taken mostly on federal level.

Vorarlberg in cooperation with the University of Amsterdam is the first in Austria carrying out a geotop (Earth-science sites) inventory (KRIEG 1988).

In Sankt Gallen in Switzerland inventories on munincipal level have been started (SCHLEGEL 1987). Promotion on national level is initiated by the Nature Conservation (Society: Schweizerischer Bund für Naturschutz) (HANTKE 1986).

In several states of the Federal Republic of Germany inventories have been carried out by the Geological Surveys or Geological Institutes (GERMAN 1974, GRUBE & ROSS 1982).

Outside of Europe New Zealand has a very active Earth-science conservation movement led and inspired by the various earth-science societies (NATURE CONSERVATION COUNCIL 1988).

The above short review is far from being complete, and data is lacking form many countries. It will serve, however, to give an impression of current activities in the field of Earth science conservation.

International Co-Operation

Until 1988, over most of the field of Earth-science conservation, there were only incidental, bilateral, international contacts, although some specialized disciplines - for example palaeontology and soil science - had conduced conservation discussions on international level for some time. In 1986, the ISSS Bulletin of the International Society of Soil Sciences included a note entitled 'Proposal for an International Register of Soil and

Vegetation Reserves' recording the discussion at the meeting in 1986 in Hamburg (MCINTOSH 1986). At a meeting in London in 1987, palaeontologists considered the necessity of protecting localities effectively (CROWTHER & WIMBLEDON 1988). Such initiatives, however, were exceptional.

Following inquiries made among Earth-science conservationists by the author in 1987, it was found that there was a great need for, and desire for, an enhanced level of international contact. Based on the results of this inquiry, the first international workshop was organized in 1988 at Leersum in The Netherlands. At this meeting, the twelve participants from Austria, Denmark, Finland, Great Britain, Ireland, Norway and The Netherlands discussed the following subjects:

- legislation; conservation policy; the classification, listing and selection of sites; management and education;
- the establishment of an international working group;
- the production of a newsletter;
- the execution of international projects.

During the meeting, it became clear that, in the participating countries, Earth-science conservation has been treated more or less as a step-child in comparison to 'biological' conservation, although the legal provisions make Earth-science conservation possible. There was confidence that this situation would be improved if an active working group were set up to operate on a national and international level. Great importance was attached to Informing national Earth science organizations and authorities, and international bodies such as the Council of Europe, EEC, IUCN, and UNESCO and, of course, the international scientific organizations, about Earth science conservation. But besides informing officials and colleagues, a very important task is to unlighten the general public on earth sciences in order to make them more earth sciences minded (GONGGRIJP 1988, DUFF, MCKIRDY & HARLEY 1985, ROBINSON 1989, SCHÖNLAUB 1988).

This first meeting which included a field trip (figure 9, 10), resulted in the establishment of the European Working Group on Earth Science Conservation with the following aims:

- exchange of information,
- mutual support,
- promotion of Earth science conservation,
- organization of annual meetings,
- production of a newsletter,
- execution of common projects.

The first common project was the preparation of an information paper on international Earth science conservation to be illustrated by examples from the different countries. It is planned that this paper will appear in Autumn 1990 in Naturopa, a journal on nature conservation published by the European Information Centre for Nature Conservation of the Council of Europe. A wide variety of sites and areas of Earth science significance in the different countries will be discussed to give attention to this special branch of nature conservation.

A second project discussed in the last meeting (included Switzerland) in June 1989 in Bregenz in Austria is the publication of a manual on earth-science conservation. This will be concerned with items such as legislation, conservation policies and procedures, considerations of access etc. in the various countries and is intended to stimulate standardization. A framework for this manual should be set up by the end of 1989.

A third project, the inventory of type sites suggested by the author at the first meeting, has been developed by Dr. W.A. WIMBLEDON of the Nature Conservancy Council on the Bregenz meeting. It implies the selection of Europe's most important sites: Eurogeosites?

WIMBLEDON suggested by labelling of these 'Eurogeosites' we could:

1.   Add support to local or national initiatives to protect sites.

2.   Submit finalised European lists to the EEC, Council of Europe, UNESCO, etc. for use in their work in the wider protection of geological, geomorphological or landscape features.

3.   Gain added status for sites which, although already recognized locally, deserve wider recognition.

4.   Publizise the existence of such sites, in order to heighten public and government awareness of all earth science sites, be they tiny fossil sites or enormous wilderness areas.

The sites that should be considered for such status, could be categorized as follows:

-    Best sites: that are the best examples of a particular category.

-    Unique sites: localities with international renown for the nature of their geology, be it rocks, minerals, fossils or landforms.

-    Firsts: the localities where the first recognition of a depositional or erosional process took place, where a major time unit was defined, or an orogenic or stratigraphic event of vital step in organic evolution was identified. These are all of the highest historical interest.

-    Patterns: the commonest category of sites in most classification systems, the sites which demonstrate the salient or significant features, be they hard or soft rock or landform, which occur in, or typify an area, large or small.

To carry out this project, classifications of all kind of features should be made and, of course, a standard set of criteria for judging these features in a European context should be developed. The last two operations can become successful only if they are fully supported by a wide range of organizations and individuals from all European countries working in the field of Earth sciences, and also if these operations receive financial backing.

Up to now various countries are not represented, their experts are welcome on our next meeting in 1990 in Norway.

The first steps in European earth science conservation have been made, but there is still a long way to go!

References

BJÖRKLAND, G. (1987): Geomorphological evaluation in international perspective. - UNGI Rapp., **67**: 1-66, Uppsala.

BRANCA, W. (1915): Schutz den geologischen Naturdenkmälern.- Naturdenkmäler, Vortr. u. Aufs. **9/10**: 1-82, Berlin.

CROWTHER, P.R. & WIMBLEDON, W.A. (1988): The use and conservation of palae-ontological sites.- Spec. Pap. Palaeont. **40**: 1-200, London.

DALY, D. (1988): Earth Science Conservation in the republic of Ireland.- Workshop Pap. intern. Meet. on Earth Sci. Conserv., Research Inst. Nat. Management: 37-40, Leersum.

DUFF, K.L., MCKIRDY, A.P. & HARLEY, M.J. (1985): New sites for Old.- A student's guide to the geology of the east Mendips: 1-192, Nature Conservancy Council Peterborough.

ERIKSTAD, L. (1984): Registration and conservation of sites and areas with geological significance in Norway.- Norsk. geogr. Tidskr. **38**: 199-204, Oslo.

[FREDNINGSSTYRELSEN] (1984): Fredningsplan lägning og Geologi, Nationale geo-logiske interesse områder.- Fredningsplanorientering **4**, København.

GERMAN, R. (1974): Das mittelfristige Programm zum Schutz geologisch besonders wichtiger Naturdenkmale in Baden-Württemberg.- Veröff. Landesst. Natursch. Land-schaftspfl. Baden-Württ. **42**: 85-92, Ludwigsburg.

GONGGRIJP, G.P. & BOEKSCHOTEN, G.J. (1981): Earth-science conservation: no science without conservation.- In: [VAN LOON, A.J.]: Quaternary geology: a farewell to A.J. WIGGERS. Geol. en Mijnb. **60**: 433-445, Amsterdam.

GONGGRIJP, G.P. (1988): Earth-science conservation: the Gea project.- Ann. Rep. Res. Inst. Nature Managem. **1987**: 93-101, Arnheim/Leersum/Texel.

GONGGRIJP, G.P. (1989): Nederland in vorm.- Ministerie van Landbouw, Natuurbeheer en Visserij: 1-141, 's-Gravenhage.

GORDON, J.E. (1987): Conservation of geomorphological sites in Britain.- In: [GARD NER, V.]: International Geomorphology 1986, Part II: 583-591, London.

GRUBE, F. & ROSS, P.H. (1982): Schutz geologischer Naturdenkmale.- Die Heimat, Z. Natur- u. Landesk. Schleswig-Holst. u. Hamb. **89**: 37-48, Neumünster.

HANTKE, R. (1986): Erdgeschichtliche Naturdenkmäler.- In: [WILDERMUTH, H.]: Natur als Aufgabe: 230-240, Basel (Schweiz. Bund Natursch.).

[INSTITUTO GEOLOGICO Y MINERO DE ESPANA] (1985): Puntos de interes geologico de Asturias.- 132 pp., Madrid.

JAHN, A. (1976): Geomorphological Modelling and Nature Protection in Arctic and Subarctic Environments.- Geoforum **7**: 121-137, Oxford.

JANSEN, I.J. (1986): Kvartärgeologi, Jord og landskap i Telemark gjennom 11.000 ar.- 87 pp., Oslo (Instituut for Naturanalyse).

JØSANG, O. (1979): Landoversikt over verneverdige naturtyper og forekomster innen geologi og geomorfologi.- Unpublished report, Oslo (Ministry of Environment).

LARSEN, G. (1987): Quaternary geology and nature conservation.- Boreas **16**: 405-410, Oslo.

KRIEG, W. (1988): Earth-science conservation in Austria.- Workshop Pap. intern. Meet. on Earth Sci. Conserv., Research Inst. Nat. Management: 14-16, Leersum.

MCINTOSH, P.D. (1986): Proposal for an International Register of Soil and Vegetation Reserves.- Bull. intern. Soc. of Soil Sc. **69**: 35-36, Wageningen.

[MILJØVERNDEPARTEMENTET] (1983): Utkast til verneplan for mineral-forekomster; Sør-Norge.- 67 pp., Oslo.

[NATURE CONSERVATION COUNCIL] (1976): Shetland, localities of Geological and Geomorphological Importance.- 67 pp., Newbury.

[NATURE CONSERVATION COUNCIL] (1988): Landforms and Geological Features, A case for preservation.- Inform. Bookl. **28**: 1-17, Wellington.

MUNTHE, H. (1920): Strandgrottor och närstående geologiska fenomen i Sverige. - Naturskyddsutredning, SGU Ser. C **302**: **1-67**, Uppsala.

NIELSEN, A.V. (1988): Nature Conservation and Geology in Danmark.- Workshop Pap. intern. Meet. on Earth Sci. Conserv., Research Inst. Nat. Management: 24-31, Leersum.

ROBINSON, E. (1989): European Geological Conservation.- Terra **1**: 113-118, Oxford.

SCHLEGEL, H. (1987): Geotopinventar des St. Gallischen Bezirks Werdenberg.- Ber. Bot. Zool. Ges. Liechtenstein-Sarganz-Werdenberg **16**: 133-184, Vaduz.

SCHÖNLAUB, H.P. (1988): Vom Urknall zum Gailtal - 500 Millionen Jahre Erdgeschichte in der Karnischen Region.- 168 pp., Hermagor.

SOYEZ, D. (1971): Geomorfologisk kartering av nordvästra Dalarna.- Forskningsrapport **11**: 1-130, Stockholm (Natur Geogr. Inst.).

SOYEZ, D. (1973): Geowissenschaften und Naturschutz in Schweden, Rückblick und Entwicklungstendenzen.- Erdkunde, Arch. für wissensch. Geogr. 27: 140-146, Bonn.

[STATENS NATURVÅRDSVERK] (1973): Råd och anvisningar för Naturinventering och Naturvårdsplanning.- SNV, PM **398**: 1-149, Stockholm.

Figure 1: Levelling for agricultural reasons of a small coversand hill, an interesting but very vulnerable geomorphological phenomenon (prov. of Overijssel).

Figure 2: Regulation of a part of the river Dinkel, one of the most unspoiled river-systems in The Netherlands, carried out in the frame of a convention with the Federal Republic of Germany, which owns the upper course (prov. of Overijssel).

Figure 3:   The installation of a golf-course in a very important and vulnerable geomorphological area. In the back-ground an original border between two coversand
deposits forming a low step, effected by the construction of the course (prov.
of Utrecht)

Figure 4:   Unique exposure of Tiglian humic clay threatened by waste disposal (prov.
of Limburg).

Figure 5:   Scientific research on the type section of the Usselo layer, a Late-Glacial podsolic soil possible passing into an Allerød peat (prov. of Limburg).

Figure 6:   Cleaning in the Cottesen quarry in Carboniferous sediments as part of a geological conservation project (prov. of Limburg).

Figure 7:   Geological site in a sand pit in a push-moraine showing ice-pushed river-sediments which are nowadays seldom exposured (prov. of Gelderland).

Figure 8:   Cleaned, protected exposure in Cretaceous chalk layers showing two different formations (prov. of Limburg).

Figure 9: The geological monument 'de Zândkoele' founded in 1984. The exposure in the back-ground shows Saalian boulder clay covered with Weichselian coversands containing periglacial features. For educational purposes a 'boulder map' of Scandinavia was constructed on the floor of the pit to provide visitors with information about ice ages and erratic boulders (prov. of Overijssel).

Figure 10: Several members of the European Working Group on Earth-science Conservation examining a Scandinavian limestone boulder erected in the geological monument 'de Zândkoele', observed by the local mayor and official (prov. of Overijssel).

Introduction

Mining, together with agricultural and livestock activities and large urban agglomerations, gives rise to the anthropic landscapes most characteristic of human activity because of the large transformations involved with respect to the pre-existing medium and the contrast to the natural environment.

From the time flint stones from outcrops of river beds were used to make tools during the Lower Palaeolithic up to current mining activities involving advanced mining, mineralogical and metallurgical techniques, Man has carried out an uncontrolled and systematic exploitation of his mining resources. Particularly during the 19th and the beginning of the 20th century, mining activity achieved an unheard of productive scale, with little consideration of integral resource management. One can speak of predatory use of both mining and other natural resources, which has been the result of successive technological revolutions taking place since the second half of the 18th century.

As in other industrial activities, mining activity has not remained untouched by the conservationist movements that began in the 1950s and 60s. Very differing social sectors called for the protection and conservation of nature, as well as for a more controlled economic development more in harmony with the surrounding environment. The relationship between education and the quality of life and the environment, along with the depletion of many mining resources, has given rise to the current process of geographical transfer of environmental problems from the Industrialized World to the Third World. It is there that mining companies are destroying the environment at alarming rates that threaten to defy international boundaries, even more so when this degradation takes place in very fragile ecosystems such as in tropical forests.

The current situation is formed by an attempt to rationalize exploitation of unrenewable natural resources, compelling the recovery of the medium degraded by extractive activities and promoting territorial planning that awards land-use according to

Reclaiming Areas Degraded by Mining Operations

by ERNESTO GALLEGO VALCARCE & LUCAS VADILLO FERNÁNDEZ [*]

Abstract

The current situation of reclaiming land affected by strip mining is discussed and the main environmental changes caused are mentioned. The various European laws and their most notable differences are compared. A cost estimate is made for reclaiming several mining operations in accordance with the type of operation and deposit characteristics and their effect on operation costs. Measures needed to control the environmental impact of abandoned mines, their technical possibilities and economic feasibility and a **Basic Program** for the possible execution thereof, are specified.

In conclusion, the territorial planning studies in which strip mining should be integrated are considered, making a distinction as to whether they are of a preventive or corrective nature.

[*] Authors' address: E. GALLEGO VALCARCE, Instituto Technológico GeoMinero de España (ITGE), Area de Ingeniería Geoambietal, Rios Rosas, 46, 28003 Madrid, España.

its natural form. The most generalized technical-administrative instrument used worldwide is the environmental impact study supported by environmental standards for each type of activity and government. This includes recommendations from different international organizations (FAO, EEC, etc.).

The potential negative repercussions of mining activity can be summarized as follows:

- Contamination of surface and underground water.
- Variation of phreatic levels.
- Deviation or damming of river courses; change in flows.
- Induced flooding and appearance of surface seepage.
- Transformations in aquifer refilling areas.
- Emission of gases, dust and particles.
- Vibration and noise.
- Land loss and pollution; alteration of phreatic levels.
- Induced erosion and compacting.
- Decrease in productivity.
- Loss of associated flora and fauna; alteration of natural ecosystems.
- Morphological alterations with generation of hollows, clearings and slag heaps.
- Induced instability of natural and artificial slopes.
- Improper land-use after the activity is abandoned.
- Visual impact on the landscape, with changes in shape, volume and color.
- Changes in the ways of life of the affected or adjacent human population.

This brief list summarizes the most characteristic impacts of this activity and has been compiled from multiple impact identification sources developed to date. This list could be subdivided in accordance with the work phases (research, exploitation and abandonment) or the impact producing operations (land movement, vehicle traffic, building, treatment plants, etc.).

Many of the above-mentioned impacts disappear upon termination of the activity; the desolate aspect presented by the majority of mining zones when abandoned is, however, characteristic.

The disappearance of land, vegetation, fauna and flora, and the appearance of anthropic morphologies, in connection with dry climates such as the Mediterranean, does not favor natural restoration of the medium, but it also converts these areas into a center of extension of erosive and polluting processes for adjoining zones.

In view of this situation, the governments of many countries have made it obligatory to restore zones subjected to new extraction and to favor new policies and investment for restoring abandoned mining areas of historical nature. Reconditioning land affected by mining can be carried out by pursuing different objectives:

- Simply encouraging natural recovery when such conditions are present. This involves making the ground fit using various forms of treatments. The proper name would be **reclamation**.
- Restoring the preoperative aspects and features to zones in which conditions are not at all conductive to natural recovery. This always involves great technical, human and economic effort. Its proper name would be **restoration**.
- Encouraging ground uses that are compatible with the new situation in the zone, consistent with the general nature of the environment: **rehabilitation**.

Although the "ideal" strategy would be to "leave things as they were", this should not be considered the only solution, nor should restoration be considered an added cost to existing costs. Possible use of these zones for suitable projects and construction, such as for sports, recreational and leisure activities or agricultural and forest use, have proved to have subsequent economic profitability.

They may also be good places for disposing of urban waste, thus demonstrating an interest in environmental management and also encouraging the morphological recovery of the old operation with these activities. Likewise, siting bothersome industries or installations in quarries and large hollows has also favoured a landscape protection policy.

Finally, the use of old mines as a storage site for toxic, dangerous and radioactive waste or natural gas is perhaps the most noteworthy application.

Current Situation

Restoration of areas degraded by mining was not begun in Spain until 1981, when Law 12/1981 was enacted by the Generalitat de Cataluña regarding **protection of places of natural interest affected by extractive activities**. Subsequently, Royal Decree 2994/82 was enacted regarding **restoration of natural locations affected by mining activities** binding the rest of the nation. In other words, Spain belatedly entered into reclamation of areas affected by mining activities, especially compared with the other countries of the EEC. Even after enactment of R.D. 2994/82, the participation of mine operators in land restoration has at best been scant excluding public firms and some private coal firms (with their own legislation; R.D. 1116/84. Due to the lack of precedents and of technical and human resources of the Mining Services, this discipline has been taken on very slowly and with great reservation on the part of these operators who see restoration as only one more expense in their operating costs. The very ambiguity of Royal Decree 2994/82, which in practice is not being carried out, makes it difficult to put it into effect, since one of its main faults is the **absence of reclamation followup**, with the exception of the legislation on coal mining reclamation patterned on British legislation. In Great Britain, the Town and Country Planning (Minerals) Act, 1981, lays down the followup and care of revegetation for a period of 5 years, with a special emphasis on compacting fill, improving draining, ground structure and ground fertility. Since a large portion of Spanish territory involves semiarid or arid climates, this legal aspect is very important because the healing rate is very low and, thus, endangers maximum revegetation when left without care or protection.

On the other hand, there has been no territorial planning infrastructure to enable the evaluation of different natural resources that can come into conflict, mainly with respect to ground classification from the standpoint of fertility, susceptibility to erosion, landscape, etc.

There have also not been any initiatives by the government and corporations to date to create funds to provide for the restoration of abandoned operations or ones that for some reason have not been able to be restored, as for example, the fund in France to the **steering committee of the parafiscal tax on granulated material**, which helps to reclaim abandoned aggregate exploitations, and provide funding for investigation of resources, replacement materials, etc.

Another important aspect is the loophole to the Mining Law tacked on by the M.O.P.U. (Ministry of Public Works and Urban Development) which frees railroad and highway leasing works (Recourses of Section A), directly performed by the M.O.P.U., from requiring an exploitation permit and thus from carrying out restoration in this type of quarry. The enactment of Legislative R.D. 1302/1986 on **Evaluation of Environmental Impact (E.I.A.)**, and R.D. 1131/1988, which clarifies the former, can and should improve integral management of natural resources, whereby Spain follows the EEC Guideline 85/337/CEE on the **Evaluation of the Environmental Impact of Certain Public and Private Works**. By way of example, it can be indicated that while the average price of reclamation in exploitations of natural aggregate is 600,000 pesetas/hectare with an incidence of 0.5 % on tonne sold; in the case of coal it is 1,800,000 ptas./hectare and 800,000 ptas./hectare, depending on whether or not there is hydrosowing, with an average incidence of 0.6 % on MT sold. These costs are not insignificant and contradict those who consider restoration to be only a measure that runs counter to development in the mining industry.

Reclamation Engineering; Implementation Programs

The impact of mining activites on the environment depends on the interrelationship of several factors, such as size of the operation, location, physical characteristics of the environment, type of mineral, operating method, transformation processes, technology used, infrastructure created, socioeconomic site aspects, etc. The most detrimental impacts on life in the area are those that directly affect the trophic chains, such as physical and chemical water pollution, the former with its increase in turbidity of water courses that afffect fish life and the latter with the resulting acidity and toxicity of the heavy metals, mainly metallic, that are inherent in mining. The effect produced by chemical pollution in the soil due to leaching produced by intermittent water courses or slag water does not seem to be as serious, due to the absorption effect of clay restricting mobility of metallic ions. Therefore, it has less of an effect in drier climates such as in most of Spain.

The mining repercussions that indirectly affect the trophic chains and directly cause morphological changes would be the effects of sedimentation in river channels and erosion processes.

Reclamation of water courses affected by these two mining processes is slow but technically feasible. The solution is based on locating the centers of pollution and reclaiming them, by treating and sealing off the focus of pollution, and on-site, subsequent follow-up using water, sludge and aquatic plant samples. In some cases, it may be necessary to clean sludge loaded with heavy metals or install hydraulic works that could be large engineering projects, for they could affect the filling up of bays with changes in coastal dynamics.

In the case of erosion-sedimentation processes caused by deforestation, changes in natural drainage, etc., reclamation could be carried out by means of ground treatment techniques, reforestation techniques, etc.

Other impacts, such as scenic changes and their integration into the natural landscape, depend on the type of mining, and here we can focus on two principal cases: quarries with large hollow formations and metal and coal mining.

In the majority of strip mining cases and mainly in quarries, where there is a higher rate of technical and economic insolvency of the operators, the absence of a planned operating project and at times the inexistence of minimum research projects and economic profitability studies and disregard for drilling and blasting techniques, have at times resulted in designs with only one very high hollow, in some cases 80-100 m of height in hard rock. This in no case aids in reclaiming these quarries.

In metal and coal mining, with their large slag heaps both from felling and from interior and panning areas, the main problem lies in the large volumes to be moved to reconcile and reclaim these landscapes.

In the first case of quarries with hollow formations, the adjustement of slopes with replanting of vegetable species, mainly trees, to minimize the scenic impact, involves retrenchment of the front of the quarry. This would entail very high drilling and blasting expenses, in addition to increasing the impact on the landscape, replanting of suitable vegetable species and maintenance thereof, at least for the first several years, and the risk of being too difficult. The easiest and cheapest solution is to use these sites, following a geomechanical and stability study, as rubbish dumps or recreational areas with a road infrastructure and some scenic restoration of slopes with climbing and hanging plant species. These measures could be undertaken at a provincial and even municipal level.

In the second case of metal and coal mining with large slag heaps, studies made by the ITGE estimate that land movement could involve costs amounting to 14,800,000,000 pesetas just in the province of León alone. This does not take into account the reclaiming of rivers, streams and lakes polluted by metal and coal mining.

These budgets can only be undertaken by the regional or the federal government.

In both cases, a program in which the following basic objectives are undertaken would be necessary:

1. **Inventory of operations and mining waste**

    (Prior to Royal Decree 2994/82) (Partially performed by the Inventory of Pools and Slag Heaps. ITGE),

2. **Environment reclamation plans**

    -   Current situation of water courses and reclamation thereof.
    -   Integration of mining shafts within the master plans for inert, solid urban toxic and dangerous waste,
    -   Integration of mining structures (quarries, slag heaps, sludge pools, installations, etc.) within the territorial planning plans.

3. **Creation of public organizations**

    By means of a tax on each tonne sold, these organizations would undertake the task of investigating mining reserves, restoring old, abandoned zones and researching the reuse and new applications of mining waste and quarry barrens.

4. **Fiscal deductions**

    These would be offered to those mining firms that undertake the reclamation of old, abandoned zones in the concessions they request.

Mining and Environmental Reclamation Studies Applied to Territorial Planning

The rational exploitation of natural resources recommended by several international organizations and the minimization of environmental impacts occuring is best aided on many occasions by application studies on territorial planning analyzing all those parameters (resource, environmental impact, economic, social) inherent to any extractive activity, as well as the subsequent restoration of the medium.

In general, these studies can be divided into two large groups:

A) Those of corrective nature -- applied to historical mining zones that are totally or partially abandoned to which the different restorative laws in force are not applied. As a last resort, their purpose is to encourage investment programs that can be undertaken by different governing bodies on the basis of precise knowledge of the specific environmental situation of this area.

B) Those of preventive nature -- they are to be applied to zones of great mining interest in the research or initial operating phase, especially if the current land-use or natural form is obviously incompatible with extractive activities. In this case, the ultimate objective of these studies will be activity planning aimed at maximum, compatible exploitation of the different resources, minimizing the negative or undesirable effects that could arise.

The following is a schematic description of the phases and contents of the two types of studies under consideration.

Corrective Studies

In Spain these would be applied to historical mining zones abandoned before 1982, which, therefore, do not fall within the scope of the Royal Decree 2994/1.982 on reclamation of natural locations affected by mining activities, or Royal Decree 1302/1.986 regarding environmental impact evaluation.

- Objectives: Define the technical basis for an environmental reclamation program of an area affected by old extraction.
  Identification of scope: In the last resort, this will depend on the governing body in charge, normally a regional or autonomous one involving several munincipalities.
  Delimitation of boundaries of maximum extraction. These are easy to identify with existing maps of extractive activities, with cartographic delimitation of lithostratigraphic units of mining interest, or with photographic interpretation.

- Environmental inventory diagnosis. In this phase, detailed data on each exploitation is obtained. The most common technique is to use data support cards, especially if computer support is available. In addition to general administrative data, these cards will contain the characteristics of the operation and its surroundings, identification of impacts, evaluation of these impacts, corrective measures and recommended uses. The diagnosis phase is extended to the social, economic and environmental features of the area involved in the extractive activity, and a sufficiently extensive environmental inventory is prepared to characterize all important parameters on different scales (geology, geomorphology, soil and vegetation, land uses, unique areas of natural interest, urbanistic qualification, etc.).

- Definition of Typologies. In a study of regional scope, it is essential to define typologies to help analyze investment and priorities.
  These typologies can be defined by the type of operation, identification of basic impacts or possibilities of restoration. Experience shows that in the close relationship between the three possibilities, size is the only distinguishing factor within similar types of operations involving similar impacts and reclamation processes.

- Definition of basic reclamation criteria. An analysis according to typologies of reclamation possibilities, parameters and basic techniques allows for a greater approximation of the economic cost evaluation and the definition of priorities within a program.
  A feasibility analysis of the reclamation, the priority plan and an approximation of the actual process costs will make it possible to define a long or medium-term environmental reclamation program.

Preventive Studies

In the case of Spain, these studies, sponsored by local or regional governments, have the legal support of the mining legislation proper, of the land law that acknowledges the preparation of special plans whose purpose is to "protect the landscape, means of communication, land, and urban, rural and natural environment", and of several autonomous regulations referring to matters of protection and conservation of the environment.

- Objectives: Planning of extractive activity and drafting of basic reclamation criteria aimed at maintaining existing ecosystems or proper land-use once the extractive activity has been completed.

- Study of mining feasibility and exploitation quality. Delimitation of zones with different levels of mining interest. It is logical to assume that the planning study will be applied to an area with an important background of mining research, deemed positive in that a quality and quantity exist that justify its exploitation, and also that it is necessary to delimit the area more precisely according to levels of interest. These levels of interest should refer not only to intrinsic resource quality criteria, but also to economic extraction cost criteria.

- Diagnosis. Environmental inventory. Study of restrictive elements of the extractive activity. Conflict maps:
  The diagnostic phase prior to the evaluation of the anticipated environmental impacts can be developed by following now classic criteria such as those set forth in Guideline 85/337/CEE on evaluation of environmental impact: *"description of environmental elements liable to be notably affected by the proposed project"* (population, flora, fauna, land, air, climate, etc.).
  Knowledge of the area's elements and the uses most consistent with the natural form of the land will make it possible to know the factors limiting extractive activity and to make conflict maps. These maps should be understood as the cartographic representation of zones in the territory whose natural form may be contradictory to other proposed uses, or that the uses are simply incompatible a priori and come into conflict with each other.

- Guideline maps for land use: This phase corresponds to a more exact definition of the zones that should not be used for extractive activity, zones recommended for extractive activity, and alternative or standby zones for subsequent extraction phases. Finally, the criteria and basis for the reclamation to be carried out in the region shall be indicated, following some guiding standards that prevent imbalances from one zone to another.

It is logical to assume that this map preparation may respond to purely technical criteria of the team making them and may have a certain amount of subjectivity or a combination of technical and political criteria, the latter being those that take priority in the end. The foregoing in no way distorts the performance of these application studies in territorial planning; on the contrary, it permits more extensive knowledge of the zone and the proposed activity, it facilitates discussions between the interested parties regarding the scientific basis, or at least the most objective bases, and in the end it aids in the integration of environmental variables in the policy decision-making process.

References

[AYALA, F.J.] (1989): Legislación Ambietal aplicable a la Minería. Nacional, Autonómica y Comunitaria.- 307 pp., Madrid (ITGE).

AYALA, F.J. et al. (1986a): Bases para la Ordenación Minera y Ambietal de las extracciones de Picón en Las Canarias.- 81 pp., Madrid (ITGE).

AYALA, F.J. et al. (1986b): Estudio geoambietal para la Restauración del Espacio Natural afectado por las explotaciones de carbón en las cuencas palentinas.- 65 pp., Madrid (ITGE).

AYALA, F.J. et al. (1987): Criterios Geoambietales para la restauración de canteras graveras y explotaciones a cielo abierto en la Comunidad de Madrid.- 86 pp., Madrid (ITGE).

AYALA, F.J. et al. (1988): Programa Nacional de Estudios Geoambietales aplicados a la minería. Provincia de León.- 234 pp., Madrid (ITGE).

AYALA, F.J. et al. (1989a): Programa Nacional de Estudios Geoambietales aplicados a la minería. Comunidad Autónoma de Valencia.- 204 pp., Madrid (ITGE).

AYALA, F.J. et al. (1989b): Programa Nacional de Estudios Geoambietales aplicados a la minería. Comunidad Autónoma de Navarra.- 170 pp., Madrid (ITGE).

AYALA, F.J. et al. (1989c): Manual de Restauración de Terrenos y Evaluación de Impactos Ambientales en Minería.- 321 pp., Madrid (ITGE).

BASCONES, M. & GALLEGO, E. (1986): Análisis y criterios de restauración para áreas degradadas por antiguas explotaciones en la Comunidad de Madrid.- In: MOPU-CEDEX: Curso sobre Geología Ambiental.- 16 pp., Madrid.

[CAIRNEY, T.] (1987): Reclaiming contaminated land.- 260 pp., London (Blackie & Son Ltd.).

DIAZ DE TERAN, J.R. & GONZALEZ LASTRA, J.R. (1980): Un método de evaluación y de jerarquización de los afloramientos de rocas industriales.- I Reunión Grupo Nacional de Geología Ambietal y Ordenación del Territorio, Vol. Comunicaciones: 1-20, Santander.

FERNANDEZ, R., FERNANDEZ, S. & ARLEGUI, J.E. (1986): Abandono de minas. Impacto hidrogeológico.- 267 pp., Madrid (IGME + ETSIM).

GARCIA, J.J., MURAIS, J. & OSBORNE, J. (1985): Guía para la restauración del Medio Natural afectado por explotaciones de canteras.- 117 pp., Madrid (IGME).

[GENERALITAT DE CATALUNYA] (1987): Recomenaciones tecniques per la restauración i condicionamento de los espais afectatas per activitatos extractives. - 421 pp., Barcelona (Dir. Pol. Terr.).

GRIGG, C.F.J. (1988): Landscaping Techniques and Restoration.- Rev. Ming. Mag. **1988**: 492-497, London.

GURS, P. (1988): Vert-Le-Grand: Les dechets entrent dans le carrière.- Rev. Min. Carrières **70**: 36-40, Paris.

[HARRIS, J.R.] (1983): Approval of State and Indian Reclamation Program Grants Under Title IV of the Surface Mining Control and Reclamation Act of 1977.- 268 pp., Washington D.C. (US Dept. Int.).

HYDRO, M. (1987): Essai d'aquaculture dans une carrière en eau: cas de l'experimentation du Teoula en région Midi-Pyrénneés.- Industr. min., Mines Carrières **69**: 261-264, Paris.

[INSTITUTO GEOLOGICO Y MINERO DE ESPAÑA] (1985): Guía para la restauración del medio natural afectado por explotación de camteras. - 117 pp., Madrid (IGME).

--- (1986 a): Estudio geoambiental para la restauración del espacio natural afectado por las explotaciones de carbon en las cuencas palentinas. - 65 pp., Madrid (IGME).

--- (1987): Criterios geoambientales para la restauración de canteras, graveras y explotaciones a cielo abierto en la Comunidad de Madrid. - 86 pp., Madrid (IGME).

--- (1988 a): Minería y medio ambiente. - 19 pp., Madrid (IGME).

--- (1988 b): Investigación de factores ambientales en labores mineras. Diseño de experiencias piloto de revegetación en canteras y sus escombreras. - 255 pp., Madrid (IGME).

[INSTITUTO TECNICO Y GEOLOGICO DE ESPAÑA] (1989 a): Programa nacional de estudios geoambientales aplicados a la minería. Comunidad Autónoma de Valencia. - 204 pp., Madrid (ITGE).

--- (1989 b): Programa nacional de estudios geoambientales aplicados a la minería. Comunidad Autónoma de Navarra. - 170 pp., Madrid (ITGE).

--- (1989 c): Legislación ambiental aplicable a la minería nacional, autonómica y comunitaria. - 307 pp., Madrid (ITGE).

--- (1989 d): Programa nacional de estudios geoambientales aplicados a la minería. Provincia de León. - 234 pp., Madrid (ITGE).

LYLE, E.S. (1987): Surface mine reclamation manual.- 268 pp., Amsterdam (Elsevier).

LYLE, V.A.S., YAZICIGIL, H. & CARLSON, C.L. (1983): Surface Mining. Environmental Monitoring and Reclamation Handbook.- 750 pp., Amsterdam (Elsevier).

LÜTTIG, G. (1986): Ressources et reserves en sables et gravières en Europe.- Bull. Assoc. intern. Géol. de l'Ing. **33**: 27-30, Paris.

[LOPEZ JIMENO, G.] (1985): II Curso sobre las alteraciones en el Medio Embiente y la restauración de terrenos en minería a cielo abierto.- 339 pp., Madrid (Fundación Gómez Pardo).

ORDOÑEZ, S. et al. (1986): Curso sobre el Impacto Ambietal del aprovechamiento de los recursos minerales.- 214 pp., Madrid (Asociación de Geólogos Españoles).

ORDOÑEZ, S. & CALVO, J.P. (1981): Impactos derivados de las explotaciones de recursos, minería de minas metálicas y rocas industriales.- In: MOPU-CEOTMA: Geología y Medio Ambiente: 341-354, Madrid.

RAMOS, F. et al. (1983): Tratamiento funcional y paisajístico de taludes artificiales.- 269 pp., Madrid (Escuela Técnica Superior de Ingenieros de Montes).

READY MIX CONCRETE (1987): A practical guide to restoration.- 83 pp., Middlesex, U.K. (R.M.C. Group).

[RELEA I GENES, F.] (1987): Recomanacions tècniques per a la restauració i condicionament dels espais afectats per activitats extractives.- 421 pp., Barcelona (Generalitat de Catalunya. Direcció General de política Territorial).

SIONNEAU, J.M. (1987): Les potentialités ecologiques des carrières.- Industr. min., Mines Carrières **69**: 224-232, Paris.

VADILLO, L. (1987): Restauración de canteras y Planes de Restaración.- In: MOPU-CEDEX: Curso sobre Geología Ambiental, 62 pp., Madrid.

VADILLO, L. (1989): Evaluación y corrección de impactos en minería.- In: [ITGE]: Geología Ambiental: 187-197, Madrid.

VADILLO, L., AYALA, F.J., GAZAPO, C. & ALONSO, F. (1988): Minería y Medio Ambiente.- 19 pp., Madrid (IGME).

Land-use Planning and Management in Brazil:
A Brief Review of Present Experience
and Environmental Problems

HELIO M. PENHA [*]

Introduction

This paper tries to give a broad image of land-use planning in Brazil, taking into acount technical and political aspects and also indicating the main environmental problems at the moment.

The large size of the country and the irregular distribution of its population originate great contrasts in the environmental and socio-economic conditions, producing a very complex situation for planning and usage. There is also an important cultural factor, due to the fact that large sectors of the population have a cultural standard characteristic of Third World countries which have been colonies, directly or indirectly, until recently. Land-use practices have often been primitive, with no organized system of land and resource planning and management. Thus, grave environmental problems are being created, principally in metropolitan regions, where 70% of the population live; at the present time this being nearly 140 million inhabitants. Accelerating erosion in agricultural areas, many of them with large gullies, desertification, floods, slides, surface and groundwater contamination, landscape deterioration are some of the many problems Brazil has to face today due to bad land-use planning.

[*] **Author's address:** Prof.Dr. H.M. PENHA, Depto. de Geologia, Universidade Federal do Rio de Janeiro, Brazil.

Present situation

To approach land-use planning in Brazil, the diversity of the country's physical environment, as well as the ethnical and cultural characteristics and the differences in the degree of development in the various regions, should be taken into consideration.

In the 8.5 million km of continental area of Brazil there are important geological, morphological and climatic variations. Some 4 million km are occupied by a humid, ecuatorial forest, the Amazonia. This relatively untouched forest reserve, so important for the country and for the whole World, has a density of population of 3 inhabitants/km , but it already shows serious problems of environmental degradation, mostly due the burning of large areas and to river pollution by Hg from the "garimpos" (gold-digging operations).

The SE region, principally in the area of Rio de Janeiro - Sao Paulo, the two largest cities in Brazil, has the highest human agglomeration in the country. Nearly 40% of the country's population live here, thus exerting a strong pressure on land and water resources and creating many problems which are difficult to solve.

Over the years, the development of the large Brazilian cities has been carried out with an almost complete lack of concern for the natural environment, with short, medium and long term consequences for the living conditions of the population.

In the last decades, the policy of development followed by the different governments has caused a serious degradation of natural environment. Enormous human agglomerations have been created in the metropolitan areas, and this growing urbanization, due to socio-economic reasons, has resulted in a disorderly use of the land in problematic areas, such as unstable slopes, flood prone areas, terrains subject to compaction and subsidence, and even in areas considered as protected by the existing legislation.

The present natural environment and landscape of Brazil show important changes and are entering a stage of acute degradation. The devastation due to uncontrolled occupation and use of the land has also originated hazards of a level of economic significance similar to the one arising from geological and climatic processes such as earthquakes, volcanic eruptions, cyclones of tidal waves, which do not exist in Brazil.

In the southern part of the country, bad agricultural practices are provoking intense erosion in lands that were once fertile and productive. Large areas of agricultural land are becoming deserts, for example in Alagrete, in the state of Rio Grande do Sul. Also in the southern state of Paraná, bad land utilization during 1950-1970 has almost completely destroyed the forest cover through indiscriminate deforestation and "queimadas" (man-provoked fires), this being similar to the degradation seen in the Amazonia, particulary in the Rondônia, Mato Grosso and Pará states.

In the state of São Paolo (SE) more than 60 municipalities are affected by the accelerated erosion of the rural and urban areas, with annual losses of nearly 200 million tons of fertile soil (CHIOSSI, 1982). The main rivers in the states of São Paolo, Minas Gerais and Rio de Janeiro are heavily polluted by urban waste and physical and chemical products from industries and mines.

A large number of roads have been projected and constructed without proper land-use planning nor due consideration to geological and geotechnical factors, producing irreversible damages to the landscape, even altering important touristic sites, or triggering fast erosion processes and landslides in adjacent slopes. A good example of this is the road from Rio to Santos, constructed 15 years ago in one of the most beautyful parts of the Brazilian coast.

This alarming state of affairs is indicative of the almost complete lack (with rare exceptions) of adequate land analysis and evaluation studies oriented towards land-use planning and management, at the municipal, state and federal levels of the administration.

Legislation

In Brazil, there are numerous laws and decrees refering to land-use planning and environmental management, even including a chapter in the new constitution of 5, October, 1988 which refers to the environment. This constitutional innovation probably makes Brazil the only country in the World with a constitution which contains a whole chapter devoted to environmental matters. The present constitution considers that any harm against the environment is a crime and it states that the duty of the law is "*to preserve and restore the essential ecological processes*". It also establishes the need for environmental impact studies prior to any activity which may produce any significant degradation of the environment. With this new constitution, the Brazilian Amazon Forest, the "Mata Atlantica" (Atlantic Forest), the "Pantanal de Mato Grosso", the "Serra do Mar", and the coastal areas are part of the national patrimony and their usage has to be carried out in such way that the preservation of environmental quality and the sustained use of the natural resources are assured.

Curiously, this same constitution favors the wild gold-digging operations carried out by the "garimpeiros", a predatory and highly pollutive activity, even giving them priority rights for the investigation and extraction of minerals.

On the other hand, the National Environmental Policy Act, established in 1981, has other criteria for the organization of the use of soil, sub-soil, water, air and biota, considering them from the point of view of protection of ecosystems or of the preservation of representative areas. The highest organization in the National System for the Environment is the National Council (CONAMA), whose function is to advise the

Presidency on all matters of the CONAMA include ministers, secretaries of the environment from the different states, and representatives of different social groups, such as scientific interest societies, associations of environmentalists and societies for the protection of animals.

Certainly, if the existing laws were respected, most of the problems described, due to bad land-use practices, would be lessened or even eliminated, but there is a wide lack of knowledge about them among the population and often even a lack of interst from the part of the authorities in their actual enforcement when it comes to the drafting of development plans.

The most apparent response to this state of affairs from society is expressed in the form of preservationist ideas coming from the activities of conservation groups. Often these groups, although well-meaning, lack the necessary information and scientific basis and, consequently, make extremist proposals, difficult to put into practice. As in many other countries, but perhaps more so in Brazil, due to the cultural characteristics referred to above, there are political groups that make a demagogical use of environmental issues, through pre-election promises which are not fulfilled afterwards. This sort of attitude leads, for instance, to the occupation of hazard zones by lower income sectors of the population, after promisses of development and urbanization of the areas for new housing schemes. The result of this un-planned occupation of marginal areas is a growing number of deaths and damages with every new geomorphological event.

Example: The Metropolitan Area of Rio de Janeiro

The situation in the country can be illustrated through the analysis of land-use planning and management in the metropolitan area of Rio de Janeiro.

The large Metropolitan Area of Rio de Janeiro (RMRJ) constitutes a typical example of the processes of uncontrolled urban expansion and the problems that arise because of it. This area covers 646,400 hectares, with a population at the moment of nearly 7 million inhabitants, and with an expected population of 17 million for the year 2000, distributed between the 14 municipalities of the region. Nearly 41% of the area of the RMRJ is subject to some form of legal protection and preservation and 21% is still covered by forests.

The climate is tropical, warm and fairly damp, with a rainfall of over 1500 mm/year. The relief is very varied, with pronounced slopes and high altitudinal variations. The geology is predominantly made up of Pre-Cambrian granites and gneisses.

The highest concentration of population is located in the coastal plain, mainly in the city of Rio de Janeiro itself, squeezed between the sea and the mountains, the Maciço de Tijuca (Tijuca Massif). The growth of the city took place through the succesive occupation of lagoons, swamps and near-shore areas by large-scale filling and reclamation. This has destroyed most of the coastal ecosystems which regulate floods, both due to surface run-off and to the invasion of marine waters. The large mangle swamps which act as natural filters for the sediments transported by run-off, and thus prevent the silting up of the rivers and lagoons, have disappeared to a very great extent. Also, the forest cover on the slopes surrounding the city, which is the natural protection against erosion and surficial landslides, has been greatly reduced as a consequence of the urban-demographical explosion of the last decades.

The modification of the natural environment through human intervention, principally because of uncontrolled land occupation and use in the small and complex available space, has provoked numerous disasters with large losses of human life and property. The frequency of landslides and flooding in the metropolitan area has increased clearly and places which until now have been considered as safe have experienced these processes in recent times. Only three days of intensive rain, in March of 1988, provoked a tragedy with 278 dead, 735 injured and 17,330 homeless.

Every year the problems increase in number and intensity, and more money is needed to cover the cost of stabilizing slopes, dredging of rivers and channels, etc.; that is, mainly corrective measures. The lack of planning is reflected in the general tendency to repair the consequences (rather, part of them) of the processes, directing the efforts to the effects without attacking the causes. There is an urgent necessity to analyse the dynamics of the different processes active in the area, to assess the existing hazards and to evaluate potential risks in order to take preventive steps for the reduction of damages, to draft plans for the restoration of the environment and for the prevention of further degradation, and to re-organize the settlements as well as to plan new developments with due consideration to the constitution and activity of the zone. Of foremost importance is the immediate enforcement of the existing environmental laws (PENHA, 1988).

An example of the problems derived from uncontrolled urban expansion, with the occupation of the land in a disorderly and caotic way are the "favelas" (slums), numbering 180 in the Rio de Janeiro area and housing nearly 40% of the population of the city. These "favelas" are responsible for the degradation of many slopes, the destruction of the vegetation cover in forest areas protected by law and the activation of geological hazards. The problem is thus not only scientific or technical, but to a great extent social.

In this sort of situation, one of the first needs is to gather the basic scientific information on the environment (and to use the already existing knowledge), in order to take appropriate planning decisions. Of foremost importance is the preparation of different types of geoscientific maps, mainly the ones on geomorphological and geotechnical features, which are still not available for the Metropolitan Area of Rio de Janeiro. This is an essential input for the enforcement of land-use planning laws.

Existing actions

The technical-scientific instrument initially proposed for the solution of some of the problems described is the geological-geotechnical map, as a basis for the understanding of the phenomena resulting from the interaction between the geological materials and structures and the activities which the land surface has been subject to. Earlier experiences, derived from the analysis of historical documents which are already stored in a data bank (BARROSO et al., 1986), are also being taken into account. The preparation of geological-geotechnical maps for the Metropolitan Area of Rio de Janeiro is one of the lines of research being carried out in the Department of Geology of the Federal University of Rio de Janeiro (UFRJ).

According to BARROSO et al. (1986) the first attempt to elaborate a geological-geotechnical map of Rio was made in 1966 in the School of Geology of UFRJ, because of the terrible destruction caused by the torrential rains in January of that year. A preliminary map was made, scale 1:5.000, which included a few districts of the city. The work was not finished and was only started again in the middle seventies, when the Dept. of Geology of UFRJ decided to include the geological-geotechnical map of the RMRJ is being made at a scale of 1:50.000, and is based on the systematic geological mapping which has been carried out in the state of Rio de Janeiro in the present decade.

In the elaboration of these semi-detailed maps geological formations with the same origin are identified and mapped. They are grouped forming categories with similar litological characteristics and associated pedogenetic units. These homogeneous units, with similar geological-geotechnical features, are the basic units for mapping.

The collection of samples of the different mapping units is carried out with the aim of performing the standard tests of soil and rock mechanics, in order to obtain a better evaluation of the units and of their likely behaviour under different conditions.

At the present time, investigations have already started in order to determine the areas of present and potential hazards, on the basis of the geological-geotechnical map and on the known impacts derived from actions which have been undertaken in the area. This work is being carried out particulary in the "favelas" (slum areas) of the Tijuca Mountains, and in the flood prone areas of the coastal plain.

With reference to land-use planning in the urban areas and the suburbs which are being mapped at the moment, many examples of inadequate use of the physical environment have been found. Badly planned developments are thus provoking catastrophes, mainly, as should be expected, during heavy rain periods.

Among the main problems which are found in the area, flooding and associated silting up as well as slope instability are the most prominent (BARROSO et al., 1986).

Flooding and silting up happen more and more frequently in the coastal plain between the "Serra do Mar" and the Bay of Guanabara. This region, called "Baixada Fluminense" (Fluminian Plain), has experienced an enourmous demographic explosion, never before known in Brazil, with an irrational and caotic land occupation, mostly on the flood plains of the main rivers and on the shores of the lagoons. These same rivers are being intensely silted up because of deforestation of the nearby mountain slopes, and they also receive sewage effluents and all sorts of wastes from the surrounding towns. As a consequence of this situation, the Guanabara Bay is being rapidly filled up with sediments and many of its intertidal and marginal zones are changing into extensive swamps.

The rock massifs of the RMRJ are made up mainly of granitic and gneissic formations of Pre-Cambrian age, covered by a thick layer of regolith, produced as a ressult of intense chemical weathering under warm and humid conditions. The bedrock crops out practically only in scarps, steep slopes and in the river beds. The large, flat lowland areas are made up of residual soils and coluvial materials deposited at the foot of the slopes.

BARROSO et al. (1986) consider that rock instability problems in the area are always associated with discontinuities in the rock massif, represented by different states of primary alteration and weathering; joints, schistosity, spheroidal weathering structures, displacements of termal origin and lithological heterogeneities are the commonly determinant factors, either isolated or associated. In the slopes of the Rio de Janeiro area, where podzolic soils are predominant, the problems of instability show a close correlation with the mineral composition, texture, structure and thickness of the C horizon of the soil. Pedologically mature soils are more stable and when they are present the areas affected by natural instability or by erosion and gullying are rare and of limited extent.

Surficial and deep slides, affecting soil and/or bedrock are frequent in the migmatitic massifs, with deep blankets of weathered materials and crossed by granitic dykes which form large boulders on the deforested slopes. These slopes are subject to sheet run-off during torrential rainfalls.

The experience obtained in Rio de Janeiro about areas subject to rapid mass movements, in which soils and rocks are displaced, shows that the lithological heterogeneities of the subsoil are reflected in the weathering layer. Therefore, a good knowledge of the bedrock geology as well as of the physical and chemical properties of the soils derived from it is of great importance for the delimination of hazard zones.

The problems briefly described, landslides, erosion, silting up and floods are closely linked to the environmental degradation and disequilibria produced by the progressive deforestation of areas formerly covered by the "Mata Atlantica" (Atlantic Forest) especially in the mountain zones with strong slopes, like the Tijuca Massif, near the cities of Rio de Janeiro and Petrópolis.

Conclusions

The situation briefly described here shows the consequences which arise from rapid urban growth in areas with complex environmental features, in which the existing delicate balance can easily be disturbed as a result of ill-planned development. Many parts of Brazil, particulary the metropolitan regions in the southeastern part of the country, suffer the types of environmental problems commented above. In areas with low density of population, such as parts of the Amazonia, the degradation is also intense and the destruction of the forests is bringing about important modifications in the climatic, geomorphological and geochemical processes.

Although the existing legislation is, in theory, adequate, the actual application and enforcement of the laws and regulations leaves much to be desired. Profound cultural changes, both at the level of the authorities and of the general population, are needed to bring about a more careful approach in the use of the environment.

Any action aimed at correcting this kind of situation must be preceeded by careful planning, on the basis of a detailed knowledge of the constitution and structure of the natural systems and of the different active processes operating in it. In this respect, the consideration of the ethnical and cultural characteristics of each region, particularly the ones relating to traditional land-use practices by the aboriginal population, is important. Prior knowledge of the geographical area to be used and of its capability to support different types of human activities is an essential pre-requisite for planning. One of the first steps to be taken in this direction is the preparation of geological, geomorphological and geotechnical maps conceived as tools for planning.

Because of the multiplicity and complexity of the problems to be considered for land-use planning, and taking into account the degree and the rate of degradation presently existing in Brazil and the lack of basic studies about most of the country, it is of foremost importance to undertake a program of mapping, inventory and evaluation of the basic environmental features, at least in the areas where occupation and development are more intense. The undertaking of studies about the most expressive and determinant environmental components (geology, geomorphology, surficial dynamics, soils, vegetation, resources, hazards, etc.) and about their modification by humans, should lead to the elaboration of different thematic maps, and of maps showing the advisable land-uses and the possible restrictions on the use of the territory.

References

BARROSO, J.A:, CABRAL, S., LINO, G.R.S. & PEDROTO; A.E.S. (1986): O mapeamento geológico-geotécnico da Grande Região Metropolitana do Rio de Janeiro; uma síntese explicativa. Mesa Redonda sobre Aspectos Geotécnicos de Encostas. - 25 pp, Univ. Federal do Rio de Janeiro.

CHIOSSI, N.J. (1982): Ocupaçao do sole e impacto ambiental. - Rev. bras. tecnol., 13 (5): 44-49. Brasilia.

PENHA, H.M. (1989): Degradaçao das áreas florestadas e suas consequências; Região Metropolitana do Rio de Janeiro. - Ciclo de Mesas Redondas sobre: Grande Rio; impropiedades no uso do solo. 5-20, Univ. Federal do Rio de Janeiro/FINEP.

Problems and Approaches in Environmental Geology in Latin America: The Colombian Experience

by MICHEL HERMELIN *

Abstract

Colombia as a region with a tremendous amount of natural hazards and social problems is to be considered as a key region for environmental geology methods. The author aims his contribution to an increase of the use of geological programs in this field.

Introduction

Latin America may appear at first glance as an homogeneous region, where problems related to the geological environment are similar. A closer observtion enables the distinction of several well characterized areas which could be defined as a function of the occurence of specific natural hazards. For instance active volcanism is limited to Andean countries, Central America and Lesser Antillas. Seismicity includes the same areas and the entire Caribbean Basin. Hazards related to exogenous processes, as high floodings and mass movements cover practically the whole continent. Main volcanic events are listed in Table 1; important earthquakes which have occured during the present century appear in Table 2.

* Author's address: Prof. Dr. M. HERMELIN, U. EAFIT, U. Nacional, Medellin, Colombia.

Another important aspect to be taken into account is population. Table 3 lists the population by country in 1960 and in 1983. Table 4 gives the tendency of the population growth. Only the Caribbean countries and three countries from the Southern Cone (Argentinia, Chile and Uruguay) show a relatively low increase, in opposition to the trend followed by the others. The impacts caused by such a population increase are numerous and can be observed everywhere. Actually, many of the problems related to environmental geology are produced because people are increasingly living in unsafe areas or because areas which were initially secure are now facing hazards induced by human activity.

The present paper will focus on the author's own country, which offers a great variety of geographical and geological situations: crossed by the equator, with an area slightly above one million of km , it includes domains as different as cordilleras topped by ice-covered summits, savannas, Amazonian forest and coasts both on the Caribbean Sea and on the Pacific Ocean.

Table 3 lists the population by country in 1960 and in 1983. Table 4 gives the tendency of the population growth. The Caribbean countries and three countries from the Southern Cone (Argentinia, Chile and Uruguay) show a relatively low increase, in opposition to the trend followed by the others.

Natural Hazards in an Andean Country

Young, high moutain chains located near the equator mean for a large variety of climates: from dry to very humid lowlands to snow-capped peaks. Average temperatures decrease with altitude, and precipitation is often controlled by mountain barriers.

Main external processes are:

-- floods, which affect both mountainous areas and plains. They may be produced by high precipitation but also by the rupture of natural dams formed by slope movements or by lahars. Substitution of original forests by pasture or urban development greatly contribute to increase peak discharges.

-- Mass movements are very common in most slopes. They may be triggered by strong rains but also by river or artificial cutting and by earthquakes.

-- Surface erosion and sedimentation due to poor land management may be important even in areas with well distributed precipitation.

-- Subsidence, natural or produced by mining and destruction due to expansive clays may occur locally.

Endogenous hazards include earthquakes, tsunamis and volcanism, as expected in a country located at the intersection of three tectonic plates with two subduction zones.

Social Aspects

The population of the continent has triplicated in about four decades. Only few countries escaped from this explosion, which produced several consequences:

-- an enormous increase of urban population. In Colombia, for instance, rural population was 73 % at the end of the 50's and is now about 34 %. The country has now more than 30 cities with more than 100,000 inhabitants. The process is due to rural migration only partially. Growth rates are proportional to the size of the cities. The evolution of the population of Medellin area, second largest in Colombia, is given in Table 5. This increase in the size of cities often happens through the occupation, legal or not, of areas which are not suitable for urban uses: too steep, unstable, subjet to flooding, etc. The result is that a growing number of persons is exposed to geologic risks. Many are competely aware of this insane situation, but have developed a dangerous sense of resignation.

-- Local economic and political influences, render in many case very difficult the starting of a zonification program and its enforcement.

In rural areas, even if a large part of the increasing population migrated to the cities, deforestation of steep terrains has been an accelerated process for the last decades, producing soil destruction and river siltation. The long term climatic consequences of this practice are now subject to international discussion.

Taking into account the present population trends, the pressure on natural resources has apparently no immediate possibility to decrease.

Legal and Institutional Aspects

The adoption of rules regarding environmental management may take place due to the influence of organized groups, the foresight of individuals or the reaction of government to the occurence of catastrophes. A short recount of the measures taken during the last two decades in Colombia is probably a sample of what has been done on a continental level:

1974:   Adoption of the National Code on Natural Resources and Environment. The enforcement has been difficult for lack of sufficient financial resources. (Republica de Colombia, 1974).

1979:   INGEOMINAS, the geological survey, opens its Division of Environmental Geology.

1983:   As a consequence of Popayan Holy Thurday Earthquake, the government adopts the Anti-Seismic Building Code, which apparently has been successfully applied since then.

1985: Following the Nevado de Ruiz volcanic disaster, the government opens the National Office for Emergency Attention, now called the National Office for Disaster Attention. The National Volcanological observatory is also founded.

1989: Adoption of the Urban Reform Laws, direct consequence of the recent law related to popular election of mayors, a major change in the Colombian traditional administration. It states that cities and villages must prepare a Development Plan, which necessarily has to include a set of regulations on urban land uses and the identification of natural risks. This is a very important breakthrough toward the security of urban population against geological hazards.

1989: Creation of the National System for Prevention of and Attention to Disasters which includes several entities, both at national and regional levels, and comprises national institutions as Ministries and technical offices (Hydrology, Meteorology, Communications, Geological Survey, National Resources, Red Cross, etc.).

The National Office for Disaster Attention, coordinator of the national system, has many projects, including a national seismic net, seismic microzonations for cities, flood hazard maps and warning systems, tsunami hazard prevention system, hazard mapping projects for several cities, hydroelectric dam control system, etc.

A Province Approach

Only the province in Colombia has so far decided to carry on an integral action to enforce the Urban Reform: the Department of Risaralda, located in central Colombia, through its regional corporation, CARDER. The study is being carried out for the capital city (Pereira, 350,000 inhabitants) and 12 towns (population between 5,000 and 20,000). Using previous experiments destinated to test the method, the following approach was used:

-- Photointerpretation at scale between 1:10,000 and 1:20,000, using airphotos taken at different times when available.

-- Field work, including identification of lithological units, slopes, active and recent processes; conversation with inhabitants and revision of historical documents. The information was plotted on enlarged airphotes with scale 1:5,000.

-- analysis of the information, to produce a map of aptitude for urban use at scale 1:5,000.

The final product is a map with units easy to use for people with no previous training in geology. Typical units could be as follows:

A. Urban areas

A-1. No observed geological problems.

A-2. Geological problems which must be corrected.

A-3. Areas which must be relocated.

B. Non-urban areas (surroundings)

B-1. Areas with no apparent restrictions for urbanization.

B-2. Areas which may eventually urbanized after detailed technical study and treatment.

B-3. Areas which may retain their present use (agriculture, coffee growing, pastures, etc.)

B-4. Areas which must receive conservation or recuperation treatments but should not be urbanized in the future.

Furthermore, reports offer detailed comments on each of the units, with specific reference to local conditions; a general geomorphic analysis of the watershed of the rivers which directly affect the urban area; recommendations about the location of a sanitary landfill; suggestions about places where building materials can be obtained and about conservation practices to be taken in the watershed which provides drinkable water.

This method permits the obtention of a level of information on the geological risks which does not pretend to be complete: it would correspond to the first stage, following BRINK's classification (BRINK et al. 1982).

However, it permits a very useful and rapid approach to land evaluation for planning and management. The average time necessary to study each town is about 4 months (2 months for a 2 geologist-team under the supervision of an expert). The study benefited by the fact that previous knowledge of the Quaternary stratigraphy of the area was partially available through tephrochronology. Final reports are being concluded; a special document of 15-20 pages will be written for each town, with illustrations, to be distributed in high schools and amongst selected people. It is aimed to explain the dangers which arise from a careless management of the town terrains, and to convince the community to collaborate with authorities by avoiding bad practices and giving notice of mismanagement or of apparition of new processes.

The regional corporation will on the other hand oblige mayors to enforce recommendations arising from zonification and will carry on, in collaboration with munincipal authorities, the necessary relocalizations, structural interventions and conservation projects.

Several drawbacks exist as associated with this kind of projects:

-- Results cannot be considered definitive: more detailed studies are needed from the standpoint of soil mechanics, hydrology and seismic hazards, to mention only the more relevant.
-- Results can only be applied for the welfare of the inhabitants of a strong willingness exists among regional and local authorities to enforce the recommendations.
-- Annual re-evaluation of the corrections and of the general situation should be carried out.

Suggestions for the Future

To be successful, an environmental geology program has to go farther than the identification of natural hazards and human risks. It must also include a permanent monitoring of processes; the implementation and enforcement of laws and rules on land use, both in urban and rural areas, and on building standards (seismic code); the installament of alert nets; the construction of structures and relocation of building the education and training of people living in hazardous areas.

Many of these tasks go much beyond what is normally considered a geologist's activity. However, geologists are an essential ingredient of this type of program, even taking into account that in many cases they have not been trained in this field.

On the other hand, it is impossible to recommend a specific organization: nobody would discuss the necessity of a national coordination body, but many of the activities could be executed more efficiently on a regional or even local level.

From the educational standpoint, several aspects may be proposed:

a)   The preparation of specialists in hazard identification should be improved offering university graduate programs in environmental geology, which could include subjects as Quaternary geology, geomorphology, soil science, remote sensing, volcanology, seismology, hydrology/climatology, coastal studies, geotechnics, communications, etc. One or several schools specialized in these topics with a high level staff carrying on regional basic research on these topics should be established in Latin America.

b)   Basic courses in environmental geology should be compulsory in all the geology programs; in a perhaps less technical version, they should be offered to other professions as civil engineering, architecture, urbanism, public administration, etc.

c)   Simplyfied versions might be prepared for high school students.

d)   Printed and adiovisual documents should be prepared for the general public.

Adoption of new techniques must include remote sensing and computer operated geographic information systems and data bases. Automatic measurement, storing and transmission of data should also be considered as important developments.

Geological research should focus the determination of hazards: team studies including archaelogists and historians could be very useful in some regions. A strong emphasis should be given to the use and development of methods for Quaternary and particularly Holocene stratigraphy.

Communications remains an important part of the success; geologists able to convince other people of the necessity to apply environmental geology will probably be very important to communities in the next years.

References

ACEVEDO, P. (1986): Mapa generalizado de riesgos volcanicos potenciales de Colombia.- Rev. Centro interam. Fotointerpret. **11** (1): 53-54, Bogota.

ALVARADO, G. (1986): Relacion entre la neotectonica y el vulcanismo en Costa Rica.- Rev. Centro interam. Fotointerpret. **11**: 246-264, Bogota.

BASTIDAS, H. & PUIGDOMENECH, H. (1986): Evaluacion sismotectonica y estudio del fallamiento cuaternario del noroeste argentino.- Rev. Centro interam. Fotointerpret. **11**: 341, Bogota.

BERMUDEZ, A. (1985): Estudio de la sismicidad del noroeste de Suramerica y areas vecinas.- Mem. VI Congr. latinoam. Geol. **2**: 292-293, Bogota.

BOLT, B.A. (1988): Earthquakes.- 282 pp., New York (Freeman).

BRINK, A.B.A., PARTRIDGE, T.C. & WILLIAMS, A.A.B. (1982): Soil survey for engineering.- 378 pp., Oxford (Claredon Press).

BUSTAMANTE, M. & ECHEVERRY, L.M. (1984): Inventario de desastres recientes de origon geologico en el Valle de Aburra.- Mem. prim. Conf. Riesg. geol. Valle de Aburra, 21 pp., Medellin.

BUSTAMANTE, M. (1988): Los desastres en Medellin, naturales? - Mem. Conf. Riesg. geol. Valle de Aburra, 21 pp., Medellin.

BUSTAMANTE, M. & HERMELIN, M. (1988): Aplicacion de la geologia en el plan de ordenamiento de la zona sur del Valle de Aburra.- Mem. V Congr. colomb. Geol. 1: 496-515, Bucaramanga.

BUSTAMANTE, M. & VELASQUEZ, A. (1984): Actividad tectonica cuaternaria y recomendaciones para la evaluacion de la amenaza sismica en el Valle de Aburra.- Mem. prim. Conf. Riesg. geol. Valle de Aburra, 16 pp., Medellin.

CABALLERO, H. & MEJIA, I. (1988): Comentarios acerca del evento torrencial de la Quebrada Ayura (Envigado) del 14-04-88 y sus implicaciones en la evaluacion de amenaza al munincipio.- Mem. seg. Conf. Riesg. geol. Calle de Aburra, 23 pp., Medellin.

CARRACEDO, J.C. (1987): El Riesho Volcanico.- In: Riesgos Geologicos: 83-97, Madrid (IGME).

CARDONA, O. (1986): Vulnerability and seismic risk assessment.- Rev. Centro interam. Fotointerpret. 11: 158-177, Bogota.

CEPEDA, H. et al. (1984): La difusionde los conceptos geologicos a la poblacion, un metodo de mitigar los danos por catastrofes naturales.- Mem. Conf. Riesg. geol. Valle de Aburra, 19 pp., Medellin.

CEPEDA, H. et al. (1987): Mapa preliminar de amenaza volcanica en Colombia, escala 1:3.000.000.- Rev. Centro interam. Fotointerpret. 11: 179-188, Bogota.

CEPEDA, H. (1989): Mapa preliminar de amenaza potencial del Nevado del Tolima, Colombia.- Mem. V Congr. colomb. Geol. 1: 442-472, Bucaramanga.

DAVILA, S. (1986): Zonificacion por riesgos naturales, metodologia y aplicaciones.- Rev. Centro interam. Fotointerpret. 11: 472-473, Bogota.

DEEB, A. & ORDONEZ (1987): Zonificacion de riesgos hidraulicos en la zona de influencia del Nevado de Ruiz.- Rev. Centro interam. Fotointerpret. 11: 305-324, Bogota.

DE GRAFF, J.V. et al. (1989): Landslides: extent and economic significance in Caribbean.- Abstracts, 28th intern. geol. Congr. 1: 1.382-1.383, Washington.

DELGADO, A. (1987): Los desastres naturales en el peru y el banco de datos del

CISMID.- I Simp. per.Prevenc. y Mitigac. Desastr. nat.: 89-93, Lima.

ELIZALDE, L. (1985): Implicaciones geologicas de la sismicidad actual en el Ecuador.-
Mem. VI Congr. latinoam. Geol. 2: 314-335, Bogota.

ERICKSON, G.E. et al. (1989): Landslide hazards in Southern Andes.- Abstr. 28th intern.
geol. Congr. 1: 1.456, Washington.

ESPINOSA, A., GARCIA, L.E. & SARRIA, A. (1985): Riesgo sismico en Colombia.-
Mem. VI Congr. latinoam. Geol. 1: 206-243, Bogota.

FORERO, E. (1986): Plan de desarrollo de Ambalema.- Rev. Centro interam. Forointerpret.
11: 560-571, Bogota.

GALLEGO, I. (1988): Infraestructura de ingenieria para el control de la erosion:
tratamientos practicos implementados por CRAMSA.- Mem. seg. Conf. Riesg. geol.
Valle de Aburra, 13 pp., Medellin.

GARCIA, T. (1987): Microzonificacion de la ciudad del Cusco.- I Simp. per. Prev. Mitig.
Desastr. nat.: 275-285, Lima.

GIESECKE, A. (1987): Programa para la mitigacion de terremotos en la zona andina -
proyecto SISRI-CERESIS.- I Simp. per. Prev. Mitig. Desastr. nat.: 287-296, Lima.

GUARDADO, R. (1985): La geologia aplicada al planeamiento construccion y
remodelacion de la ciudad de Santiago de Cuba.- Mem. VI Congr. latinoam. Geol.
1: 491-492, Bogota.

HERMELIN, M. (1984): Riesgos geologicos en el Valle de Aburra.- Mem. prim. Conf.
Riesg. geol. Valle de Aburra, 22 pp., Medellin.

HERMELIN, M. (1988): Aspectos geologicos del plan de ordenamiento territorial de la
zona norte del Valle de Aburra.- Mem. seg. Conf. Riesg. geol. Valle de Aburra, 20
pp., Medellin.

HERMELIN, M., & VELASQUEZ, A. (1984): Evolucion reciente de la cuenca de la
Quebrada La Iguana.- Mem. prim. Conf. Riesg. geol. Valle de Aburra, 24 pp.,
Medellin.

ISAZA, J. (1988): Informacion geografica del medio natural en el area metropolitana del
Valle de Aburra.- Mem. seg. Conf. Riesg. geol. Valle de Aburra, 26 pp., Medellin.

JAMES, M. (1987): Evaluacion de las amenazas geologicas en la cuenca del Rio Otun.-
Rev. Centro interam. Forointerpret. 11: 327-329, Bogota.

JAMES, M. (1987): Riesgo sismico en el area del Viejo Caldas.- Rev. Centro interam.
Fotointerpret. 11: 266-268, Bogota.

JAMES, M. & LONDONO, E. (1987): Programa de vigilancia del Volcan Nevado de Santa Isabel: cuenca del Rio Otun.- Rev. Centro interam. Forointerpret. **11**: 396-397, Bogota.

JIMENEZ, G. (1984): Riesgo sismico del area metropolitana de Medellin.- Mem. prim. Conf. Riesg. geol. Valle de Aburra, 42 pp., Medellin.

JOSHUA, D. (1982): Development planning at the interphase on mountain and plain: a Venezuela case study.- Mountain Res. Developm. **2**: 1-30, Boulder.

KUROIWA, J. (1987): Planeacion fisica contra desastres naturales en el Peru.- I Simp. per. Prev. Mitig. Desastr.: 297-306, Lima.

LOPEZ, F. & MEJIA, E. (1988): Riesgo geologico en el Barrio Santa Rita (Bello).- Mem. seg. Conf. Riesg. geol. Valle de Aburra, 11 pp., Medellin.

MALPARTIDO, C. (1987): Proyectros de reconstruccion de la ciudad del Cusco a raiz del sismo del 5 de abril de 1985.- I Simp. per. Prev. Mitig. Desastr.: 283-286, Lima.

MARTICORENA DE, G.D. (1987): Estudio hidrogeodinamico de huaycos en la cuenca del Rio Rimac.- I Simp. per. Prev. Mitig. Desastr.: 233-236, Lima.

MEYER, H. et al. (1988): Haciendo el Observatorio Sismologico del Suroccidente.- Mem. seg. Conf. Riesg. geol. Valle de Aburra, 14 pp., Medellin.

MORENO, H. (1986): Problemas de Riesgo volcanico en los Andes del Sur de Chile.- Rev. Centro interam. Fotointerpret. **11**: 139-152, Bogota.

MORA, S. (1987): Analisis preliminar de la amenaza y vulnerabilidad potenciales generados por el Rio Reventado y el deslizamiento de San Blas, Costa Rica.- Tecnol. en Marcha **9** (1): 19-37, San Jose/Costa Rica.

MUNOZ, J. (1984): Bases y desarrollo del Codigo Colombiano de Construcciones Sismo-resistentes.- Mem. prim. Conf. Riesg. geol. Valle de Aburra, 36 pp., Medellin.

MURCIA, A: & VERGARA, H. (1987): Riesgos geologicos potenciales en la cuidad de Ibague, Tolima, Colombia.- Rev. Centro interam. Fotointerpret. **11**: 330-347, Bogota.

PANIAGUA, S. & SOTO, G. (1986): Reconocimiento de los riesgos volcanicos potenciales de la Cordillera Central de Costa Rica.- Rev. Centro interam. Fotointerpret. **11**: 178-199, Bogota.

PARRA, L.N., MEJIA, J.I. & CABRERA, K. (1988): Amenaza sismica en Medellin.- Mem. seg. Conf. Riesg. geol. Valle de Aburra, 14 pp., Medellin.

RAMIREZ, J.E. (1975): Historia de los terremotos en Colombia.- 150 pp., Bogota (Inst. Geogr. A. Cardazzi).

REGAIRAZ, A. (1986): La geomorfologia aplicada a problemas vinculados al vulcanismo y a estudios de neotectonica en zonas seleccionadas para el emplazamiento de un repositorio nuclear.- Rev. Centro interam. Fotointerpret. **11**: 300-318, Bogota.

[REPUBLICA DE COLOMBIA] (1974): Codigo Nacional de Recursos Naturales, Modificaciones al Codico Civil.- 86 pp., Bogota (INDERENA).

[REPUBLICA DE COLOMBIA] (1984): Codigo Colombiano de Construcciones Sismo-resistentes, Decreto 1400, Junio 7 de 1984.- 307 pp., Bogota (Asoc. Col. de Ingen. sis.).

[REPUBLICA DE COLOMBIA] (1989): Reforma Urbana, Ley 09 del 11 de Enero de 1989.- Bogota.

[REPUBLICA DE COLOMBIA] (1989): Sistema Nacional para la Prevencion y Atencion de Desastres de Colombia, Decreto 919, Mayo 1 de 1989.- Bogota.

RODRIGUEZ, C. (1986): Implicaciones hidrogeologicas de dos proyectos de gran impacto ambiental en Colombia.- Mem. seg. Simp. Hidrogeol., 210 pp., Bogota.

ROMERO, J.A. & NIETO, A.H. (1986): El Observatorio Vulcanologico Nacional.- Rev. Centro interam. Fotointerpret. **11**: 154, Bogota.

SARRIA, A. (1985): Incorporacion de los efectos de los terremotos en la normativa sismo-resistente.- Mem. VI Congr. latinoam. Geol. **1**: 244-268, Bogota.

TANAHASHI, I. (1987): Establecimiento del Banco de Datos para investigaciones en ingenieria antisismica y planeamiento de Desastres.- I Simp. per. Prev. Mitig. Desastr. nat.: 95-99, Lima.

TORRES, A. (1989): Peruvian Andenes as an alternative of land space planning in hillsides.- Abstr. 27th intern. geol. Congr. **3**: 245-246, Washington.

VARGAS, G. & MENDOZA, H. (1988): Riesgos geologicos en el munincipio de Cepita, Santander.- Mem. seg. Conf. Riesg. geol. Valle de Aburra, 39 pp., Medellin.

VIECO, B. & SALVA, P.J. (1988): Zonificacion geotecnica de las laderas de El Poblado y Envigado.- Mem. seg. Conf. Riesg. geol. Valle de Aburra, 15 pp., Medellin.

VELASQUEZ, A. (1987): Macrozonificacion de riesgos naturales en el munincipio de marsella, Risaralda.- Rev. Centro interam. Fotointerpret. **11**: 284-286, Bogota.

VELIZ, J. (1987): Problemas de la Geodinamica en el Valle del Rimac.- I Simp. per. Prev. Mitig. Desastr. nat.: 205-206, Lima.

Year	Volcano	Country	Victims
1741	Cotopaxi	Ecuador	1000
1845	Ruiz	Colombia	1000
1877	Cotopaxi	Ecuador	1000
1902	Soufriere	St Vicent	1565
1902	Mt Pelee	Martinique	29,000
1902	Sta Maria	Guatemala	6000
1982	Chichon	Mexico	5000
1985	Ruiz	Colombia	23,000

Table 1:   Recent Desastrous Volcanic Events in Latin America
(adapted from CARRACEDO, 1987)

Year	Date	Place	Magni.	Remarks
Mexico - Guatemala				
1806	March 25	Jalisco		many killed
1845	March 7	Acapulco	7	Tsunami
1845	March 15	Guatemala		
1875	February	Zapopan	7.5	
1902	April 19	Quezaltenango	8.3	many killed
1911	June 7	Jalisco	8	45 killed
1918	January 3	Guatemala		several killed
1943	February	Guerrero	7.5	Damage in Mexico City
1973	January 30	Michoacan-Colima	7.7	56 killed
1973	August 28	Puebla-Veracruz	7.1	500 killed
1976	February 4	Guatemala	7.6	27,000 killed
1985	September 19	Ixtapa	8.1	9,000 killed
Central America				
1820	October 19	Honduras, San Pedro Sula		
1841	September 2	Central Costa Rica		
1847	July 31	Nicaragua		
1859	July 31	Nicaragua		
1859	December 8	El Salvador		
1882	September 7	Panama		
1904	December 20	Costa Rica, Panama	8.3	
1951	May 6	El Salvador, Jucuapa	6.5	400 killed
1972	December 23	Nicaragua, Managua	6.5	about 10,000 killed
1986	October 10	El Salvador		ca. 1,000 killed, 200,000 homeless
Caribbean and Venezuela				
1812	March 26	Venezuela, Caracas		destruction
1831	August 11	Barbados		
1842	May 7	North of Haiti		
1852	August 20	Cuba, Santiago		
1874	September 26	Antigua		
1900	June 21	Cayman Islands	7.9	
1907	January 14	Jamaica, Kingston		1,000 killed
1923	December 22	Colombia, Boyaca	6.9	Tsunami
1946	August 4	North of Santo Dom.	8.1	
1967	July 29	Venezuela, Caracas	6.5	250 killed

Table 2: Important Earthquakes, Latin America
(Modified from BOLT, 1988)

Year	Date	Place	Magni.	Remarks
Colombia and Ecuador, Peru-Bolivia				
1805	July 16	Colombia, Honda		
1838	March 30	Peru, Callao		
1859	March 22	Ecuador Quito		
1868	March 22	Ecuador-Ibarra		70,000 killed
1875	March 22	Colombia, Cucuta		destruction
1906	January 31	North of Ecuador	8.9	
1942	August 24	Peru, Alazca	8.6	
1946	November 10	Peru, Ancash	7.8	1,500 killed
1949	August 5	Ecuador, Ambato	6.8	6,000 killed
1967	July 29	Colombia, Santander		
1970	May 31	Peru, Ancash	7.8	67 killed
1979	December	Colombia, Tumaco	8.2	200 killed, tsunami
1983	March	Colombia, Pon-ayan		200 killed, 1,000 million damage
Chile and Argentina				
1819	April 3, 4, 11	Chile, Copiago	8	Tsunami
1835	February 20	Chile, Concepcion	8.5	Tsunami
1861	March 21	Argentina, Mendoza		18,000 killed
1839	January 25	Chile, Chillan	8.3	30,000 killed
1944	January 15	Argentina, San Juan		5,000 killed
1960	May 22	Chile, Valdivia	8.4	
1966	December 28	Chile, Taltal	7.5	
1971	July 8	Chile, La Ligua	7.5	
1977	September 23	Argentina, San Juan	7.4	
1985	March 3	Chile, Algarrobo	7.4	176 killed

Table 2 (cont.):   Important Earthquakes, Latin America
(Modified from BOLT, 1988)

Country	1960	1983
Argentina	20,616	29,627
Bahamas	113	258
Barbados	231	272
Belice	92	175
Bolivia	3,428	6,034
Brazil	72,594	129,766
Colombia	15,538	27,518
Costa Rica	1,236	2,470
Cuba	7,029	9,906
Chile	7,609	11,665
Dominica	59	87
Ecuador	4,413	8,857
El Salvador	2,574	5,232
Granada	89	115
Guatemala	3,964	7,524
Guyana	538	941
Haiti	3,723	6,258
Honduras	1,943	4,092
Jamaica	1,629	2,288
Mexico	37,073	75,108
Nicaragua	1,493	3,058
Panama	1,105	2,088
Paraguay	1,778	3,472
Peru	9,931	18,707
Republica Dominica	3,224	5,962
Santa Lucia	93	123
Suriname	290	419
Trinidad and Tobago	843	1,218
Uruguay	2,538	2,968
Venezuela	7,502	16,394

Table 3:   Total Population [1,000 inhabitants]

Source: Anuario Estadistico de America Latina/Cepal. Santiago de Chile: Comision
Economica para America Latina y el caribe, 1984.- 765 pp.

Country	1950 1955	1955 1960	1960 1965	1965 1970	1970 1975	1975 1980	1980 1985
Argentina	2.0	1.7	1.6	1.5	1.7	1.6	1.6
Bahamas	2.9	4.4	4.8	4.4	2.9	3.4	2.1
Barbados	0.9	0.9	0.3	0.3	0.5	1.4	1.0
Belice	3.1	3.4	2.9	2.5	3.1	3.0	2.6
Bolivia	2.1	2.2	2.3	2.4	2.5	2.6	2.7
Brazil	3.2	3.0	3.0	2.6	2.4	2.3	2.3
Colombia	2.9	3.0	3.1	2.8	2.2	2.2	2.2
Costa Rica	3.6	3.8	3.7	3.2	2.6	3.0	2.7
Cuba	1.9	1.8	2.1	1.9	1.7	0.8	0.6
Chile	2.1	2.4	2.4	2.0	1.8	1.5	1.6
Dominica	2.2	0.7	2.0	1.8	1.4	1.8	1.4
Ecuador	2.8	3.0	3.2	3.2	3.1	2.9	2.9
El Salvador	2.7	3.0	3.1	3.6	3.0	3.0	3.0
Granada	2.3	0.9	0.7	0.4	2.2	1.1	1.2
Guatemala	2.9	2.9	2.9	2.8	2.8	2.8	2.9
Guyana	3.6	3.8	2.9	2.7	2.2	2.2	2.1
Haiti	1.7	2.0	2.1	2.2	2.3	2.4	2.5
Honduras	3.3	3.4	3.5	2.8	3.2	3.6	3.4
Jamaica	1.9	1.1	1.6	1.2	1.8	1.4	1.5
Mexico	3.0	3.2	3.2	3.3	3.3	2.9	2.6
Nicaragua	3.1	3.2	3.2	3.2	3.2	2.8	3.4
Panama	2.6	2.9	3.0	3.0	2.8	2.8	2.2
Paraguay	2.7	2.6	2.6	2.6	3.2	3.4	3.0
Peru	2.6	2.7	2.9	2.8	2.8	2.7	2.6
Republica Dominica	2.8	3.1	3.0	2.8	2.9	2.4	2.4
Santa Lucia	1.9	1.3	0.6	1.0	2.1	1.0	1.5
Suriname	3.1	3.0	2.7	2.2	-0.4	1.3	2.8
Trinidad and Tobago	2.5	3.2	3.0	1.0	1.0	1.5	1.4
Uruguay	1.2	1.4	1.2	0.8	1.0	0.6	0.7
Venezuela	4.2	4.1	3.6	3.4	3.6	3.5	2.9

Table 4: Total population growth (annual average rate per 100 inhabitants)

Source: Anuario Estadistico de America Latina/Cepal. Santiago de Chile. Comision Economica para America Latina y el Caribe, 1984.- 765 pp.

1951	499,756	(1)
1964	1,084,660	(1)
1973	1,601,804	(1)
1981	1,963,882	(2)
1988	2,316,709	(2)

Table 5:   Population of Medellin Area

(1)   Result of population census, DANE, Censos de poblacion de 1951 y 1964, ajuste censo 1973-1981.

(2)   Estimation, DAPM, Anuario, Medellin, 1987.

Chapter 4:

Organisational and Institutional Topics

The Integration of Environmental Geology Methods and Concepts within the Framework of a Regional Administration; the Case of the Comunidad Valenciana, Spain

C. AURENHEIMER *

Abstract

The process of transfer of jurisdictions on environmental matters from the central administration to an autonomous regional government is described. The main approaches taken by different autonomous regions in Spain, to deal with the problems of setting up an environmental administrative organization are also briefly commented here. Finally, the specific structure of the environmental administration in the Comunidad Valenciana and the main guidelines of its environmental policies are explained. It is concluded that earth science can provide a very useful basis for the drafting of these policies in the region, because of the specific nature of the problems to be dealt with, related mainly to water management, soil conservation, land occupation, natural hazards and preservation of natural areas.

* Author's address: Dr. C. AURENHEIMER, Agencia de Medio Ambiente, Comunidad Valenciana, Bailia 1, 46030 Valencia, Spain.

Introduction

During recent years there has been, worldwide, a spectacular increase in public awareness and concern about environmental problems. Governmental authorities have reacted in different ways in different countries as a result of the prevalent political ideology (the differences between Conservatives, Christian Democrats, Liberals, Social Democrats, Socialists or Communists in their approach to environmental matters are well known) and of the pecularities that these problems have in each country and region. These political differences, as well as the diversity of the "environmental problem" in each country, has led to a great variety in the organization of administrative structures and in the assignment of economic resources.

In the case of Europe, where most countries have some sort of de-centralized administration, the jurisdiction over many environmental matters rests to a great extent on the state/regional/provincial governments. Unfortunately, a common fact in almost every country is the distribution of jurisdictions, both horizontally, among different ministerial departments, and vertically among the different levels of the administration: central, state or regional governments, provinces or countries, city councils, communities, etc.

The case of the autonomous communities or regions [1] in Spain is particularly interesting, because they represent a case of transformation from a centralized to a highly de-centralized administration, during a period of general environmental awareness. These changes took place at the same time as other important changes were affecting the Spanish society, such as the transition from dictatorship to democracy and the joining of the European Economic Community.

The way in which the administration of environmental matters has been arranged

---

[1] These regions are in a way similar to the "states" in the U.S.A. or to the "Länder" in the F.R.G. they all have their own executive, legislative and judicial powers and considerable budgets that they administer with total independence from the central government.

within the regional government of one of the main autonomous regions in the country and the role of environmental geology in the planning and management process represent an interesting case study about the incorporation of earth science into environmental decision-making at the level of regional administration.

From the initial stages in the transformation of the country from a centralized to a de-centralized state, the regional governments displayed greater sensitivity towards environmental problems than the central governments. This is probably due to the fact that having to administer a much smaller territory, they know it better and are closer to its citizens; thus they are highly interested in dealing directly with matters which are of concern to the majority of their potential voters. Therefore, the jurisdiction over most environmental matters was assigned to the regional governments when the by-laws of the different autonomous regions were established.

However, the rush to complete the transfer of power from the central to the regional governments prevented an orderly organization of environmental matters within the new legal framework. Thus, the environmental responsibilities which were distributed amongst the different ministerial departments of the central administration, were often assigned in an equally disordered way to different departments of the regional administrations.

Environmental Geology and Environmental Management

The development of environmental geology (understanding this term in its widest sense) within an administration should be intimately linked with other environmental problems, and is concerned particularly with the integrated evaluation and management of land and other natural resources. Therefore, given the diversity of environmental problems in different regions, the role of environmental geology will vary considerably from areas where air or noise pollution are the main concerns, to others where land occupation and degradation or geological hazards are important. The role that environmental geology must play in a specific regional administration should be greater where the problems of territorial character, especially those derived from different land occupation alternatives, are more serious.

The techniques and methods contributed by environmental geology are means or instruments that can be interesting as a support for the management of environmetnal matters. However, it must be pointed out that they are not an end in themselves and thus they should not be a specific aim in the confines of environmental management. The utilization of environmental geology concepts, methods and techniques should not be limited to the management of "geological problems", but on the contrary, they can and should be used in fields as diverse as agriculture, solid and liquid waste management, landscape protection, etc.

Thus, the way to approach the problem is not to decide how a regional administration must manage environmental geology, but rather where, when, how, and at what level environmental geological methods must be applied in environmental management.

The Process of Assuming Environmental Jurisdictions in the Comunidad Valenciana

The Comunidad Valenciana is a region with a total area of 23,300 km and about 4 mio inhabitants, distributed among three provinces (Castellós, Valencia and Alicante) along the Mediterranean coast of SE Spain. It has one of the highest rates of economic growth in the country, centered in the industrial and service sectors, in which tourism plays an important role.

In 1980, responsibilities regarding legislation, planning and management of most environmental matters were transferred from the central to the autonomous regional government ("Generalitat Valenciana"). The jurisdictions over these matters were distributed among the following departments: Public Works, Urban Planning and Transport; Agriculture and Fisheries; Public Health and Consumer Affairs; Industry, Commerce and Tourism.

This process of transfer was undertaken at the same time as new national laws regarding the environment were passed at the national parliament. Also, laws were passed at the regional legislature, defining with greater detail and broadening the scope of environmental management within the region. The list of legal norms in the appendix of this paper gives an idea of the high dispersion of environmental jurisdictions between national and regional administration and among departments of the latter. This situation reflects that in fact the concept of environment was never clearly defined from the point of view of planning and management, in part probably due to its complexity, thus preventing the establishment of a unified management of environmental matters.

Among the different departments of the Generalitat Valenciana with responsibilities for environmental affairs, the Department of Public Works, Urban Planning and Transport proved to be the most active and innovative in this field. In particular, with regards to environmental geological matters, certain programs which had been initiated by the Diputación Provincia de Valencia (the administration of the province of Valencia), were extended to the whole region, covering the other two provinces. These include the master plan for management of solid waste and the elaboration of geoscientific maps as a basis for land-use planning. With this department, an "Office for Land-use Planning and Environmental Affairs" (GOTMA) was created, which constituted the nucleus of the present Environmental Protection Agency (Agencia de Medio Ambiente), now attached to the Department of Public Administration Affairs.

The Departments of the Environment in the Autonomous Regions

The serious problem represented by the distributon of jurisdictions in environmental matters in almost every country has led to efforts aimed at achieving the unification of management for environmental affairs. The models adopted vary considerably from one country to another, to the point that it can be said that no two models are exactly the same.

In the case of Spain, at the level of the central administration environmental affairs are attached to a General Directorate of the Ministry of Public Works and Urban Planning. However, some of the responsibilities are scattered among other ministries: Agriculture and Fisheries, Industry, Public Health, etc.

The process of transfer of jurisdictions from the central government to the autonomous regions or communities runs parallel to the efforts by these to unify the management of environmental affairs in a single organism. However, the development of this process was quite diverse in each region. So, some communities did not create general directorates and, finally, in others the process has lead to the creation of "Agencias del Medio Ambiente" (environmental protection agencies or agencies for the environment). This is the case for Andalucia, Asturias, Madrid, Murcia and the Comunidad Valenciana.

This "Agencia de Medio Ambiente" model represents an attempt to group the disperse environmental jurisdictions, so that an autonomous, unified environmental policy can be implemented within the region. The environmental programs and actions can thus be projected "horizontally" to other departments of the regional government. Nevertheless, although they have the same name, not all "Agencias de Medio Ambiente" have the same power. Some are offices with limited decision-making or mandatory capacities, whereas others have a vice-ministry status and are directly linked to the presidency of the autonomous community.

The Agencia del Medio Ambiente in the Comunidad Valenciana

The Environment Protection Agency of the Comunidad Valenciana, in its present form, was created by decree on January 16th, 1989, following a resolution of the regional parliament, on October 5th, 1988. The agency is attached to the Department of Public Administration Affairs, in order to make it independent from other departments which carry out investment programs which might affect the environment.

The activities of the agency follow two main lines. On the one hand, it coordinates, advises and controls policies and programs in other departments of the regional government in matters of environmental significance. On the other hand, it lays down policies and designs programs which it executes from start to end.

The first line of action referres to:

-- Coordinating the actions of all departments and services of the Generalitat Valenciana (regional government) with regards to the environment.
-- Informing and supervising master plans for urban and rural areas.
-- Making proposals for the redistribution of jurisdictions on environmental matters and for the assignment of offices to the Environmental Protection Agency.

The decree explains that it is planned to progressively increase the powers of the agency and its mandatory capacities in other departments of the Generalitat Valenciana. Its status will enable the agency to lay down norms and regulations which must be followed by both private individuals and corporations at regional administration departments.

The second line includes the legislation and programs of the Comunidad Valenciana on natural parks, special activities and environmental impact mitigation, as well as the responsibilities for management of solid and liquid waste, environmental education, environmental health, reclamation of derelict or degraded land, etc.

At present the agency has a moderate budget of about 700 million pesetas per year, of which 270 million is devoted to specialized staff. A program of steady growth has been laid down for the next three years.

Implementation of Actions on Environmental Geology in the Comunidad Valenciana

The Office for Land-use Planning and Environmental Affairs (GOTMA), precursor of the present Environmental Protection Agency, continued and extended, from 1985 onwards, some actions in the field of environmental geology which had been initiated by the administration of the provinces, particularly that of Valencia, the most active of the three in this area. So, the geoscientific mapping and evaluation of the provinces of Castellón and Alicante, for land-use planning at a general level, was undertaken following a methodology similar to the one used for the 1:200,000 map of Valencia, carried out by the Diputación Provincial in cooperation with the universities of Valencia and Cantabria. Master plans for the management of urban and industrial waste were another important part of environmental geology programs. The one in Valencia had been carried out by the Diputación Provincial in cooperation with the Generalitat Valenciana and the Ministry of Industry of the central government. Others were implemented in the other two provinces directly by the regional administration.

The Environmental Protection Agency has also undertaken a program of restoration of degraded lands, including the development of management plans of certain areas of special interest. These areas are defined as parts of the territory in which the said plans can bring about a significant improvement in: flood prevention and control; aquifer recharge; soil conservation; increase in faunal abundance and/or diversity; protection and/or social use of scientific, cultural or recreational resources.

Another area of priority interest has been the application of geoscientific knowledge to the drafting of regional development policies, through specific plans. These plans include water resources management, mangement and restoration of woodlands, management plans for special interest zones, and the like. Among these plans, emphasis will be placed on the following in the intermediate future:

-- Integrated water resource management with special attention to conservation measures (this is particularly important in this region with high density of population and industries, intensive agriculture and tourism, depending to a great extent on groundwater).
-- Woodland regeneration as a means to improve aquifer recharge, to reduce flood frequency and intensity and to curtail soil erosion.
-- Determination of the vocation and carrying capacity of sensitive areas, areas subject to hazards or areas of natural interest. Setting up a network of natural and recreational parks and educational trails.

The recommendations of these plans will be applied directly or through their incorporation into municipal and other master plans for town and country planning. It is expected that in this way protection of natural resources and use of the territory can be made compatible, gradually establishing sound, integrated planning and management schemes.

Final Comments

In this type of region, in which the main environmental problems are related to groundwater protection and management, soil conservation, land occupation, geological hazards and preservation of natural areas, environmental geology plays an important role at the level of the regional administration. Other matters, such as the ones related to air quality or urban environment, rest mostly on the national and municipal administrations, respectively.

Thus, the assessment of geological factors and the application of geological know-how constitute an essential basis for the drafting of regional environmental policies. However, earth science input must be integrated with inputs from other disciplines and should never be considered as the only, or even the main, factor to be taken into account for decision-making. The economic and social consequences of environmental policies can be very great indeed and it would be too simplistic to lay down such policies just on the basis of geological or, more generally, natural constraints and potentialities.

Appendix

Recent environmental legislation applicable in the Comunidad Valenciana [2]

1.1. Decrees (R.D.) regulating the transfer of jurisdictions from the central government to the autonomous regional government.

-   R.D. 278/1980; Jan. 25, on potentially unpleasant, unhealthy, dangerous or harmful activities.
-   R.D. 2595/1982; Jul. 24, on industry and energy (includes industrial environment and urban and industrial waste).
-   R.D. 2835/1983; Oct. 5, on land-use planning and environmental studies.
-   R.D. 3411/1983; Nov. 2, on environmental matters.
-   R.D. 3411/1983; Nov. 9, on coastal zone planning and disposal of waste in the sea.
-   R.D. 2365/1984; Feb. 8, on nature conservation and natural parks.
-   R.D. 1974/1985; Sept. 11, on land reform and agricultural development (includes the

[2] Environmental regulations from the European Economic Community also apply.

regulation of natural resources, agricultural exploitation, groundwater, waste-water and actions against salination of soil and water).

- R.D. 1871/1985; Sept. 11, on water supply, waste-water treatment, regulation and protection of river margins.

1.2. National legislation affecting the Comunidad Valenciana

- R.D. 1654/1985; July 3, on the organization of the Ministry of Public Works and Urban Planning.
- National Water Act, 29/1985, Aug. 2.
- R.D. 849/1986; Apr. 2, regulating the application of the Nat. Water Act.
- National Act on Toxic and Hazardous Waste, 20/1986, May 14.
- R.D. 1163/1986; June 13, modifying national act 42/1975 on Solid Urban waste.
- R.D. 1302/1986; June 28, on environmental impact assessment.
- National Coastal Zone Act, 22/1988; July 28.
- National Act 4/1989; March 27, on the conservation of natural areas, fauna and flora.

1.3. Regional legislation

- Decree of Nov. 5, 1982, on the assignment of jurisdiction over industry and energy to the Dept. of Economy, Industry and Energy.
- Decree of Dec. 1, 1983, assigning to the Dpt. of Public Works, Urban and Land-use Planning (COPUT) the jurisdictions and functions defined in the R.D. 2835/1983 (studies on land-use and environment).
- Decree of March 1, 1984, assigning to the COPUT the jurisdictions and functions defined in the R.D. 3411/1983 (environment).
- Decree of March 8, 1984, assigning to the COPUT the jurisdictions and functions defined in the R.D. 310/1983 (coastal zone planning and disposal of waste in the sea).

- Decree of April 2, 1984, on the organization and functions of the Dept. of Public Health and Consumer Affairs.
- Decree of April 2, 1984, on the organization and functions of the Dept. of Agriculture and Fisheries (CAP).
- Decree of April 2, 1984, on the organization and functions of the Dept. of Industry, Commerce and Tourism.
- Decree of Feb. 7, 1985, assigning to the CAP the jurisdictions and functions defined in the R.D. 2365/1984 (nature conservation and natural parks).
- Decree of Nov. 15, 1985, assigning to the COPUT the jurisdictions and functions defined in the R.D. 1871/1985 (water supply, waste-water treatment, regulation and protection of river margins).
- Decree of 25/1985, assigning jurisdictions and functions to different departments (includes potentially unpleasant, unhealthy, dangerous or harmful activities, assigned to the COPUT).
- Decree of Oct. 31, 1985, assigning jurisdictions and functions on land reform and agricultural development to the CAP.
- Decree of Aug. 23, 1985, establishing guide-lines for the coordination of functions to be carried out by the provincial administration, on matters of interest for the whole autonomous region (environmental matters are regulated through the COPUT and the CAP).
- Decree of Nov. 11, 1985, modifying the structure and organization of the Dept. of Industry, Commerce and Tourism.
- Decree of 193/1985, modifying the structure and organization of the Dept. of Agriculture and Fisheries.
- Order of June 18, 1985, of the CAP, on the creation of provincial forestry councils.
- Decree of Jan. 13, 1986, modifying the structure, organization and functions of the Dept. of Public Health and Consumer Affairs.
- Decree of Mar. 21, 1986, on primary health care within the Comunidad Valenciana.
- Decree of 52/1986, modifying article 45 of decree Jan. 13, 1986, on the organization of the COPUT.

- Decree of July 8, 1986, on the legal status of the natural park of L'Albufera.
- Decree of June 19, 1987, creating the Natural Park of the Penyal d'Ifach.
- Decree of Mar. 16, 1987, creating the Natural Park of El Montgó.
- Decree of Apr. 13, 1987, creating the Natural Park of La Font Roja.
- Decree of Jan. 25, 1988, creating the Natural Park of the Columbretes Islands.
- Decree of Dec. 12, 1988, declaring El Hondo as a Natural Site of the Comunidad Valenciana.
- Decree of Dec. 12, 1988, declaring the area Cabanes-Torreblanca as Natural Sites of the Comunidad Valenciana.
- Decree of Dec. 12, 1988, declaring the lakes of La Mata and Torrevieja as Natural Sites of the Comunidad Valenciana.
- Decree of Dec. 12, 1988, declaring the Salinas de Santa Pola as a natural Site of the Comunidad Valenciana.
- Decree of Jan. 16, 1989, creating the Environmental Protection Agency (Agencia de Medio Ambiente).
- Regional Act 5/1988 (June 24), on the regulation of Natural Sites of the Comunidad Valenciana.
- Regional Act 2/1989 (March 3), on environmental impact.
- Regional Act 3/1989 (May 2), on special activities.
- Regional Act 6/1989 (July 7), on land-use planning in the Comunidad Valenciana.

Environmental Geology in the United States: Present Practice and Future Training Needs

LAWRENCE LUNDGREN *

Abstract

Environmental geology as practiced in the United States confronts issues in three large areas: Threats to human society from geologic phenomena (geologic hazards); impacts of human activities on natural systems (environmental impact), and natural-resource management. This paper illustrates present U.S. practice in environmental geology by sampling the work of 7 of the 50 state geological surveys and of the United States Geological Survey as well. Study of the work of these agencies provides a basis for identifying avenues for the training of those who will deal with environmental issues in the future. This training must deal not only with the subdisciplines of geology but with education to cope with the ethical, interdisciplinary, and public-communication aspects of the work of the environmental geologist.

* **Author's address:** Prof. Dr. L. LUNDGREN, Dept. of Geological Sciences, University of Rochester, Rochester, N.Y. 14627, USA.

Introduction

In 1970, FLAWN published the first textbook to bear the title **Environmental Geology** (FLAWN, 1970). Other textbooks bearing similar titles followed, leading ALEXANDER (1983) to observe, in a review of these books, that a new sub-discipline of geology had been established in North America. This sub-discipline was defined in one of the most popular of these textbooks as follows (KELLER, 1982, p. 1):

"Environmental geology is applied geology. Specifically, it is the application of geologic information to solving conflicts, minimizing possible adverse environmental degradation, or maximizing possible advantageous conditions resulting from our use of the natural and modified environment."

This definition suggests that the environmental geologist is expected to make judgments concerning the degree to which human use of a particular environment adversely affects that environment and is also expected to resolve conflicts. Since this paper is concerned with the present and future practice of environmental geology, the author first examines the nature of environmental geology. This provides a perspective from which to review present practice in environmental geology within the United States. This review shows what kinds of work are being done, how public policy helps to account for the distribution to effort, and what kinds of new demands are being placed upon geologists. This leads in turn to a discussion of needs for the future.

The Nature of Environmental Geology

Environmental geology is an applied field. Its subject matter is defined by the problems that arise out of the relationships between humans and the environments in which they live, work and play. The radon-exposure hazard is a current example of the subject matter of environmental geology (KRIMSKY & PLOUGH, 1988). This hazard was made a matter for national concern when the U.S. Environmental Protection Agency and the U.S. Surgeon General recommended on 12 September 1988 that most homes in the U.S. be tested for radon (SHABECOFF, 1988). The geological roots of the radon hazard are to be found in the differing rates of radon generation in rocks in different areas and in the factors that govern radon migration. The societal roots of the problem are to be found in our land-use and building-design practices.

The radon-exposure hazard illustrates the view of BURTON, KATES & WHITE (1978) that natural hazards arise out of, and can only be understood in terms of, interactions between natural systems and social systems. It follows that the assessment of and management of almost any environmental problem demand analysis by, and perhaps interaction between, natural scientists and social scientists.

ALEXANDER reminded geologists of this view when he wrote (1983, p. 127) "... that the principal defect of environmental geology is that it is too geological."

Although ALEXANDER made this observation in a review of environmental-geology textbooks, the statement was aimed at professional environmental geologists, who, according to ALEXANDER, should become more effective by obtaining a better understanding of the way that nature (geology) interacts with society (human decisions) to create hazards and environmental problems. The author rephrases the question posed by ALEXANDER as follows: Can the contributions of some environmental problems be improved by a better understanding of the non-geologic dimension of these very problems? If so, which environmental geologist can benefit most from this understanding, and how shall they obtain it?

The author will not answer this question decisevely, but shall examine it. In order to do this, first the geologic content of environmental geology is considered and then the actual practice of environmental geology in the United States examined.

Geologic Content of Environmental Geology

The purely geologic content of environmental geology has two principal parts:

1.) The description of past and present states of a particular environment, and
2.) The forecasting of possible futures for that environment.

This may be illustrated by studies of Mount St. Helens volcano. CRANDELL & MULLINEAUX (1978) studied the stratigraphy of volcanic and sedimentary deposits around Mount St. Helens to reconstruct the behavior of Mount St. Helens over the past 4500 years. They were able to show what kinds of events had occurred during this time interval and were also able to show how large an area had been affected by each of the different events.

These geologic studies of the past provided them with a basis for making statements about future eruptions and future environmental effects of these different eruptions (see LUNDGREN, 1986, p. 27 for a tabulation of these statements). These studies found unexpected application when Mount St. Helens became active in March 1980. Their work provided a framework in which to place the observations obtained by real-time monitoring from March 1980 to the present. The synthesis of all this work provides a basis for understanding the day-to-day behavior of the volcano and for improving the ability to predict future activity at Mount St. Helens and at other volcanos (LIPMAN & MULLI-NEAUX, 1981).

The accounts of studies done at Mount St. Helens presented in LIPMAN & MULLI-NEAUX (1981) illustrate the breadth of the geologic research necessary to deal with the problems of environmental geology. The author believes that the proper geologic study of environmental problems must contain three elements represented by work at Mount St. Helens and elsewhere - study of the record of past behavior, good understanding of the present state of a system, and study of the dynamics of processes now in operation. This is the geologic content of environmental geology.

Current Practice in Environmental Geology in the United States
Sources of information

There are three readily available sources of information from which to draw samples of work being done in environmental geology within the United States. The Yearbooks of the United States Geological Survey (USGS) (see U.S. Geological Survey, 1987, for an example) provide an overview of work at the federal level. The annual reports and annual lists of publications of the state geological surveys provide a view of work at the state level. The third source comprises the annual reviews of activity in geology that are published each February by the American Geological Institute in Geotimes. All these sources have been drawn up, but the reports of 7 state geological surveys (California, Louisiana, Maine, Minnesota, New York, Oregon, and Texas) provide most of the information cited.

Categories of Environmental-Geology Activity

Guides to the major areas of environmental-geology activity are provided by the tables of contents of textbooks of environmental geology. Virtually all such books deal with geologic hazards and with the environmental impacts of human activity (see KELLER, 1982 and LUNDGREN, 1986); many, but not all, also examine the role of geology in assessing and managing natural resources. A complete review of all of these areas is not possible in a short paper. Work that falls under the following headings has been surveyed:

1)   Coastal-Zone Problems
2)   Hazards
3)   Hydrogeology
4)   Environmental Mapping and Geographic Information Systems
5)   Education and Information.

An examination of activity in these areas illuminates all of the issues faced by environmental geologists working in the three major areas - hazards, environmental impact, resources - named above.

Each of the 7 states examined here delegates responsibility for environmental-geology activities to two or more state agencies. In New York State, for example, environmental geology is practiced within at least three agencies: the New York State Geological Survey, the Department of Environmental Conservation, and the Office of Parks, Recreation, and Historic Preservation. The author has only reviewed activities in the geological agencies listed in SOCOLOW (1988). A review of the work done in these agencies identifies all of the different types of environmental-geologic work being done in the United States even if it does not identify all of the agencies in which environmental geology is being practiced.

The environmental issues faced by the geological agencies are identified in the three sections titled: Coastal Zone Problems, Hazards, and Hydrogeology. Most of the work done on these issues is ultimately transformed into map form. Thus the fourth subsection deals with Environmental Mapping and Geographic Information Systems. The last subsection deals with Education and Information.

Coastal-Zone Problems

Coastal-zone problems are important in all six of the coastal states sampled, but not all of the geologic agencies in these states have responsibility for dealing with these problems. The major problems confronted in the six coastal states include shore-line change, loss of wetlands, pollutant loading in bays and estuaries, and flood hazard (OFFICE OF COASTAL ZONE MANAGEMENT, 1976).

Shore-line change is experienced as barrier-island migration on the Atlantic and Gulf coasts, as bluff and sea-cliff erosion on the Great Lakes and on the Pacific coast, and as beach erosion and change on all coasts (U.S. ARMY CORPS OF ENGINEERS, 1971). The loss of coastal wetlands and the transformation of estuarine and near-shore environments due to pollutant loading are problems experienced in greater or lesser degree by the six coastal states sampled here (HORWITZ, 1978). Development of the coastal zone in all six states combines with natural factors to create substantial flood hazard problems as well (OFFICE OF COASTAL ZONE MANAGEMENT, 1976).

Although these problems affect all of the coastal states sampled, few of the state geological agencies are given responsibility to deal with all of them. The Maine Geological Survey has publication series dealing with coastal hazards, marine environments, and neotectonics (MAINE GEOLOGICAL SURVEY, 1989). The California, Oregon, and New York state geological agencies do not currently have comprehensive responsibility for addressing coastal problems. Since 1983, the Louisiana Geological Survey has devoted much of its resources to geological-biological research on the causes of coastal land loss cooperating with the U.S. Geological Survey in this research (CHARLES GROAT - Louisiana Geological Survey, personal communication 13 September 1989). Coastal research by the Texas Bureau of Economic Geology is represented in two major map series, one the Environmental Geologic Atlas of the Texas Coastal Zone (see e.g., FISHER et al., 1972), the other a series dealing with geology-biology of the Submerged Lands of Texas (see e.g., WHITE et al., 1985).

Hazards

A major activity of most geological agencies is dealing with geologic hazards. These hazards include purely natural hazards - earthquake and volcano hazards for example - and purely technical hazards - nuclear and toxic-waste hazards. They also include hazards that are neither purely natural nor purely technical.

In this section, four hazards are considered: radon-exposure, earthquake, volcano, and landslide. Additional hazards are considered in the section on hydrogeology, and flood hazard is noted in the section on mapping.

Radon-exposure: The radon-exposure hazard has only recently been recognized as important. This recognition reflects the accomplishments of an interdisciplinary approach based on studies of radon generation and migration in geologic materials, building studies documenting the factors that control the concentration of radon within structures, and epidemiologic studies of the health hazard posed by radon (KRIMSKY & PLOUGH, 1988). The role of environmental geology in dealing with this hazard is well documented by the response of the New York State Geological Survey (NYSGS). The NYSGS and Northeastern Science Foundation sponsored a conference on Radon and Geology in June,

1988. The papers presented at this conference appear in a special issue of Northeastern Environmental Science (7, 1, 1988). The NYSGS does not itself carry out research on radon but provides laboratory facilities to scientists of the New York State Department of Health (NYSDOH) as part of a cooperative project with NYSDOH. The Maine Geological Survey is investigating the relationship between bedrock type, fracture systems, and radon in ground water (WALTER ANDERSON, personal communication, 12 February 1990). The Minnesota Geological Survey (RICHARD LIVELY, personal communication, 13 February 1990) has completed a pilot study documenting a positive relationship between bedrock lithology and radon in ground water.

Earthquake hazard: Earthquake hazard is a concern in many states, but it is the State of California that has the greatest experience with this hazard and therefore devotes the most attention to it. The earthquake-hazard activities of the California Division of Mines and Geology (CDMG) illustrate particulary well how the environmental-geology activities of a state geological agency are influenced by directives from the state legislature. For example, the Alquist-Priolo Special Studies Zones Act of 1972, passed by the California State Legislature after the 1971 San Fernando earthquake, directed the CDMG to map faults displaying evidence of activity on differing time scales, and the CDMG has prepared maps of 3700 km of fault lines as result. All maps prepared since 1977 under the directive of this act display only those faults on which motion is believed to have taken place in the past 11,000 years (CALIFORNIA DIVISION OF MINES AND GEOLOGY, 1988).

The CDMG has also addressed the hazard due to earthquake shaking, notably through the establishment of the first of a projected series of test sites where the dependence of shaking on local geologic conditions will be studied (CALIFORNIA DIVISION OF MINES AND GEOLOGY, 1988). The initial site is located at Parkfield, which is itself located on that segment of the San Andreas fault for which the U.S. Geological Survey has issued an official prediction of a magnitude 6+ earthquake before 1993 (BAKUN, 1987).

The CDMG also began a program in 1988 to determine the feasibility of a long-term

Urban Seismic Hazard Mapping Program, a program that could lead to publication of an Urban Seismic Hazard Map Series (CALIFORNIA DIVISION OF MINES AND GEOLOGY, 1988). Such information would be directed especially at the insurance industry, which might employ these maps in the same way that the insurance industry uses the FIRM series prepared under the U.S. National Flood Insurance Program.

Following the 1971 San Fernando earthquake, the California State Legislature also passed legislation creating a Strong Motion Instrumentation Program (SMIP) operated by the CDMG and funded by a tax of $0.07 per $1,000 on all new constructions within the state (CALIFORNIA DIVISION OF MINES AND GEOLOGY, 1987).

Oregon and New York have both undertaken studies of earthquake hazard with support from the Federal Emergency Management Administration (FEMA) as part of the National Earthquake Hazards Reduction Program (FAKUNDINY, 1988; DONALD HULL - Oregon Department of Geology and Mineral Industries, personal communication, 28 August 1989).

Volcano hazard: Volcano hazard is a concern in California and in Oregon. The statement of Missions and Programs: 1985-1991 issued by the Oregon Department of Geology and Mineral Industries, makes no mention of volcano hazard. The CDMG, in contrast to Oregon, has instituted a Volcanic Hazards Monotoring Program. It has issued disaster-planning materials for the Mammoth Lakes area as part of this program, and it collaborates with the USGS to monitor the potential for volcanic activity at Mammoth Lakes (CALIFORNIA DIVISION OF MINES AND GEOLOGY, 1988).

Landslide hazard: Landslide hazard is a major concern in California and is a significant problem in some of the other states sampled. The U.S. Geological Survey supports a program in which the individual state geological agencies prepare statewide landslide-inventory maps (see, e.g., FICKIES, 1989). In Oregon, responsibility for landslide investigation is divided among many agencies, both federal and state. The greatest effort devoted to landslides is in California, where both the CDMG and the USGS are active (see section on Mapping).

The work of the CDMG illustrates some of the major approaches to landslide hazard and also illustrates how legislative directives affect these approaches. A long-standing program of landslide mapping carried out by CDMG was terminated in 1978 by California legislation called Proposition 13. In 1980, the legislature authorized CDMG to carry out the first site-specific slope-stability project at the Baldwin Hills in Los Angeles. Then in 1983, the legislature passed the Landslide Hazard Identification Act of 1983, which called upon the CDMG to map landslide hazards in areas already developed or undergoing development. This legislation led to the preparation between 1983 and 1987 of landslide hazard maps covering 3100 square kilometers of urban area (CALIFORNIA DIVISION OF MINES AND GEOLOGY, 1988).

Hydrogeology

The decade 1980-1989 is surely be thought of by most environmental geologists in the world as the Hydrogeology Decade. The rapid growth of hydrogeology is docmented in the National Water Summary annual monographs published annually by the U.S. Geological Survey since 1984 (U.S. GEOLOGICAL SURVEY, 1984). These annual summaries provide state-by-state coverage. Even more dramatic evidence is provided by the increase in membership of the Association of Ground-Water Scientists and Engineers (AGWSE) and the Geological Society of America's Hydrogeology Division (Fig. 1). The AGWSE data represent members with a variety of backgrounds, but they are compatible with evidence of increased demand for bachelor's and master's-degree holders in geology seen in any department of geology during this period.

Some of the initial concerns with ground-water contamination in the United States arose as one uncontrolled hazardous-waste disposal site after another was identified. Love Canal, in the city of Niagara Falls, New York, one of the best known of these sites, illustrates the problems confronted at hundreds of other sites in the United States (see

LUNDGREN, 1986, for a review and references). Additional concerns were raised by the identification if actual or possible migration of radionuclides from sites where nuclear wastes were stored. The West Valley, New York low-level nuclear-waste storage site is an example (PRUDIC, 1986).

Concern with hazardous-waste sites led to the passage of federal legislation commonly referred to as RCRA and CERCLA. The Resource Conservation and Recovery Act of 1976 (RCRA) is intended to control the disposal of toxic wastes. The Comprehensive Environmental Response, Compensation, and Liability Act of 1980 (CERCLA) provided for the cleanup of existing hazardous waste sites such as Love Canal. CERCLA provides a means for funding such cleanup through a fund generally known as Superfund. Both these laws helped to create an extraordinary demand for hydrogeologists.

This national concern with ground-water contamination was matched by concern at the state and local level focused especially on existing and planned solid-waste landfills. State legislation regulating the siting and design of landfills added another increment to the demand for hydrogeologists.

At the same time that these problems were being faced, the federal government was confronted with the problem of storing high-level nuclear waste (including spent fuel) produced by military operations and commercial nuclear power plants. The U.S. Nuclear Waste Policy Act of 1982 specified that the President of the United States would identify the first site for storage of spent fuel elements by 31 March 1987. This law led to intense activity by hydrogeologists, but the deadline was not met.

The Low Level Radioactive Waste Policy Act of 1980 addressed the problem of low-level nuclear waste (LLNW). This act required each state (or group of states) to identify a site which the low-level nuclear waste generated by that state (or group of states) could be stored. These sites were supposed to be identified by 1 January 1986; that deadline was extended to 1 January 1993. This act also increases the demand for hydrogeologists.

All of these concerns with observed and potential ground-water contamination put pressure on the states to map aquifers to be protected from contamination and also led to calls for federal and state ground-water protection legislation. As such legislation is enacted, it will further increase the demand for hydrogeologists.

The initial search for high-level nuclear waste led to major activities in each state identified as having the potential to become a repository site. Texas, Minnesota, and Maine were among the original candidate sites, and at one time all had programs to address this issue. The Texas Bureau of Economic Geology carried out major hydrogeology research under the auspices of the Texas Nuclear Waste Program. A number of research reports in the Palo Duro basin were issued (see e.g., GUSTAVSON et al., 1983), and a final synthesis of all hydrogeologic work on nuclear waste sites in Texas was published in the final report of the Texas Nuclear Waste Commission (BUREAU OF ECONOMIC GEOLOGY, 1989). All three states were relieved of responsibility for further work in site characterization when the U.S. Congress directed the U.S. Department of Energy in December 1987 to focus all work in one site, Yucca Mountain in Nevada.

The search for LLNW sites has been handled very differently in the different states. The Texas Bureau of Economic Geology conducts its own hydrogeology investigations of proposed sites, and it also coordinates other studies by university researchers. In Maine, the State Geologist is a member of the LLNW Authority. The approach of this Authority is discussed in the section on Future Needs. The New York State Geological Survey plays only an advisory role; the major responsibility for site studies is delegated to the New York State Low-Level Radioactive Waste Siting Commission.

Aquifer protection is a major responsibility of many state geological agencies. The Maine Geological Survey published maps and reports in significant aquifers (see e.g., WEDDLE, 1988) and has several other publication series in hydrogeology (MAINE GEOLOGICAL SURVEY, 1989). The Minnesota Geological Survey publishes hydrogeologic information in the County Geologic Atlas series (see e.g., BALABAN, 1989).

Mapping and Geographic Information Systems (GIS)

The single most common method of presenting information on the environmental problems discussed in this paper is to employ maps. The presentation of geologic data and hypotheses on geologic maps is fundamental in geology, and it may seem that environmental geologic maps simply represent a continuation of long-established practices. This is not true.

An environmental geology map is prepared by a geologist or a team for use by people whose geologic training is not at all comparable with that of the geologist preparer. The nature of the environmental problem determines what geologic information is presented on the map. The purpose of the map is to aid in the management of a specific environmental problem. The criteria for evaluating an environmental geology map are societal as well as geologic.

These differences between a conventional geologic map and an environmental geologic map commonly demand that the preparer of the environmental map understand dimensions of environmental problems that go beyond what is considered in any purely geologic program of training. The preparer synthesizes different kinds of information and selects information with societal goals in mind. The synthesis may be accomplished by manually combining various types of spatially related data, an approach familiar to any field-oriented geologist. The synthesis may also be accomplished by employing the techniques of Geographic Information Systems or GIS. Whatever the techniques of synthesis, it is the principles that guide the synthesis that are important.

In the next section, a brief overview of some different kinds of maps produced by environmental geologists and their collaborators will be provided, and then landslide maps more carefully considered, illustrating the role of Geographic Information Systems in this treatment.

Types of Environmental Maps: Many types of environmental geologic maps are produced in the United States. Flood-insurance rate maps (FIRM) produced under the National Flood Insurance Program show subzones within the area (river or coastal) expected to be inundated by a flood having recurrence interval R = 100 years (see LUNDGREN, 1986 for a sample of FIRM). Any environmental geologist interested in risk mapping can benefit from an examination of these maps, which illustrate how geologic (hydrologic) information can be combined with economic (damage) information to illustrate levels of risk. Each such map shows the area indundated by a flood of recurrence interval R = 100 years but also shows many subzones. These subzones provide the basis for determing the rate that is to be charged for the purchase of flood insurance covering structures within any subzone.

Many other types of environmental maps have been prepared for the coastal zone. Many of these maps combine geological and biological information. The processes of coastal erosion, sedimentation, wetland and estuarine evolution, and even coastal flooding demand this synthesis. Excellent samples are to found in 63 maps that constitute the Environmental Atlas of the Texas Coastal Zone issued by the Texas Bureau between 1972 and 1980 (FISHER et al., 1972). Publication of a comparable series of maps of the submerged coastal lands and associated wetlands was begun by the Texas Bureau in 1983 (WHITE et al., 1985). The MAINE GEOLOGICAL SURVEY (1989) also publishes a Coastal Marine Environmental Maps, Geology and Flood Zone Maps, and Coastal Hazards Maps and a coastal Neotectonics series. The Louisiana Geological Survey publishes a Coastal Geology Map Series that consists of aerial videotapes of coastal Louisiana (LOUISIANA GEOLOGICAL SURVEY, 1989).

Hydrogeologic maps have become an important part of the output of many state geological agencies. The Geological Survey of Minnesota's County Geology Atlas Series provides excellent examples (BALABAN, 1989). The Hennepin County Atlas (BALABAN, 1989) comprises 9 maps with text; the maps in this atlas include surficial and bedrock-geology maps, and four hydrogeology maps (Bedrock Hydrogeology, Quaternary Hydrogeology, Sensitivity of Ground-water Systems to Pollution, and Geology and Well

Construction). A Data Base map shows the location of all wells from which information was obtained.

The MAINE GEOLOGICAL SURVEY (1989) has produced a series of maps showing the extend of sand and gravel aquifers and significant aquifers. These maps are the first step in the development of any aquifer-protection program.

Landslide mapping: Landslide mapping is a major activity of state geological surveys and of the United States Geological Survey (USGS). The USGS cooperates with individual states that then prepare statewide landslide-inventory maps (see FICKIES, 1989, for an example). The USGS has created a Regional Landslide Analysis Group (RLAG) within the USGS Branch of Geologic Risk Assessment. EARL BRABB, who is a member of the RLAG, has written extensively on landslide mapping and serves, along with WILLIAM COTTON, as editor of the International Landslide Research Group Newsletter in which numerous reports on landslide mapping may be found (COTTON & BRABB, 1989).

The USGS landslide-hazard mapping projects began in the 1970's as part of the San Francisco Bay Region Environment and Resources Planning Study, which was intended to show how geologic information might be brought to bear on land-use decisions and planning. A synthesis of work on landslides done under this program is available (NILSEN et al., 1979).

This program produced maps showing the distribution of landslides in California; it also produced landslide-susceptibility maps. The landslide maps were objective representations of existing landslides; the susceptibility maps were interpretive maps intended to show which areas not already affected by landsliding were most likely to be so affected in the future.

NILSEN & BRABB (1972) made a direct effort to show the importance of landslide maps for those who make land-use decisions in California. In this report, they presented side-by-side maps of the Penitencia Creek landslide area in Santa Clara County, California.

One map shows geologic information; the other shows non-geologic (damage) information. Taken together, the maps illustrate how the interaction between nature (landslides) and land-use choices determine later costs to society. At the time of mapping, a major land-use decision was under coonsideration, a decision concerning construction of a $10 million water-treatment plant to be built on the Penitencia Creek landslide. BRABB (1987 b) has since reported on this decision; the water treatment plant was built on the landslide that NILSEN & BRABB had shown on their 1972 maps. The landslide continues to be active. This case could provide the basis for a multi-disciplinary analysis of approaches to landslide mapping and risk assessment, and BRABB (1987 b) has solicited on on this or similar situations.

Since performing that early work on landslide risk mapping, BRABB has reviewed other approaches to landslide mapping (BRABB, 1984). This in turn has led to a pilot landslide mapping project employing the techniques of Geographic Information Systems (GIS). This project illustrates the potential of GIS and points to some of the issues raised at the beginning of this paper. BRABB's 1984 review begins with a specific recommendation:

"Landslide hazard and risk maps should provide information needed to judge the impact that the hazard would have on people and structures at risk."

Thus BRABB seems to be recommending that the environmental geologist incorporate significant non-geologic information in some of the maps. One model for such an approach is provided by the FIRM maps produced under the National Flood Insurance Program maps. Another model is provided by the land-use capability maps produced by a multi-disciplinary USGS team as part of the San Francisco Bay study (LAIRD et al., 1979). Both the FIRM and the land-use capability maps seem to meet the specifications in BRABB's recommendation cited above.

BRABB also argues that the geologist should prepare landslide maps that are predictive. Such maps might express prediction in terms of the future probability of landsliding, or they might express predictions in terms of contingency (see LUNDGREN,

1986, a treatment of different types of prediction). BRABB raised a number of questions about the forms that landslide mapping might take in the future. These questions amount to a challenge to geologists to rethink the whole approach to creating and evaluating environmental geology maps. Some of these questions have been circulated internationally in a questionaire (COTTON & BRABB, 1989), and responses have been tabulated. The responses show a great diversity of viewpoints.

An initial report on the USGS-RLAG pilot project (BRABB, 1987 a) shows how the GIS techniques have been used to prepare maps that could not previously have been prepared. This project has created maps that show the predicted distribution of debris flows contingent on specific rainfall thresholds being exceeded. These maps have made it possible to create a system for issuing debris-flow warnings to the public at the same time that flash-flood warnings are issued by the U.S. Weather Service.

This GIS-based mapping is innovative; it demands a very large data base, a multi-disciplinary team to design the system for combining information, and a team to evaluate these maps.

Education and Information

Most of the geological agencies surveyed report that they bear a much greater responsibility for educating and informing a broad public now than in the past. Perhaps in the past this public information role could be met by producing brochures and maps presenting simplified presentations of geologic findings that had few societal implications. Now, given the intense interest that can be generated around almost any environmental issue, the public-information role clearly is demanding much greater attention to the needs of the public.

The maps published in the County Geologic Atlas series of the Minnesota Geological Survey illustrate one way to provide this information (see e.g., BALABAN, 1989). Hydrogeologic maps show aquifer distribution and the susceptibility of aquifers to contamination. These atlases were a direct response to requests from the counties, and those that have been produced to date are said to be much in demand. These maps are accompanied by practical information on well construction. The Minnesota Survey also publishes a model bulletin on the use of ground-water data in its educational series (OLSEN, MOHRING & BLOOMGREN, 1987). This publication and others like it deserve the attention of all environmental geologists interested in improving communication and public levels of understanding.

The need for improving the ability to serve the public is not a matter of providing material in printed form. PETERSON (1988) makes a strong and effective case for improved person-to-person communication between geologists and the public. Those agencies, such as the Maine Geological Survey, that have begun to develop Geographic Information Systems and computerized data bases are creating another base for improving the ability to provide information.

Future Training of Environmental Geologists

Some environmental geologists do fundamental research that will be brought to bear others on environmental problems. Some of these researchers carefully maintain a separation between themselves and the public. The training of such researchers will presumably proceed as in the past; doctoral and post-doctoral disciplinary research will define that training.

Others engaged in environmental geology -- perhaps the majority -- have responsibilities that do not allow them to keep their distance from the issues and from those people who confront these issues directly. Most of what follows bears upon this group of environmental geologists whose training may range from bachelor's degree to doctoral degree. In this concluding section, some conclusions or propositions will be stated so that comment that bears upon the future training of environmental geologists will be provided. In most instances, the author's examples are drawn from hydrogeology.

1) Problems confronted by environmental geologists are interdisciplinary; they can neither be defined nor understood in terms entirely familiar to any one specialist. Some environmental geologists will fit into this interdisciplinary picture by contributing geologic expertise without regard for the non-geological dimensions of the problem. Others, especially those whose responsibilities demand management of the larger problems, will function best if they understand ALEXANDER's point (quoted on page ...) that environmental geology may be *too geological*. Students and their teachers might consider the following example: The State Geologist of Maine, Dr. WALTER ANDERSON, has observed (personal communication, 19 September 1989) that the Maine LLNW Authority, of which he is a member, recognized at an early stage that the LLNW Authority would at an early stage need not only the contributions of geologists and engineers but also contributions from a social science perspective. The social scientists would examine ways to involve the public, and, by so doing, decrease the chance of confrontation of the Authority by the public, confrontation well known to those in New York State in 1989 where a Commission in Low-Level Radioactive Waste is also trying to find an LLNW site. Agencies that have not appreciated the need for a social-science perspective have sometimes found it difficult to do the geologic and other work that must be carried out. In other words, it may be worth training environmental geologists to understand that no environmental problem is purely a technical problem manageable by natural science alone and to have them devote some serious study to the larger practice of "environmental management".

2) Environmental geology is a dynamic and exciting hybrid field that has grown very rapidly in the past decade; the subfield of hydrogeology has shown the most dramatic

growth of all. Departments of Geology and universities struggle to find ways to provide effective instruction in such new subareas of geology as hydrogeology. Some of the problems to be met and overcome are described in an analysis by WARREN WOOD, the Geological Society of America Birdsall Lecturer for 1989. He observes that hydrogeologists in the United States have not yet convinced other geologists, the National Academy of Sciences, and the National Science Foundation of the viability, importance, and future of hydrogeology as a scientific discipline within the larger field of geology (WOOD, 1989). Thus, although the demand for hydrogeologists has increased dramatically, opportunities for students to be trained or retrainied are relatively limited. WOOD notes that departments of geology that cannot meet this or other comparable environmental challenges will relinquish important opportunities in the future.

3) Environmental geology is a high-visibility activity for many who practice it. U.S. Geological Survey geologists learned this while working at Mount St. Helens (PETERSON, 1988; SAARINNE & SELL, 1985). PETERSON (1988, p. 4161) states: *"We must apply the same degree of creativity and innovation to improving public understanding of volcanic hazards as we apply to the problems of volcanic processes. Only then will our full obligation to society be satisfied."*
Environmental-geology agencies should consider building groups and training individuals to deal with situations like those descibed by PETERSON.

4) The environmental geologist, like any other natural scientist, understands the merits of scientific detachment. DAVID BREW (1974), of the U.S. Geological Survey has expressed the need for avoiding bias in handling the data and the interpretations might also seem to demand avoidance of any involvement in the environmental issues themselves. Evidence of substantial and effective involvement on the part of any number of respected environmental geologists suggests an alternative position must be considered. An essay by the respected hydrogeologists R. ALLAN FREEZE and JOHN A. CHERRY on U.S. ground-water legislation (FREEZE & CHERRY, 1989) illuminates many facets of this alternative position. The training of environmental

geologists should at least recognize the inevitability of involvement in the issues. Having recorded this, training or retraining of many geologists should provide an understanding of ethical, economic, and societal issues and should introduce or re-introduce these subjects periodically by concrete example.

5) The problems of environmental geology differ in a fundamental way from those of conventional geology. Conventional geology has almost always been concerned with the past, and prediction in conventional geology generally means "prediction in the past". Environmental geology builds on conventional geology's ability to understand the past, but almost always employs this understanding to make statements about the future. Thinking about the future is not a part of the training of most geologists. It must become a part of that training. Part of this training should consider the societal impact of published geologically based predictions. In this training, the geologist should consider the issues that must be faced if a prediction (e.g., a landslide contingency map) is issued where there is no public policy in place that is designed to deal with the consequences of the prediction.

6) The handling of various data bases through the techniques of Geographic Information Systems will become an essential part of work in environmental geology. Every effort should be made to incorporate work with computers in every course and to require appropriate courses as well.

7) Geologists have important contributions to make on a broad range of environmental problems. In order to insure that these contributions are made, some geologists will have to concern themselves with the role of geology and the geologist in the multi-disciplinary field of environmental management.

Acknowledgments

The author thanks the following people from the state geological agencies who helped him greatly: WALTER ANDERSON, ALAN DUTTON, ROBERT FICKIES, CHARLES GROAT, DONALD HULL, DAVID SOUTHWICK, WOODROW THOMPSON, and BRIAN TUCKER. Given their schedules and the author's, they have not had the opportunity to review the manuscript. All responsibility for errors of misplaced emphasis is with the author.

References

ALEXANDER, DAVID (1983): Environmental geology: A hazard in its own right? Environmen. Managem. **7**, 2: 125-128.

BAKUN, WILLIAM H. (1987): Predicting the next major earthquake in the Parkfield area of California. U.S. geol. Surv. Yearbook fiscal Year **1986**: 12-15, Washington, D.C.

BALABAN, N.H. (1989): Geologic Atlas -- Hennepin County, Minnesota: Minnesota geol. Surv. County Atl. Ser. Atlas C-4. St. Paul, Minnesota (University of Minnesota).

BRABB, EARL E. (1984): Innovative approaches to landslide hazard and risk mapping. 4th Intern. Landslide Sympos. Proc., **1**: 307-323, Toronto, Canada.

--- (1987 a): Analyzing and portraying geologic and cartographic information for land-use planning, emergency response, and decisionmaking in San Mateo County, California. GIS (Geographic Information Systems) '87 - San Francisco: 362-374, Falls Church, Virginia (American Society for Photogrammetry and Remote Sensing).

--- (1987 b): The Penitencia Creek landslide, Santa Clara County, California: A brief description and a request for input from readers: Intern. Landslide Res. Group Newsl. **1**, 2: 2-4. Los Gatos, California.

BREW, DAVID (1974) Environmental Impact Analysis: The Example of the Proposed Alaska Pipeline. U.S. geol. Surv. Circ. **695**: 1-16, Washington D.C.

[BUREAU OF ECONOMIC GEOLOGY] (1988) Annual Report. Austin, Texas (Bureau of Economic Geology).

[BUREAU OF ECONOMIC GEOLOGY] (1989) List of Publications (1989). Austin, Texas, (Bureau of Economic Geology).

BURTON, I., KATES, R.W., & WHITE, G.F. (1978): The Environment as Hazard. 240 pp. London (Oxford University Press).

[CALIFORNIA DIVISION OF MINES AND GEOLOGY] (1988): 80th Report of the State Geologist. 22 pp. Sacramento, California (Division of Mines and Geology).

COTTON, E. & BRABB, E. E. (1989): Priorities for the International decade of disaster reduction. Intern. Landslide Res. Group Newsl. 2, 3: 1-7, Los Gatos, California.

CRANDELL, D.R. & MULLINEAUX, D.R. (1978): Potential Hazards from Future Eruptions of Mount St. Helens Volcano. U.S. geol. Surv. Bull. 1383-C: 1-26, Washington D.C.

FAKUNDINY, ROBERT H. (1989): New York State Landslide Inventory Map. New York State Mus. Science Service Chart Ser., Albany, New York (New York State Museum and Science Service).

FLAWN, PETER T. (1970): Environmental Geology: Conservation, Land-use Planning and Resource Management. 313 pp. New York (Harper and Row).

FREEZE, R.A. & CHERRY, J.A. (1989): What has gone wrong? Ground Water, 27: 458-466, Dublin, Ohio.

GUSTAVSON, T.C. et al. (1983): Geology and Geohydrology of the Palo Duro Basin, Texas Panhandle, A Report on the Progress of Nuclear Waste Isolation Feasibility Studies (1982). Bureau of Economic Geology geological Circ. GC 83-4: 1-156, Austin, Texas.

HORWITZ, ELAINE L. (1978): Our Nation's Wetlands. 70 pp. Washington D.C., (U.S: Govt. Printing Office).

KELLER, EDWARD A. (1982): Environmental Geology. 3d ed., 526 pp. Columbus, Ohio, (Charles E. Merill Publishing Co.).

KRIMSKY, S. & PLOUGH, A. (1988): Environmental Hazards: Communicating Risks as a Social Process. 333 pp. Dover, Massachusetts, (Auburn House).

LAIRD, RAYMOND T. et al. (1979). Quantitative Land-Capability Analysis: U.S. geol. Surv. prof. Paper 945, 1-115, Washington D.C.

[LIPMAN, P.W. & MULLINEAUX, D.R.] (1981): The 1980 Eruptions of Mount St. Helens, Washington. U.S. geological Survey prof. Paper **1250**: 1-844, Washington D.C.

[LOUISIANA GEOLOGICAL SURVEY] (1989): Publications of the Louisiana Geological Survey 1989. Baton Rouge, Louisiana, (Geol. Survey).

LUNDGREN, LAWRENCE (1986): Environmental Geology. 576 pp. Englewood Cliffs, New Jersey (Prentice-Hall).

[MAINE GEOLOGICAL SURVEY] (1989): Publications. Augusta, Maine (Geological Survey).

NILSEN, T.H. & BRABB, E.E. (1972): Preliminary Photointerpretation and Damage Maps of Landslide and other Surficial Deposits in Northeastern San Jose, Santa Clara County, California. U.S. geol. Surv. misc. Field Studies Map **MF-361**, Washington, D.C.

NILSEN, T.H., WRIGHT, R.H., VLASTIC, T.C. & SPANGLE, W.E. (1979): Relative Slope Stability and Land Use Planning in the San Francisco Bay Region, California. U.S. geol. Surv. prof. Paper **944**: 1-96, Washington, D.C.

[OFFICE OF COASTAL ZONE MANAGEMENT] (1976): Natural Hazard Management in Coastal Areas. Washington, D.C. (U.S. Dept. Commerce).

OLSEN, B.M., MOHRING, E.H. & BLOOMGREN, P.A. (1987): Using Ground Water Data for Water Planning. Minnesota geol. Surv. Educational Series - **8**: 1-24, St. Paul, Minnesota.

PETERSON, DONALD W. (1988): Volcanic hazards and public response. Jour. geophys. Res. **93**, B5: 4161-4170, Baltimore.

PRUDIC, DAVID E. (1986): Ground-Water Hydrology and Subsurface Migration of Radionuclides at a Commercial Radioactive-Waste Burial Site, West Valley, Cattaraugus County, New York. U.S: geol. Surv. prof. Paper **1325**: 1-83, Washington, D.C.

SAARINEN, THOMAS F. & SELL, J.L. (1985): Warning and Response to the Mount St, Helens Eruption. 240 pp. Albany, New York (State University Press).

SHABECOFF, PHILIP (1988): Major radon peril is declared by U.S. in call for tests. p. Al, A20. New York, (New York Times, 13 September 1988).

[SOCOLOW, ARTHUR A.] (1988): The State Geological Surveys -- A History: American Assoc. of State Geologists. 499 pp. Tuscaloosa, Alabama (Geol. Gurv.).

[U.S. ARMY CORPS OF ENGINEERS] (1971): National Shoreline Study. Washington, D.C. (U.S. Army Corps of Egineers).

[U.S. GEOLOGICAL SURVEY] (1984): National Water Summary 1983 --- Hydrologic Events and Issues. U.S. geol. Surv. Water Supply Pap. **2250**: 1-243, Washington, D.C.

[U.S. GEOLOGICAL SURVEY] (1987): United States Geological Survey Yearbook Fiscal Year 1987. 174 pp. Washington, D.C.

WEDDLE, T.K. et al. (1988): Hydrogeology and Water Quality of significant Sand and Gravel Aquifers in Parts of Hancock, Penobscot, and Washington Counties, Maine. August, Maine. Maine geol. Surv. Open-File Rept. **OF88-7a**.

WHITE, W.A. et al. (1985): Submerged Lands of Texas, Galveston-Houston Area: Sediments, Geochemistry, Benthic Macroinvertebrates, and Associated Wetlands. 145 pp. Austin, Texas (Bureau of Economic Geology).

WOOD, WARREN W. (1989): Birdsall lecture report 1989. Ground Water, **27**: 729-730, Baltimore.

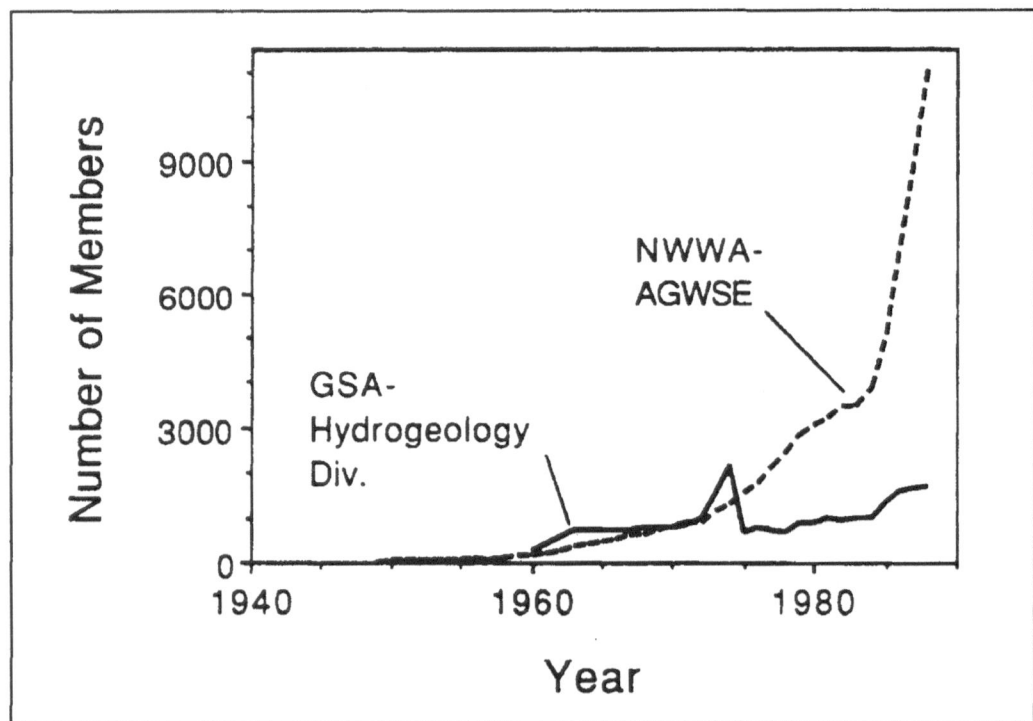

Figure 1:   Growth in membership of the Assoc. of Ground-Water Scientists and Engineers (AGWSE) and Hydrogeology Division of the Geol. Society of America

(courtesy of A. DUTTON, Texas Bureau of Economic Geology).

Standards in Environmental Geology

ERIK STENESTAD [*]

Abstract

Environmental problems are worldwide and increasing and extensive programmes related to environmental geology have been launched all over the world. Data and experience from these programmes and earlier work held by geoscientific institutions should be integrated in environmental planning. A limiting factor may be the lack of common standards. Standards do exist but they are in general neither common nor are they part of the official systems of standards. The paper aims to give an idea of what already exists, what seems to be missing and how common standards might be established.

Common geoscientific standards should respect existing local standards and they should be defined according to the needs of users and data producers. They should be minimum standards necessary for the exchange and use of data and they should consider all links of the data chain from sampling and analysis to storage, transfer and presentation of data. Within the next few years most environmental geology databanks will probably become part of major data networks. This will involve some basic requirements to the data and their organization. It will be necessary to establish Indexing systems and EDP-based Catalogues for data and maps, Thesaurus defining data and types of data, standard legends for graphic representation a.o.. Types of information in data sets and thematic maps

[*] Author's address: Dr. E. STENESTAD, Miljøministeriet, Danmarks Geologiske Undersøgelse, Thoravej 8, DK-2400 N.V. København.

necessary for different issues may be defined. Standard Environmental Geology Maps and Map-Sets for certain common issues may be developed, also. This might support the development of thematic maps on geological risks, environmental impact assessment and maps combining geology and economy. The work on standards should include their promotion, also, because standardization is meaningless if the standards are not generally accepted and used.

Introduction

The increasing and global environmental problems have created an urgent need for the development of practical solutions. The BRUNDTLAND Commission Report (WORLD COMMISSION ON ENVIRONMENT AND DEVELOPMENT 1987) made its point very clear and the public attention was focused on the threats to the environment. Quite a few national scientific and monitoring environmental programmes were launched before the appearance of the BRUNDTLAND Report but in general the efforts were increased afterwards.

An important initiative is the World Bank strategy which adds an environmental aspect to the programmes.

The European Communities, early in 1989, started the planning of an European Environmental Agency and data network to be launched within the year. This agency will be an important factor in the management of the environment.

The Summit in Paris in July 1989 strongly supported the environmental thoughts of the BRUNDTLAND Commission.

The economic declaration presents environmental problems as one of the three main challenges of the world's economic situation: All countries are urged to give further impetus to scientific research on environmental issues, to develop necessary technologies and to make clear evaluations of the economic costs and benefits of environmental policies. All countries should combine their efforts in order to improve observation and monitoring on a global scale.

Directly mentioned in the economic declaration is water pollution, soil erosion and desertification, all issues which are strongly related to environmental geology. Especially interesting is the statement that *"Clear assessments of the costs, benefits and resource implications of environmental protection should help governments to take the necessary decisions on the mix of price signals (e.g. taxes or expenditures) and regulatory actions, reflecting where possible the full value of natural resources."* (SUMMIT OF THE ARCH 1989).

On the basis of the above indications it can be concluded that besides scientific research, environmental geologists should develop tools for the solution of environmental problems. Some of the tools would be monitoring, networking and cost benefit analyses of environmental solutions. This stresses the need for common standards which can assure that relevant high quality data are at hand and which can facilitate the exchange of data and knowledge.

Extensive and useful information and large collections of environmental geology data established during the last one or two centuries represent an investment of money and scientific work of gigantic dimensions. The world community is in a very lucky position to possess all these data, of which many could never be produced again. The effort and expenditure to be invested in standardization, in order to make the existing data available for international use, are small in comparison to the costs of producing new data.

Standards, once established will further the exchange of information and at the same time make it possible that new monitoring data and results are from the beginning comparable, interchangeable and ready to meet future new purposes and needs.

DEFINITIONS

Standards

Traditionally, standards refer to

1.) technical standards and specifications e.g. for machinery, products, methods, procedures etc. and

2.) to quality of performance, e.g. "up to normal standards" or "high standards" not specifying the exact meaning of the statement.

The topic of this paper is limited to technical standards since these can also be used as tools for taking care of quality aspects. Important quality characteristics of environmental geology data sets would be that they are necessary and sufficient to cover the subject, that they are clearly defined, that they can be understood and related to other relevant data and that they can be used for future purposes. In short, data should be comprehensive, clearly defined and compatible. To further the exchange of data it is essential that all links of the "data chain", from sampling and analysis to storage, transfer and presentation conform to agreed minimum standards (fig. 1).

The Environmental Geology Concept

Environmental Geology is practised by combined use of geological, physical, chemical and other basic scientific disciplines. Several definitions of the concept of Environmental Geology have been offered, based on the application of geological knowledge to the protection of human health and safety, use of natural resources, environmental planning and management of land-use (see e.g. MONRO & HULL 1986).

A practical definition of the Environmental Geology Concept might be derived from the list of topics dealt with by geological surveys, e.g. environmental issues of the U.S. Geological Survey (PECK 1987) and a tentative list of environmental issues of some European Geological Surveys (Table 1).

Needs for Common Standards in Environmental Geology

On the basis of the statements in the introduction it can be concluded that the many initiatives stress the need for work on standardization to be accelerated. Common minimum standards for topics related to environmental geology are necessary to meet user needs for many purposes, and must be developed in the coming years at both regional and international levels. The requirements of the users and the producers deserve separate consideration.

The users

The need for common standards in environmental geology relates to a diverse group of users which includes scientists, technicians, planners, administrators and politicians. A very important group of users of Environmental Geology data and maps would be the agencies and international organizations establishing national or regional data networks for monitoring the environment. Such networks are at present being established in many countries and regions -- e.g. in Denmark and the Nordic Countries -- based on geological or other data which were originally acquired for other purposes, or on data obtained from special test sites.

Another example would be the European Environment Agency and the European network for monitoring, evaluation and control of environmental data (EEC COMMIS-

SION 1989 and MINISTRY OF THE ENVIRONMENT DENMARK 1989). The agency will continue the EEC CORINE program on monitoring of the environment. The system will be based on existing national and international networks, institutions and organizations. the agency will work for better comparability of environmental data and should develop certain standards for monitoring methods, monitoring equipment, data format and others. It is suggested that the environmental geology databases held by geological institutions will be made part of the system.

Users want easy accessible and clearly presented information they can rely on

For the benefit of the users it has been recommended that geoscientific base maps or other base maps be translated into derived thematic maps (e.g. LÜTTIG 1987). Danish experiences support this approach and the national mapping of Danish raw materials is based on this principle (STENESTAD 1976). In the discussion of the scientific aspects of simplifying geological information and maps it is important to note that all user groups want quality. Information in databases and maps is used as a basis for decisions and should first of all be reliable. The scientific quality of the data or map product must be up to the highest possible standards. The "extra" dimension which the environmental geologists should add to their work is the presentation -- especially in maps -- of the scientific results in a way that helps the non-geologist user to understand better the meaning of the information and the limits for its use, which after all is what the geologist wants, also.

As for the new customer, the European Environment Agency, it is suggested that to meet the needs of the system standardization will have to be accelerated on all levels of the "data chain" and that it will include environmental geology data, methods of analysis and formats for the exchange of data. This is crucial if the overall objective is to be accomplished, which is to bring together reliable and objective data to the benefit of the environment.

In this context it is important to stress that for the benefit of future work, environmental geology experience and knowledge must be involved in the development of such standards.

Producer needs

Data producers want their results to be used: first of all for the purpose for which they are produced but also for future purposes. Data producers and data holders who want their databanks to be part of a network would like to sample, organize and maintain the data in such a way that data can be accepted for use within the network without further treatment.

The importance of standardization of sampling and measurement is illustrated by e.g. NORDKALOTT (KAUTSKY 1986) in which project the compatibility of airborne gravimetry was supported by common standards of measurements.

Since regional and global environmental problems make international cooperation on data essential it is suggested that most environmental geology databanks will become part of major networks within the next few years. To access the large amount of data which is collected and used for many different purposes, it will be necessary to agree on common definitions and standards of documentation. Only by the adoption of such a system can the data from individual datasets be integrated within local facilities many of which will employ a Geographic Information System (GIS) approach. To make such systems work, a number of basic requirements on the data and their organization will have to be met. An important part in the establishing of a GIS is the documentation which will include definitions of the data types and data in the files (a thesaurus) and standard legends for graphic representations such as maps, diagrams and sections. It will also include catalogues of maps and libraries of data bases, which again makes indexing a must. Since such systems already exist nationally and at institutional level, a major task will be to find means of having these systems work together. In this work it might be possible to benefit from experiences made by major bibliographic databases such as GEOBASE, GEO-ARCHIVE, GEOREF, GEOS, a.o.

A basic demand of a GIS is the geographic reference grid. In Denmark and elsewhere geological surveys traditionally have adopted the national ordnance grid. These should all be convertible to UTM or geographical coordinate (latitude and longitude).

Last but not least the producers needs would be to know the needs of the users, realizing that a producer normally is the main user, also. No doubt, every possible effort should and will be made to involve the users in the process of defining the goals i.e. the output and facilities of an environmental geology GIS. It should be remembered, however, that present users' needs can not account for future needs and that the environmental geologists will first of all be responsible for defining the goals to be set.

Standards and Guidelines

Many standards already exist within environmental geology. The problem is that they are in general neither common nor comprehensive. In the process of developing a system of common standards should be reviewed and discussed with the purpose of using the best of what exists instead of starting from scratch.

International Standards

ISO standards

ISO standards are formally agreed standards. They are related to products and techniques, and they are an important tool for industry and trade. They are established by comprehensive and time consume international cooperation coordinated by the ISO Central Secretariat in Geneva. Technical committees develop draft standards and amendments to existing standards. Existing standards are listed in the ISO Catalogue and standards in preparation are published in a yearly updated list. 34 handbooks bring the full text of the ISO standards within selected fields.

None of the standards relate directly to geology or to environmental geology, but some of the standards may be of interest for environmental geologists e.g. standards on graphic symbols, methods of analyses etc. A list of topics covered by ISO standards is shown in Table 2: Field identification numbers for ISO standards.

Some thematic groups within the product oriented fields are listed in Table 3. As may be seen from the ISO Catalogues, no existing or planned international standards seem at present to contribute significantly to environmental geology standards. Nevertheless, the ISO system might be a useful supplement to -- or even a part of a future environmental geology system of standards. This would especially apply for technical standards at "data chain" level one through four (fig. 1). In a recent guideline for authors the Western European Geological Surveys adopt the use of ISO 719-1:1974 through ISO 710-6:1984 on graphical symbols for use on detailed maps, plans and geological cross-sections.

UNESCO Guidelines

Very few standards related to environmental geology can be regarded as international standards, e.g. the International Legend for Hydrogeological Maps (STRUCKMEIER et al. 1983), published by UNESCO. This standard is in a class of its own since it is not part of the ISO standard system. It is suggested that the spreading of such standards is to an even higher degree that the ISO standards dependent on the acceptance by a majority of international users and national governments.

Scientific Standards

Unofficial "de facto" international standards have been established through international scientific cooperation. Some of these refer to the very basis of the "data chain", level one, primary data, e.g. the Code of Zoological Nomenclature and the Code of Botanical Nomenclature.

To some extent publications like Treatise on Invertebrate Palaeontology and Catalogue of Foraminifera set international standards, also. Guidelines for I.G.C.P. map colours and symbols also tend to be standards. International periodicals like Environmental Geology and Water Sciences, Journal of Palaeontology, Journal of Sedimentology, Journal of Geochemistry etc. have much the same function of setting professional standards for quality and output by means of editorial guidelines.

Professional Standards

Initiatives which may lead to the establishment of international standards in the field of environmental geology would be the activities of the international geological organisations such as the International Union of Geological Sciences, I.U.G.S.. During the International Geological Congress in Washington in July 1989 an ad hod committee under the I.U.G.S. was established with the purpose of preparing a Commission on Environmental Geology.

Another example of standard setting activities of professional organizations would be the guideline books made for the members practical use. Such guidelines may bridge the gap between the basic scientific standards and practice. As an example the Geological Society publication on Aggregates (COLLIS & FOX 1985) might be mentioned. One of the objectives of the book is *"to identify relevant standards and other specifications and codes of practice, to summarize and discuss some of the currently accepted limits"* and *"to explore some of the problems commonly encountered in applying specifications and codes."* The book guides on field investigations of deposits, including mapping, and on classification, description and reporting. The book also has a comprehensive chapter on sampling and testing and a glossary.

A similar book on engineering geology (a guideline for the description of geological samples to be used for engineering purposes) has recently been published in Denmark (LARSEN et al. 1988), and it is strongly believed that guideline books of this kind can be found everywhere. It is further assumed that they are generally accepted and commonly used and because of that would be an important factor in standarizing. Depending on how this factor is handled it can be a problem or a tool in standardization.

Regional Standards

European Community Initiatives

The report on Groundwater Resources of the European Community (FRIED 1982) is a good example of how this kind of information can be presented on maps. In combination with the UNESCO guideline on legends for hydrogeological maps a.o., the EEC report may be a good starting point for the standardization of such maps.

Nordic Standards

A Nordic example of an attempt of establishing regional standards would be the book on standards for characterization of environmental data published by the Nordic Council of Ministers (MUNTHE-KAAS et al. 1981). These standards cover only parts of the topics (e.g. only part of Quaternary geology). They are at present optional and are not commonly used. In connection with a planned Environmental Monitoring System for the Nordic countries, these standards or a standard system based on the principles presented on the book might be considered. The experiences of the working group reported in the book may also be of interest for other working parties on environmental standards.

Western European Geological Surveys Initiatives

The Western European Geological Survey Working Club of Directors (WEGS) has some years ago established a Standing Group on Geological Information related to the environment (WEGS-ENVI Group) and a Computer Advisory Group. These groups have as one of their tasks to further the exchange of information between the geological surveys. Two lines of action are followed. One is the establishment of a common EDP-network, which at present is being elaborated based on existing VAX mainframes at some of the geological surveys. The other line is the establishment of common minimum standards for data and maps. These standards are also in preparation. Preliminary experiences from the work are mentioned in the paragraphes below.

National Standards

Official Technical Standards a.o.

Examples of national standards are the German DIN standards, the Danish DS standards a.o. None of the Danish DS standards are directly related to geology or environmental geology and it is suggested that this is probably the case for most of the national standards. Like the ISO standards they are first of all related to industry and trade and not to science and administration. The standards may to some extent relate to level I of the "data chain" (fig. 1), the primary data level, i.e. techniques and equipment for analysis and measurement and standards for products.

The situation will presumably be somewhat different in the future because of the EEC procedure for the establishment of standards. Safety regulations a.o. will be given in directives in a rather general way while technical standards will be provided by CEN/CENELEC. This means that the central position of the national organizations responsible for technical standards will be strengthened and that environmental geology standards in many cases should be coordinated with the national/EEC standards system.

Institutional Standards

Some institutional standards present interesting solutions to common problems related to standards within environmental geology and tend to be adopted by other institutions as good professional standards. As an example the "Guide to Authors -- A Guide for Preparation of Geological Maps and Reports" (BLACKADAR et al. 1980) seems to have much of this quality. Much inspiration on thematic maps on environmental geology can be drawn from the well known map series of the individual geological institutes. Two map catalogues are listed in the references in which the thematic maps demonstrate how much local users needs and local geological conditions are reflected in institutional thematic map standards.

At an international level hardly none of the local standards can be adopted directly as comprehensive standards for international use. They have somehow to be combined.

On the establishment of common standards

Scope, elements, fields and level of standardization

Scope: Standards should be a help in scientific and practical work and not a hurdle to be overcome. They should further the quality and compatibility of data sets and basic data and strengthen the possibility of using the data for future purposes as yet not known.

While standards for data are in principle producer defined, standards as applied to environmental geology maps should in principle be defined by the users needs. Because of the diverse group of users and the wide range of applications, the needs of the different user groups should be carefully analysed and taken into account when standards are developed.

In the process of establishing a coherent system of environmental geology standards, the starting point will have to be the present situation and the future needs of the data producers and the different user groups. This will lead to an idea of the ideal way to meet these needs. But the development of an ideal system may not be a realistic goal to set. In defining a realistic goal the impediments to be overcome should be discussed.

One of the hurdles would be the fact that all existing database systems already do have standards of their own and that it will be prohibitively expensive to make major changes in all these systems. This means that common standards should be minimum standards and that they should be based on the existing systems so far as possible. This seems to be the only realistic approach and it has, as indicated above, been adopted by the European Commission as a basic principle for establishing the European Network for Environmental Monitoring and Control.

Elements and Fields of Standardization

Elements: It is necessary to establish minimum standards for all elements in acquiring, handling and processing of environmental data, and rules for information about local standards of common standards are not applied. This will a.o. include:

1.) Definition of the scientific content of the elements or reference to such definitions.
2.) Sampling, including sampling equipment and other equipment, sampling methods including sampling frequencies recommended for certain situations, and methods of conservation and transport of samples to laboratories.
3.) Analytical methods.
4.) Formats for data sets or types of data to be incorporated in data sets and their order in the sets.
5.) Exchange formats for data.
6.) Presentation of data on environmental maps. Certain standard map types should be defined which will cover most frequent needs. The standards will include the types of data which should be presented (necessary for the topic of the map) and the way they are presented.

Fields of standardization: As for the fields of environmental geology, the Western European Geological Surveys are working within the following nine "main fields":
(1) geology, (2) geomorphology, (3) engineering geology, (4) mining, (5) hydrogeology and hydrology, (6) mineral resources, (7) seismicity and volcanism, (8) soils and (9) geochemistry.

These main fields are being used in the discussion of the specific requirements for data and maps. In a future international system of standards for environmental geology other thematic field groupings may be applied.

Levels of standardization: Standards should be established at a minimum level which allows for flexibility and local needs. They should respect and incorporate existing standards as far as possible and they should not prohibit the necessary development of appropriate new standards. In some cases standards for the documentation on the way data are produced, processed and stored may be sufficient.

Requirements for data and data exchange

Data: Data and data sets should be reliable, relevant and comprehensive. Some types of data such as information about locality, ownership or reference numbers are the same for all topics while other data are specific for different geological fields. Besides this grouping in general or specific information, data may be classified as administrative, technical or scientific.

1.)   Administrative data would be e.g. identity of the holder of data, reference numbers, indexes, location of sampling sites, sources of data, dates of acquisition / approval / publication / copyright / last revision, etc.

2.)   Technical data should indicate methods of sampling, measurement and analyzing, sampling frequency, reliability of data, etc.

3.)   Scientific data is the record of observations and measured parameters.

Data exchange: Data producers have different data formats and consequently exchange of data calls for the development and adoption of standardized data exchange formats. As an example, the development of the nation-wide monitoring sytsem for groundwater in Denmark, based on data from 14 countries, has only been made possible through the adoption of a common data exchange format (STANDAT).

As already suggested, the European Environmental Agency and monitoring network system will probably make use of environmental geology data held by geological institutions. It can be assumed that only special types of conclusive data will be relevant for the system. Such data can be obtained from the databases by means of rather simple "search and conversion programmes", which can generate working files of the data in question from the databases and convert them into a format the system can accept. This can be easily done provided the requirements and standards of the system can be met by the individual data producer.

The establishment of an EEC network on environmental data is indeed a very important issue but it should be emphasized that regional or international networks of this kind will not solve all standardization problems related to environmental geology. At a technical level it is important that primary data and not only conclusive data can be exchanged. Thus, it is recommended that partners in environmental geology data networks should also agree on common formats for data exchange of primary data. In this context database descriptions on details of recorded information including definition of data fields is a useful tool.

Requirements for Environmental Geology Maps (EGM)

General requirements:

1.) Lay-out: Geological maps seem in general to have more or less the same lay-out which includes the map itself, the legend, possible accompanying profiles and sections, and identification, indexing and other types of information at the top and bottom margins. A generalized lay-out of a common type map sheet is shown in figure 2. The text below refers to this general lay-out.

2.) The map: maps should in general have a geographical grid (or ticks) for the definition of the geographic location of the map. Data are shown as points, lines or polygones or as symbols, codes or digits which are copied exactly and explained in the legend. Scales should be appropriate for the purpose and for the density of information. Some map producers have fixed scales for their map series, while others do not. It is recommendable that possible standard environmental geology maps (Standard EGM) are published in agreed scales and projections.

3.) Sections and profiles on maps: Stratigraphy, structural conditions, regional location or other topics are often illustrated by sections, profiles, location maps or other additional graphic representations along with the map proper. To avoid confusion the signatures and colours of the profiles and sections on a map sheet should in principle be the same as in the map.

4.) Legends, contents, signatures and colours: The legend should inform about the topic, signatures, scale and projection of the map a.o. Graphical symbols for use on detailed maps, plans and geological cross sections should follow ISO 710-1:1974 through ISO 710-6:1984.

So far possible signatures and colours should be used in accordance with normal professional practice. This would imply that e.g. in a map, profile or section showing lithology brickwork signature should be used for nothing but limestone. In a stratigraphic column Cretaceous formational units should preferably be green etc.

It is recommended that the work on standards for signatures and colours is continued at an international level and taht a list or catalogue of minimum standards be published. The use of common minimum standards for scales, projections, signatures and colours would further the readibility of the maps and facilitate the combined use of maps from different map producers e.g. across national boundaries.

5.) Top and bottom margin of map sheets: The margins of the map sheets are commonly used for administrative information such as topic and identification of the map, indexing, location, author, publisher, printer, copyright, year of publication, advice on use and non-use, etc. Minimum standards for the type of information offered at map sheet margins might add to the value of the map as a document.

Specific requirements for different EGM types:

1.) The below classification of environmental geology map types is based mainly on MONRO & HULL (1986) and LÜTTIG (1987). The comments on the maps are based on the ongoing work of the Western European Geological Surveys.

Environmental Geology Maps:

a)   Element maps (derived from geoscientific and other base maps)

b)   Combined element maps (derived from element maps)

c)   Environmental potential maps (combining EGM, land-use and other relevant information)

d)   Standard environmental geology map sets (EGM sets).

2.)   Minimum standards for environmental geology maps (EGM):

The Western European Geological Surveys have for some years been working on the basic principles and minimum criteria for environmental geology maps. Criteria for maps within the above nine "main fields" of topics have so far been maximum criteria rather than minimum criteria and they are not yet ready for publication. The basic idea is that, for the benefit of the user, a thematic map should present an appropriate minimum of information necessary for the defined purpose. An example from a draft on groundwater data and maps is shown in table 4.

3.)   Standard environmental geology maps:

Within the different categories and types of maps some maps are more frequently used than others. Very often they have practically the same content and might be developed to EGM standard maps by standardizing the thematic and graphic content. The "translation" of scientific maps into thematic maps is not very easy and it is suggested that there would be some merit in developing some standard type maps. It is further suggested that a list of EGM standards is established and published, indicating the minimum standards for their preparation.

4.)   Standard environmental geology map sets (EGM sets):

To make available the necessary basis for the work on an environmental geology related problem, it will very often be necessary to make use of a set of maps which together yield the necessary information.

A possible next step to be taken by geological surveys and other geoscientific institutions might be to agree on a series of standard map sets for different purposes. Most geological surveys have developed standard map sets for e.g. water supply and other issues. In Denmark the map set on ground water consists of three maps on:

a) Geological basic data map (1:50 000 and 1:25 000)

b) Map of the piezometric surface and transmissivity of the groundwater

c) Hydrochemical basic data map.

An example of a large set of maps would be the one mentioned as sources for the Geoscientific Map of the Natural Environment's Potential (GMNEP) (LÜTTIG 1987). The set of maps incorporates non-geoscientific data and maps produced and held by other organizations. This calls for an active inter-disciplinary and inter-institutional cooperation which may be very rewarding for geoscientists, by making geological knowledge and data known to and used by planners and decision-makers, including those responsible for the overall land-use. Many examples have shown that this group of potential users are not always interested in traditional geological maps and in some cases prefer a more conclusive type of map as a basis for their decisions. Standard EGM sets might change this situation.

It is suggested that standard EGM sets, recommendable for certain common environmental problems, are described and published as a professional guideline for geoscientists and planners working with these kind of problems.

Conclusion

Main existing research lines:

One of the main lines is the development of minimum criteria for all links in the data chain within environmental geology, with the aim of facilitating data exchange and improving data quality to the benefit of the environment. Another main line is the development of standard EGM and EGM sets on water, waste and resources and other major common environmental issues. Promising lines of work are a.o. the systematic introduction of economic aspects in relation to environmental geology and economy. The Economy Summit 1989 pointed to the need for imposing cost benefit aspects on environmental investments. This point of view has also been advocated by many governments under headings like: "More environment for less money". So, it is assumed that more emphasis should be put on risk maps and environmental impact assessment (EIA) in the future.

Topics where efforts should be concentrated would include the development of necessary thesaurus', catalogues of existing standards, indexing systems, codes and standards. The work on minimum standards for all links in the data chain and for EGM and EGM sets should be accelerated. A catalogue of environmental geology standards should be established and widely distributed.

Procedures and activities:

To establish professional agreements on types of environmental maps, scales, procedures, criteria and symbols for their preparation it is crucial that the basic principles and possible standards for data, data exchange and map types are accepted by all major groups of producers of data and maps. To have the professional standards adopted by the users it is crucial to obtain consensus on the scope of the work and of the procedures and activities. So, along with the work of the organizations and working groups which are responsible for a technical and scientific sustainable result, there should be close links to the users. This can be achieved by consensus conferences, progress reports and other well known means. Acceptance is crucial because it is difficult, time consuming and expensive to develop and introduce new standards. Provided the right procedure and strategies are followed, general acceptance of common standards should after all not be too hard to achieve. Because of the costs and difficulties related to the development of standards, there is a strong incentive for most data and map producers to adopt existing standards instead of establishing standards of their own.

One of the ways to obtain acceptance is to avoid the development of standards which are unnecessary or prohibitive for the daily work: Standards should be minimum standards. They are and should be nothing but a means of communication - words and structures of a common language.

While it can be argued that major groups of users and producers, including databank holders, will be in favour of common standards at a defined minimum level, the individual data producer, i.e. the geoscientists, may be harder to persuade. Traditionally many geoscientists are rather reluctant with standards which are not directly necessary for their own scientific work. This means that an important part of standardization work would be to promote the standards, so that the scientists do not use home made systems where generally accepted standards are in fact at hand. This is because it can be very difficult or impossible to translate such private systems to current systems, meaning that important information may be lost. And this is indeed not just an "in-house" problem of the geoscientific institutions. So, it is to be hoped that geoscientists who might disapprove with common standards will realize that such standards are the means of better communication. *"Science is organized knowledge"* as written on the walls of Library of Congress in Washington.

Acknowledgements

The members of the Western European Geological Surveys Standing Group on Geological information related to the Environment, the WEGS-ENVI Group, is cordinally thanked for fruitful discussions on standardizing of environmental geology data and maps and problems related to exchange of information.

Special thanks to Dr. BRIAN KELK of the British Geological Survey and to Dr. P. GRAVESEN and Dr. K. BINZER of the Geological Survey of Denmark, for reading the manuscript and for valuable comments and advice.

References

ANDERSEN, S., JACOBSEN, E.M., PRINTZLAU, I. & THIIM, L.N. (1985): Raw material mapping and planning in Denmark with examples from the Greater Copenhagen Region.- Striae **22**: 61-82, Uppsala.

BLACKADAR, R.G., DUMYCH, H. & GRIFFIN, P.J. (1980): Guide to authors. -- A guide for the preparation of geological maps and reports.- Geol. Surv. Canada miscell. Rep. **29**: 1-66, Ottawa.

[COLLIS, L. & FOX, R.A.] (1985): Aggregates: sand, gravel and crushed rock aggregates for construction purposes.- Geol. Soc. engin. Geol. spec. Publ. **1**: 1-220, London (printed in Belfast).

[COMMISSION OF THE EUROPEAN COMMUNITIES] (1989): Proposal for a Council Regulation (EEC) on the establishment of the European Environment agency and the European Monitoring and Information Network.- Com. 89, 303: 13 pp., Brussels.

[DANMARKS GEOLOGISKE UNDERSØGELSKE] (1988): Map catalogue.- 45 pp., København (DGU).

[FRIED, J.J.] (Coordinator of the Study) (1982): Groundwater resources of the European Community. - Synth. Rep. Comm. eur. Comm. Dir. gen. Envir. etc. Eur. **7940** EN: 1-75, Brussels.

[RIJKS GEOLOGISCHE DIENST et al.] (1986): Subsoil uncovered.- 36 pp., Haarlem (R.G.D.)

GRAVESEN, P. & FREDERICA, J. (1984): ZEUS-geodatabase system. Well data archive. Data description, code system and satellite bases.- Geol. Surv. Denm. (D) **3**: 1-259, København.

[INTERNATIONAL STANDARD ORGANIZATION] (1981): Statistical methods. ISO. Handb. **3**, 2nd Ed.: 456 pp., Geneva.

[INTERNATIONAL STANDARD ORGANIZATION] (1982a): Units of measurement. ISO. Handb. **2**, 2nd Ed.: 264 pp., Geneva.

[INTERNATIONAL STANDARD ORGANIZATION] (1982b): Technical drawings. ISO. Handb. **12**: 350 pp., Geneva.

[INTERNATIONAL STANDARD ORGANIZATION] (1983): Measurement of liquid flow in open channels. ISO. Handb. **16**: 526 pp., Geneva.

[INTERNATIONAL STANDARD ORGANIZATION] (1988): Documentation and information. ISO. Handb. **1**, 3rd Ed.: 1012 pp., Geneva.

[INTERNATIONAL STANDARD ORGANIZATION] (1989b): Technical programme 1989.- 204 pp., Geneva.

[JOHANSSON, I.] (1984): Nordic glossary of hydrology. English -- Danish -- Finnish -- Icelandic -- Norwegian -- Swedish with definition in English.- 224 pp., Stockholm (Almquist & Wiksell).

KAUTSKY, G. (1986): The Nordkalott Project. (reprint from Terra cogn. **6** (3): 564-568). In: Maps of Northern Fennoscandio.- Geol. Surv. Finl. Guide **24**: 28 pp., Espoo, Helsinki.

KORHONEN, J. (1989): Maps of Northern Fennoscandia.- Geol. Surv. Finl. Guide **24**: 28 pp., Espoo, Helsinki.

LARSEN, G. et al. (1988): Vejlending i ingeniørgeologisk prøvebeskrivelse.- Dan. geotechn. Soc. Bull. **1988**, 1: 1-144, København.

LÜTTIG, G. (1987): Large scale maps for detailed environmental planning.- Norg. geol. Unders. spec. Publ. **2**: 71-76, Trondheim.

MONRO, S.K. & HULL, J.H. (1986): Environmental geology in Great Britain. In: [BENDER, F.]: Geo-resources and environment: 107-124, Stuttgart (Schweizerbart).

[MINISTRY OF THE ENVIRONMENT; DENMARK] (1989): The European Environment Agency and the European monitoring and information network.- 48 pp., København.

MUNTHE-KAAS, H., ØSTERDAHL, L., RØNDELL, B. & BINTZER, K. (1981): Characterization of environmental data.- Handbook **1**: 123 pp., **2**: 72 pp., **3**: 105 pp., Oslo (Nordic Council of Ministers).

PECK, D. (1987): The role of the U.S. Geological Survey in meeting environmental issues.- Envir. Geol. Water Sci. **10**, 2: 63-65, New York.

STENESTAD, E. (1976): Råstofkortlaegning.- Danm. geol. Unders. (A) **1**: 1-24, København.

[STRUCKMEIER, W., MONKHOUSE, R.A., JELGERSMA, S. & GILBRICH, W.H.] (1983): International legend for hydrogeological maps.- Revised edition. 51 pp., Paris (IAH, IAHS, UNESCO).

The World economic Summit in Paris 14-16 July 1989, "Summit of the Arch" (1989): Economic declaration.- 22 pp., Paris.

[SØRENSEN; H. & NIELSEN, A.V.] (1978): Den geologisjke Kortlaegning af Danmark.- Danm. geol. Unders. (a) **2**: 1-79, København.

[WOLFF, F.C.] (1987): Geology for environmental planning.- Norg. geol. Unders. spec. Publ. **2**: 1-121, Trondheim.

[WORLD COMMISSION OF ENVIRONMENT AND DEVELOPMENT] (1987): Our common future.- 374 pp., Oxford (Univ. Press).

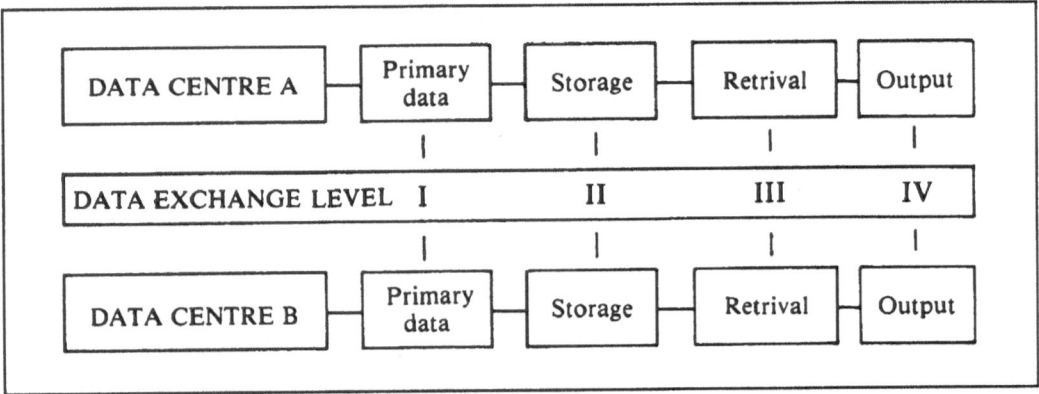

Figure 1: The "data-chain" and the exchange of data.

Information is often exchanged at Output level as maps and books, level IV in the data-chain. Common standards for this kind of output support cooperation. Even better results can be obtained if common standards are established at level I, II, and III.

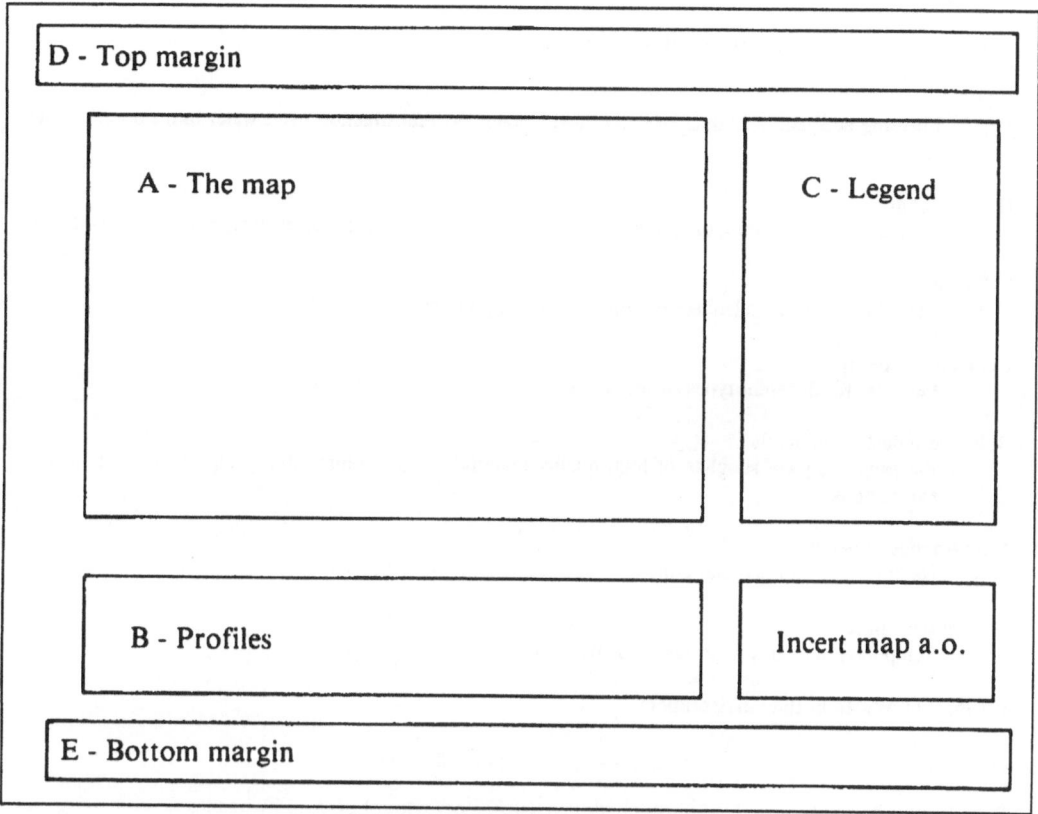

Figure 2: General lay-out of a common type map sheet.

A: The map proper with grit (or ticks) and data shown as points, lines or polygons or as symbols, codes or digits.
B: Sections or profiles, location map or other additional: graphic representations.
C: The legend giving information on topic, signatures, scale of the map a.o.
D and E: Top and bottom margin information on topic, identification, indexing, location, author(s), publisher, printer, copyright, year of publication, advice on use and non-use, etc.

Natural Resources in the Geological Environment

Water
   Ensuring adequate resources of good quality water - overabstraction - salt water intrusion in coastal areas

Hydrocarbons
   Support for the ensuring of an adequate equilibrium between resource identification and exploitation

Solid fuels
   Contributions to an adequate resource and mining policy

Geothermal energy
   Potential for different types of resources

Bulk and industrial minerals
   Ensuring adequate supplies of high quality material for the future versus impact of large surface excavations

Metalliferous minerals
   National and international policies - changing exploration strategies

Undergroud space
   Temporary storage of various substances - underground constructions

Disposal of Waste in the Environment

Surface storage of waste - Land-use versus disposal - leaching agents

Deep storage of non-radioactive industrial waste - Satisfactory control of deep hydraulics

Deep storage of radioactive waste
   High level of safety over long periods

Impact of Geological Hazards to Development

   Environmental Impact assessment EIA
   Prediction, prevention and remedial actions
   Seismic risks - Volcanic risks - Ground movements - Mudflows - Cavities - Coastal erosion - Sea bed instability - Flood hazards - Rising water table

Impact of Development on the Geological Environment

   Impact of surface mining - Compaction and subsidence - Soil contamination, erosion and restoration - Impact on hydrological cycle

The Role of Planning in Development

   Loss of resources by other developments - Land-use planning strategies - Planning the after-use

Table 1:   A tentative list of some environmental issues of European Geological Surveys - Based on WEGS ENVI Group work

005	Fundamental standards
010	Documentation and information
015	Banking and financial services
020	Metrology and measurement
025	Medical sciences
030	Environmental and safety engineering
035	Testing and analysis (in general)
040	Heat transfer engineering
045	Fluid engineering
050	Manufacturing engineering
055	Mechanical systems and compounds
060	Materials handling
065	Packaging and distribution of goods
070	Mineral extraction
075	Road vehicles
080	Shipbuilding
085	Railways
090	Aircraft and space vehicles
095	Agriculture
100	Food technology
105	Inorganic chemistry
110	Organic chemistry
115	Coal technology
120	Agricultural plant products other than foodstuffs
125	Petroleum technology
130	Glass and ceramics
135	Paint and colour industries
140	Metallurgy
145	Wood technology
150	Leather technology
155	Paper technology
160	Textile industry
165	Rubber and rubber products
170	Plastics
175	Information processing systems
180	Products of consumer interest
185	Clothing industry
190	Building and construction materials
195	Image technology

Table 2: Field identification numbers for ISO standards

Group No.	Subject
0010	Vocabulary
0020	Terminology (principles and coordination)
0030	Quantities and units
0040	Graphic symbols
0050	Technical drawings
0070	Statistical methods
0080	Quality assurance
0120	Documents in administration, commerce and industry
0200	Measurement of liquid flow in open channels
0330	Analysis of chemical products including gases
0700	Earth-moving machinery
0750	Iron ores
0760	Manganese ores
0770	Chromium ores
0780	Aluminium ores
0790	Fluorspar
0800	Mica
0810	Mining
1040	Nuclear energy
1530	Oxides
1540	Acids
1550	Hydroxides
1610	Anhydrider
1620	Alcohols
1630	Esters
1640	Aldehydes and ketones
1650	Phenols
1660	Aliphatic hydrocarbons
1670	Aromatic hydrocarbons
1680	Halyhydrocarbons
1710	Coals
1760	Crude petroleum
1770	Natural gas
1790	Liquid and liquified hydrocarbons
1810	Glass
1820	Ceramics
2240	Pesticides
2340	Character sets and information coding
2460	Graphic technology
2470	Computer graphics
2510	Building
2580	Water quality
2720	Measuring instruments
2820	Diagnostic and test equipment
2880	Concrete
2890	Gypsum
3380	Non-destructive testing
3530	Sieving
3540	Instruments for testing

Table 3: Examples of ISO Standard Groups of possible relevance for standardization within Environmental Geology

Criteria for Groundwater Data

Administrative data

Well / spring / runoff gauging station - File number - Community / region / country etc. - Drilling company - Well logging company - Owner of the well - Owners file number Name and/or registration number of the well - Map-sheet code number, printing year, etc. - X, Y coordinates for the well in the defined map - UTM-zone and coordinates - Well site elevation - Accuracy of localization - Date of completion of the well - Purpose of the well (monitoring, water-supply ..)

Technical data

Drilling method - Geophysical well logs - Diameter and total depth (TD) of borehole - Casing (material, strenght, diameter and TD) - Screen (Material, diameter, mesh/slot size) - Screened interval(s) - Gravel packing - Clay or cement plug - Duration of pumping test - Label numbers and other identification of samples - Methods of data acquisition
Field investigations: Sampling techniques including logging, exactness of observations, sampling frequency
Laboratory: Analysis techniques and equipment exactness of analysis.

Scientific data

Field observations: Rest water-level and date of observation(s) - Exactness of obervations - Flow rate of well (volume / unit time / drawdown) - Physical and chemical observations (pH, temperature, electric conductivity) - EC - Aquifer type - Hydraulic values (K, T)
Laboratory observations: Geological parameters (Lithology, facies, stratigraphy, etc.) - Geochemical parameters.

Criteria for Groundwater Maps
Tabulation based on FRIED (1982)

Aquifers: Geology and lithology - Type of aquifers - Geometry of aquifers - General information (cross sections etc.)

Groundwater Hydrology: Groundwater movement (piezometric contours, direction of groundwater flow, ..) - Particular problems - Springs - General information

Groundwater Quality: Hydrochemistry - Oxygen, temperature, pH - Pollutants (organics, heavy metals, biological activity (N, P, S), patogens ..)

Groundwater abstraction: Unit system - abstraction densities catchments - springs or groups of springs - Mine drainage - General information

Balance of resources: Unit system - Monitoring systems - Available resources for specified use - availability of resources - general information (surface waters, lakes, dams, reservoirs, urban areas, contour lines, state border lines ..)

Table 4:   Draft Minimum Standards for Environmental Geology Data and maps on Groundwater

Guidelines for the Organisation of an International Postgraduate Programme on Earth Science and Environmental Planning

ANTONIO CENDRERO *

Introduction

The many and varied problems which affect the human environment are in most cases, as it is pointed out in several of the papers included in this volume, of an inter-disciplinary nature. But not only do they cut across traditional disciplinary boundaries, they also cut across national borders, having more often than not an international dimension.

The strategies which have been proposed or implemented for the analysis and solution of those problems, both at national and international levels, do in general mention the need of an interdisciplinary approach. Nevertheless, they often fail to recognise the fact that environmental problems stem from the use of the surface of our planet by human beings and that, therefore, a proper understanding of those problems and the finding of adequate solutions for them cannot be achieved without due consideration of the constitution and dynamics of the Earth's surface (ARCHER, LÜTTIG & SNEZHKO, 1987).

In the author's opinion, this is due to two main factors. First, the number of earth scientists is in most countries relatively small as compared to other specialists. Second,

* Author's address: Prof. A. CENDRERO, DCITTYM (División de Ciencias de la Tierra), Universidad de Cantabria, Santander, Spain.

traditionally earth scientists have been involved mostly in the exploitation of mineral and energy resources and in the study of bedrock geology, devoting much less interest to surficial geology and to environmental problems. The result has been, in general, that the level of geological input into environmental programmes does not reflect the potential contribution of earth science to the solution of environmental problems. (BARAHONA et al., 1987).

With the present trends of growing social and scientific interest on environmental matters, there is an increasing need for "environmental expertise". As happens in many other fields of specialisation, a growing number of earth scientists are carrying out professional, scientific, educational or administrative tasks in relation with different environmental problems. Very often, these professionals find that their background is, on the whole, broadly satisfactory for the requirements of their jobs, but that they lack an appropriate training in certain disciplines or techniques which are relevant for an adequate grasping of the multiple aspects of environmental problems. It is also frequent to find that they lack in-depth training in some environmental geological topics.

This situation is reflected in the high number of applicants and/or participants in postgraduate courses dealing with different aspects of earth sciences and environmental analysis, planning and management. Very often, courses with an important earth science component do in fact attract as well many non-earth scientists. This is at least the author's personal experience, both at the national and international level (CENDRERO, 1989).

During the last decade or so, there has been in many countries, both industrialised and developing, a large number of short courses dealing with different environmental geological topics. However, to the author's knowledge, there are practically no comprehensive programmes aimed at providing long term training of postgraduates from different backgrounds in the field of earth science and environmental planning and management.

The replies given to evaluation questionaires in some international intesive multi-disciplinary postgraduate courses (CENDRERO, CHARLIER & SENTEN, 1989) show both a clear interest in more activities of a similar kind, and a desire to analyse with more detail and depth the problems related to each participant's field of specialisation. It seems that, without detracting from the high value of short or intensive courses, there is a major interest in longer, academic year courses, preferably leading to an academic degree.

This interest is felt not only in the industrialised countries, but even more in the developing world, whrere the implementation of land and resource use and management policies, aimed at preventing future environmental degradation, is of particular importance (ARNDT & LÜTTIG, 1987). In most developing countries there is no training scheme that can provide the new types of experts needed to meet the environmental challenge.

Main guidelines

If environmental problems are multidisciplinary and international, it is only logical to incorporate these features into any training scheme. However, there are obvious difficulties to translate this into practice. There have been many discussions about the relative merits of interdisciplinary and specialised training in the environmental sciences (STEVERS, 1989; de GROOT, 1989). The limited duration of any reasonable programme of postgraduate studies makes it impossible to cover in depth all environmentally significant topics. A possible line to follow is to train environmental generalists, with a good understanding of the different "natural" and "social" aspects of environmental problems. A second possibility is to produce environmental experts specialised in one of the traditional disciplines, according to their own background.

Both options have, of course, advantages and disadvantages, and they would provide people suitable for different types of functions. A generalist, interdisciplinary approach has the advantage of providing students with a broad understanding of the wide range of problems and factors (both physical and human) to be considered. On the other hand, the normal limitations of the human mind make in-depth analysis of the problems, and the acquisition of expertise to solve them, practically impossible.

A disciplinary approach enables students with a solid background in a given discipline to concentrate on certain matters, allowing a more profound treatment of the problems and making it possible to acquire the specialised expertise needed to solve such problems and to carry out research. The disadvantage of this approach is that the specialists thus trained might lack the general view of the complexly interwoven environmental issues.

In the author's opinion, an interdisciplinary approach is better at the introductory levels, where problem understanding is normally the main objective. At the postgraduate level, where the aim is to train professionals for the solution of problems and for carrying out reserach, a disciplinary, specialised approach seems to be more adequate. The generalist kind of training, at the postgraduate level, is probably appropriate for people who are going to fulfill managerial ot organisational tasks, especially within the public administration (BARENDESE, 1989). These persons ought to have a wide view of the various problems involved, and an ability to communicate and establish a coordination with the different specialists.

It is probably useful, when it comes to training, to make a distinction between environmental education (mostly interdisciplinary, aimed at achieving problem understanding) and environmental research (with emphasis on one discipline, for problem solving).

In the case of a postgraduate programme on earth science and environmental planning and management, it would be probably advisable to include basic courses on environmental topics normally not covered in earth science curricula, in order to provide students with a more comprehensive view of the problems. However, the emphasis should be in the teaching of subjects within the field of the earth sciences and of techniques which would enable them to analyse certain environmental problems with depth and to contribute with specific solutions to them.

When it comes to the point of internationality, it is only fitting to remember the recommendation of the BRUNDTLAND Report: "Think globally, act locally" (WORLD COMMISSION ON ENVIRONMENT AND DEVELOPMENT, 1987). Environmental experts, in any field, would benefit greatly from an exposure to the problems existing in various parts of the world, as well as to the approaches and strategies for solving them in different natural, social, cultural and economic backgrounds. This should be done, ideally, through direct contact with specialists who have professional and/or scientific experience in the study of such problems. The specific solutions to concrete, local problems and their global implications could thus be analysed and discussed.

It would be highly desirable, for a postgraduate programme intended to provide this sort of multinational approach, to be run by a multinational faculty, and to be addressed to a multinational group of students, with special attention to applicants from developing countries. The sharing of experience between students and faculty and among the students themselves would presumably be more enriching. The use of more than one language as a means of instruction would contribute to the training of specialists with a broader outlook, easier access to information sources, greater ability to communicate and more possibilities to carry out professional tasks in different cultural settings.

Approach and structure

A programme in the lines commented above should try to avoid the pitfalls of courses with a great number of subjects, often only of local significance and with limited application elsewhere. The analysis of local case studies can be extremely beneficial within an broader context, but a particular area should not be the focus of interest.

A two-block structure for the programme could be a possibility. This could consist of a core part (compulsory) and a complementary one (in which there could be a choice among several optional courses). The core part should include those subjects which are considered essential for the environmental training of graduates with an earth science background. The complementary part should be much more flexible and could focus on specific topics which might not be of general interest. For students with a limited earth science background, "bridging" courses could be offered or established as a prerequisite.

Research should be an important part of the programme, but with a level of intensity which could be different for students interested in a professional career than for those wanting to follow scientific or academic endeavours. A two-level programme (for instance, but not necessarily M.A. and Ph.D.) with the first level devoted mainly to course instruction and limited research and the second level concentrating mostly on research, might be appropriate.

The organisation of a course along the lines proposed here can best be carried out through the collaboration of several institutions (academic and non-academic, public and private) in different countries.

Organisation

The postgraduate programme should be formally established as a cooperative effort by several institutions. A programme director would be appointed, who should be assisted by a board including representatives of the institutions engaged in the programme.

A first year of study, consisting mainly of formal course instruction and only limited research, could be completed by the students in one of the institutions. This part of the progamme could be held permanently in the same place of offered placers in successive years. The courses would be taught in part by "permanent" faculty, who would remain in the institution during the whole year and supervise the research work as well, and partly by faculty from other cooperating institutions, who would teach short or specific courses. Instructions should be carried out in two languages (to be determined, among English, French and Spanish), and profficiency in both languages should be a prerequisite to be accepted in the programme. The successful completion of the first year of study could lead to the obtention of a certificate.

The second year should consist mostly of seminars and research work, and it could be completed in the appropriate departments at any of the cooperating institutions, according to the interests of the candidates and to the availability of faculty and lab facilities. This system would provide the programme with a broad research base and a variety of possible research topics, difficult to achieve in a single institution. The research would be supervised by a member of the faculty and should lead to the preparation of a dissertation. Alternatively, subject to approval by a faculty board, students could carry out their research in their institutions of origin (especially for candidates form developing countries), even if these are not formally cooperating in the programme. Successful candidates would be awarded an M.S., M. Eng. Sci. or an M.A. according to their specialization. Research of an interdisciplinary nature, including the integration of "natural" and "socio-economic" aspects should be encouraged.

Qualified students could be accepted, after approval of a thesis project by a selection committee, to pursue further work leading to a Ph.D.

It would be desirable to encourage students to undertake research and dissertation subjects which reflect their own interests and personal and professional experience, including problems in the regions of origin, so that their training could also provide some answers to such problems.

Programme contents

An initial draft of the programme, could include the following courses:

Title: Earth processes, land-use and environmental planning and management

Core

1 - Principles of planning and management, environmental protection and conservation. Assessment of economic aspects (4 credits).

2 - Ecology (5 credits).

3 - Methods and techniques of environmental impact assessment (6 credits): environmental impact, concept, types; general EIA methods, models, programme design; base-line studies; socio-economic aspects; preparation and presentation of final reports; prevention and correction of impacts.

4 - Present earth processes and natural hazard and risk assessment (5 credits): fluvial processes (erosion-sedimentation, floods); surficial processes (slopes, mass transfer); neotectonics, seismicity, volcanism; prediction, prevention and control methods.

5 - Geoscientific mapping and assessment, environmental planning and management (7 credits): descriptive parameters, measuring and representation; evaluation and integration procedures; establishment of areal management programmes.

6 - Earth resource use and exploitation, environmental effects and planning (5 credits): soils; water; low and high unitary value mineral resources; scientific, cultural and leisure natural resources; restoration of degraded lands.

7 - Biogeochemical cycles and pollution (7 credits): natural biogeochemical cycles and their modification, nutrients, trace elements and organic molecules; pollution of waters, sediments and soils; waste disposal, methods and impacts; geochemical monitoring and pollution correction.

8 - Engineering solutions to environmental problems and cost-benefit analysis of management plans (5 credits): Hazards and risks, resource exploitation, impact mitigation; comparison of alternatives.

Complementary

The following options could be offered (the courses offered could vary in different years):

1 - Computer storage, processing and display of environmental data; computer mapping; modelling

2 - Environmental legislation

3 - Population and resources

4 - Remote sensing

5 - Urban planning

6 - Economic assessment of environmental factors

7 - Techniques of visual landscape evaluation

8 - Soil capability assessment and soil conservation

9 - Geological engineering

10 - Hydrogeology

11 - Applied geomorphology

12 - Environmental planning and management in the coastal zone

13 - Environmental planning and management in mountain areas

14 - Applied "environmental science" education and communication

15 - Global change and surficial processes

16 - Specialised seminars

Students should complete a total of 60 credits, out of which a minimum of 30 from the core part (the topics to be waved would be decided after examination of the candidate's background). A credit is equivalent to 10 contact or lab hours or one day of field work. Report preparation and tests would not be counted as credit hours. The total duration of the course would be 25 weeks, with an average of 25 contact, lab or field work hours per week. The rest of time would be devoted, apart from study, to small research projects and to report and dissertation writing and preparation.

The dissertation would be presented before a jury of three faculty members. The presence of jury members from institutions not formally integrated in the programme should be encouraged. The presentation and defense of the Master's and/or Ph.D. thesis should be made in front of a five-member jury which would include external members.

Financing and sponsorship

This kind of undertaking should not be started without a reasonable guarantee of continued financial support. Some international organisations that could provide support for the programme are: UNESCO, AGID, IUGS, UNEP, FAO, OAS, CEPAL, EEC, Council of Europe. Additional financial backing could come from cooperating academic or scientific institutions (faculty, air fares, facilities, housing) and national bodies, as well as from students fees.

Final comments

Earth scientists know that the study of surficial earth's processes represents a very good basis for environmental education and training, but this is not all obvious for other professionals. It has been often mentioned that the analysis of environmental problems requires a holistic approach (SUSANNE, 1989), but no holistic view of the earth system which constitutes our environment can be obtained without analysing its behaviour in the past and the profund environmental changes it has been through, as a result of the variations in different physical and biological factors (in some cases resulting in massive extinctions of species). In recent periods of the earth's history, human influence has been added as a new factor affecting the environment and, consequently, the behaviour of the system has changed.

Recent earth history, especially the one preserved in the Quaternary sediment, fossil and geomorphological record, provides the means to learn about the response of natural systems to changes which have occured in the past (climatic changes; sea level rise; influence of vegetation destruction on erosion-sedimentation processes, on flood regimes and on geochemical cycles, etc. [TITUS, 1986]). It is through the study of these processes that meaningful predictions about some future changes can be made, in order to design strategies to cope with them. Environmental impact assessments, for instance, can be made much more precisely if empirical information about past system behaviour is available. It is, therefore, of high importance that earth science expertise (in cooperation with other disciplines) can be directed towards the solution of many contemporary environmental problems (CHAFEE, 1986), which, otherwise, will not be satisfactorily solved.

Many professionals and scientists, trained at periods when the level of environmental awareness was low, have become interested in environmental topics and re-oriented their activities, complementing their training and skills through basically individual actions. In other cases earth science has been introduced as a subject in the training of other kinds of specialists. Both kinds of training, the one directed to provide a complement to non-earth science specialists and that aimed at forming environmental earth scientists, are necessary (LÜTTIG, 1987). But a wide, adequate contribution cannot be made unless there are people of a high level of academic instruction who are specifically trained to deal with environmental problems. This training should, hopefully, avoid the main pitfalls which affect most present practioners.

Universities and other institutions devoted to the advancement of knowledge have usually played a major role in the adaption of modern societies to changing conditions, values and professional needs (UNESCO, 1977), and hopefully will continue to do so. In this particular case the cooperation of universities and societies of learning could provide the scientific background for the training of the new kind of earth science professionals necessary for the present, and even more so future, conditions. It is the authors opinion that IUGS has an essential role to play in this direction, and that the steering and/or organisation of the kind of course outlined here, with whatever improvements are considered convenient, would be an important step.

References

[ARCHER, A.A., LÜTTIG, G. & SNEZHKO, H.] (1987): Man's dependence on the Earth. - 216 p., Stuttgart, UNEP-UNESCO (E.Schweizerbart).

[ARNDT, P. & LÜTTIG, G.] (1987): Mineral resources extraction, environmental protection and land-use planning in the industrial and developing countries. - 346 p., Stuttgart (Schweizerbart).

BARAHONA, E. et al. (1987): El medio ambiente en las organizaciones internacionales. - 126 p., Madrid (MOPU).

BARENDESE, G. (1989): Postgraduate education in environmental sciences; preparing for a professional career. - In: [SUSANNE, C.]: The integration of environmental concepts in universitary teaching; UNESCO (in press).

CENDRERO, A. (1989): Environmental problems in the teaching of earth sciences at the graduate and postgraduate level. - In: [SUSANNE, C.]: The integration of environmental concepts in universitary teaching; UNESCO (in press).

CENDRERO, A., CHARLIER, R.H. & SENTEN, R. (1989): Land-use problems, planning and management in the coastal zone; analysis of an intensive course. - Int. J. environm. St. **34**: 309-313, London.

CHAFEE, J.H. (1986): Our global environment: the next challenge. - In: [TITUS, J.G.]: Effects of changes in stratospheric ozone and global climate: 59-62, Washington (US EPA-UNEP).

GROOT, T. de (1989): Teaching "environment and development" expressing a problem-oriented environmental science paradigm. - In: [SUSANNE, C.]: The integration of environmental concepts in universitary teaching; UNESCO (in press).

LÜTTIG, G. (1987): Geology versus mineral, groundwater and soils resources manage ment; approach to the public, education and training questions, outlook. - In: [ARNDT, P. & LÜTTIG, G.]: Mineral resources extraction, environmental protection and land-use planning in the industrial and developing countries: 319-331, Stuttgart (Schweizerbart).

STEVERS, R.A.M. (1989): Teaching environmental science as a compact, problem-oriented discipline. - In: [SUSANNE, C.]: The integration of environmental concepts in universitary teaching; UNESCO (in press).

[SUSANNE, C.] (1989): The integration of environmental concepts in universitary teaching; UNESCO (in press).

TITUS, J.G. (1986): Overview of the effects of changing in the atmosphere. - In: [TITUS, J.G.]: Effects of changes in stratospheric ozone and global climate: 3-19, Washington (US EPA-UNEP).

UNESCO (1977): The contribution of higher education in Europe to the development of changing societies. - 179 p., Bucharest (CEPES).

WORLD COMMISSION ON ENVIRONMENT AND DEVELOPMENT (1987): Our common future. - 400 p., Oxford (Univ. Press).

The IUGS Commission on Geosciences for Environmental Planning

FREDRIK CHR. WOLFF *

Abstract

The need for an international umbrella organization for the activities within the field of environmental geology has become evident. The International Union of Geological Sciences has proved to be the most suitable organization to host the activities carried out world-wide by geoscientists interested in geological aid to environmental planning. The present paper reviews the main international activities in this field, the preparation for a new IUGS Commission on Environmental Geology, the scope of its activities and the structure of its organization.

Introduction

The global strain on the natural environment is increasing with the rapidly growing world population and development of human technology. Long range planning has thus become essential (BRUNDTLAND 1987).

* Author's address: Dr. F. CHR. WOLFF, Geological Survey of Norway, POB. 3006, N 7002 Trondheim, Norway.

In planning, earth science information and advice have, however, commonly been neglected. Even through all our sustenance is derived from the top 5 km of the Earth. This has resulted in unforseen tragedies caused by natural disasters and economic failures, despite the efforts in many countries to make planners aware of the necessity and value of the utilization of geological information and expertise in environmental planning. To bridge the activities that are being carried out by earth scientists in this field in different parts of the world and to awake the awareness of geoscientific aid to planners, the need for an umbrella organization on an internaitonal level has become evident.

Why an IUGS commission?

The International Union of Geological Sciences (IUGS) is a voluntary professional organization, it is non-governmental, non-political and non-profit-making. This organization aims first of all to

1. encourage the study of geoscientific problems of world-wide significance,
2. facilitate international and interdisciplinary cooperation in geology and its related sciences,
3. and provide continuity to such cooperation IUGS fosters dialogue and communication among various specialists in earth sciences around the world and embraces topics from fundamental research to its economic and industrial applications, from scientific, environmental and social issues to educational and development problems (LAFFERTY 1987).

The IUGS has several standing Commissions dedicated to studying a particular geological field. They vary in size, though each is composed of a suitable number of geographically representative experts and are often subdivided into Subcommissions, Regional Committees or working groups according to the specific tasks charged to them.

On this background it was decided by a group of colleagues gathered at a symposium for Environmental Geology during the EUG-meeting in Strasbourg April 1989 to follow up the initiative of the present writer to approach the IUGS with a proposal for the creation of a new commission for Environmental Geology.

Earlier activities

To meet the demands for geoscientific information and advice much activity has been achieved both on an international and on national level over the last ten to fifteen years. Of the more significant achievements should primarily be mentioned the activities of the Commission of the Geological Map of the World (CGMW) - Sub-Commission for Maps on Environmental Geology (SC-MEG).

This sub-commission was created in 1974 to establish legends and common standards for maps of the natural environments potential within the field of geosciences.

The commission has organized a number of meetings -- Hannover / FRG 1976, Lomé / Togo 1980, Berlin / Germany 1985, Trondheim / Norway 1986 and Strasbourg / France 1989 -- where problems related to environmental geology have been discussed. Proceedings of the papers presented at this meetings have also been published (ARNDT & LÜTTIG 1987, WOLFF 1987). Compilation of maps at small scales to cover international regions and to serve as models for such compilations has been started in north western part of Central Europe and in Northern Scandinavia. The next step in the activity of the commission will be to circulate a suggestion to a legend to maps for the geological environments potential to collect world wide reaction to its ideas.

The activity of the CGMW / SC-MEG has over the years gathered a large number of colleagues world wide. These colleagues all active in different fields of environmental geology, have, however, a wider scope for their work than to produce small scale international maps.

The CGMW and its subcommissions having as its prime goal to produce international maps at small scales is consequently not suitable for the aims and prospectives of this very active international group of colleagues.

Proposal to IUGS for a new Commission for Environmental Geology

It has therefore become evident that the creation of a Commission under IUGS would be more appropriate for the activity of this group. A proposal for such a commission was consequently sent to IUGS in July 1988 and a positive answer arrived in February 1989. During the EUG V Symposium on Geology for Environmental Planning in Strasbourg in March the same year, a number of key people from European countries gathered to discuss further details for our next move towards the creation of an IUGS commission.

A new proposal was sent to IUGS in April urging its General Secretary to put in on the agenda of the forthcoming IUGS Bureau meeting in May in order to be ready for discussion on the subsequent executive committee meeting and an eventual final treatment at the IUGS Council meeting during the International Geological Congress in Washington in July 1989. A positive answer of 14th June from the IUGS General Secretary stated that the proposal for a Commission on Environmental Geology fits well in the proposed activities related to natural hazards and other environmental issues considered by IUGS. The Bureau also recommended that a Steering Committee be set up to evaluate the scope of activities of an IUGS Commission on Environmental Geology and try to work out a sensible work programme for these activities. A report on the programme should be presented to the next IUGS Executive Committee meeting in Sao Paulo, Jan 1990, for further actions.

At the 28th International Geological Congress in Washington in July 1989 a meeting of the Steering Committee chaired by the present writer and attented by a number of interested parties including the IUGS President, U. CORDANI, was organized.

Report of the Steering Committee meeting in Washington

In his speech at the opening session of the 28th International Geological Congress in Washington the IUGS President, U. CORDANI underlined the importance of geoscientific aid to environmental planning which he considered to be the most important role of the geoscientist in the next decades. Professor CORDANI said that: *"Damage to the environment, and the many related problems, are now a major world-wide concern. The challenge cut across the divides of national jurisdiction. Political decisions on the management of resources and land-use planning are crucial.*

Sustainable development will give rise to an unprecedented demand for information, advice and technologies that only an integrated approach can satisfy. In many countries the focus of the challenge ahead us shifting from protection and restoration to planning and prevention, as the possible solutions to environmental issues become more and more complex, and dependent on the cooperation of a multitude of sectors but first and foremost, that of science.

Soil degradation, desertification, deforestation, acid rain, water pollution as a result of excessive use of fertilizers and pesticides, major technological hazards in nuclear plants, or large chemical plants, mining wastes, sea water pollution from wastes and oil spills - all demonstrates that our planet is in an environmental crisis. Most of these problems are related to the growing human population, which is starting to reach saturation. Human activities now involve an annual flux of earth materials equal to that of plate tectonics. Moreover, in addition to local or regional environmental problems, there is growing concern about possible changes in the whole global ecosystem, which may have long-term effect on mankind. And finally there are the catastrophic natural hazards, earthquakes, volcanic eruptions, typhoons, floods, landslides and so on, which so frequently cause death and destruction." The IUGS President concluded by stating that: *"The geosciences have a fundamental role to play in addressing all these problems and that we as geoscientists hold a great responsibility for future generations. We must keep faith in our capacity and stay organized to accomplish our goals. We must make use of the tremendous potential of international collaboration."*

At the closing session he returned to his issue in saying that: *"We also can feel the growing interest of all us geologists toward the issues that are now the main concerns of society the environment and natural hazards."*

President CORDANI'S words created an excellent climate for the Steering Committee Meeting that was held on Thursday 13 July 1989 at the Grand Hyatt Hotel from 11.00 to 12.30 hours. Present at this meeting were:

UMBERTO CORDANI (IUGS President), Brazil; FREDRIK CHR. WOLFF (Chairman), Norway; ED DE MULDER, Netherlands; BRIAN KELK, UK; MARTIN CULSHAW, UK; HENK SCHLAKE, Netherlands; G.S. VARTANYAN, USSR; BOB HAGEMANN, Netherlands, and ERIK STENESTAD, Denmark.

After a brief introduction by the Chairman (WOLFF), the meeting was opened by the IUGS President (CORDANI). He said that with reference to his introductory speech to the 28th IGC, environmental geology was becoming a very important issue. Therefore, the IUGS was positively considering the creation of a Commission on the subject. The first opportunity to do this would arise at the next IUGS Executive Committee meeting in Moscow in early 1990. In the meantime, it was necessary to establish terms of reference for the Commission. Consequently the Steering Committee under the chairmanship of WOLFF had been formed to carry out this task.

The chairman then made a number of points:

- The solution of many environmental problems requires input from geologists.
- Many geologists, world-wide, are already working in the fields of hydrogeology, waste disposal and management, land-use and urban planning. This has been demonstrated by e.g. the SC-MEG Symposium held in Trondheim, Norway in 1986.
- There are two important factors relating the environment, geology and society. These are:
  - availability of resources, underlining those for energy,
  - the natural ability to digest the disposal of waste.
- The title of the Commission might be: "Commission on Geology for Environmental Planning".

The objectives and aims of the proposed commission might include the following:

- To promote the utilization of geoscientific data and expertise in environmental planning.
- To sponsor and encourage such activity through seminars, workshops and the publication of proceedings and manuals.
- To promote understanding of the use of geodata in planning both nationally and internationally.
- To review the state-of-the-art in environmental geology and identify fields of research and specific research projects for the future.
- A cooperation with the Subcommission of Maps on Environmental Geology (SC-MEG) of the CGMW to support the relevant activities of standardization of scales, symbols etc. for environmental geology maps.
- More specifically, some of the fields to be covered by the Commission might include:
  - Hazards assessment
  - Environmental management
  - Geochemistry in relation to the environment

- Coastal management
- Geology in favour of land-use planning and urban geology,
- data processing in environmental management,
- postgraduate and in-service training.

- A number of activities have already been undertaken in the recent past in relation to environmental geology (including as well industrialized as developing countries) via CGMW-SC-MEG in Hannover (1975), Togo (1980), Berlin (1985), Trondheim (1986); and via ESCAP in terms of meeting on Urban Geology in SE Asia (leading to the Landplan series of publications).

- The meeting of UNESCO Working Group "Geology and Land-Use Planning", CGMW-SC-MEG and INQUA (AQR) on which this volume is reporting, was held in Santander, Spain to discuss relevant topics.

A discussion then followed in which the following points were made:

- CULSHAW suggested that hazard assessment and mitigation should be added to the points made by WOLFF. He pointed out that one should rather think in broader terms than in terms of specific factors and suggested the concepts of *"resources for, and constrains on development"* were appropriate and included all necessary factors.

- It was agreed that AGID should be informed of what is proposed. DE MULDER agreed to contact the secretary/treasurer of AGID (RAU), and CULSHAW would speak to the AGID editor (STOW) to inform AGID and establish links.

- The IUGS President then explained that the ICSU had called for project proposals to be presented to its next meeting in early September 1989.

After discussion two projects were agreed:

- Training in environmental geology for developing countries.
  This was suggested by DE MULDER who agreed to draw up a provisional programme and proposal in collaboration with the Chairman and to submit it to the IUGS President before the end of the IGC.

- Preparation of a geo-socio-economic catalogue of information on environmental geology problems. This was suggested by KELK who agreed to prepare a draft to proposal and fax it to the Chairman for further completion before the end of August 1990.

DE MULDER agreed to expand a list of possible Commission members already prepared by WOLFF. The Chairman stated that it was unfortunate that more people -- particularly from developing countries -- had not been able to attend the meeting but that he was confident that direct contact through the list of names should result in better representation in the future.

Proposals to ICSU for activities related to developing countries

The International Council of Scientific Unions (ICSU) is an international non-governmental scientific organization composed of 20 international Scientific Unions like the IUGS. Since its creation in 1931 ICSU has adopted a policy of non-discrimination, affirming the rights of all scientists throughout the world -- without regard to race, religious, political or philosophical conviction, citizenship, sex or language -- to join in international activities. The Council's principal objectives is to encourage international scientific activity for the benefit of mankind. It does this in a number of ways e.g. by initiating, designing and coordinating international interdisciplinary research programmes such as

- The Internatinoal Geophysical Year 1957-1958,
- The International Biological Programme 1964-1974,
- The Upper Mantle Project 1961-1970, and
- The Global Change Programme, launched in 1986.

ICSU initiates also special studies such as those on radioactive waste disposal and on the disposal of toxic wastes and that carried out by SCOPE on the environmental consequences of nuclear war (ENUWAR).

For programmes in multi- or transdisciplinary fields which are not completely under the aegis of one of the Scientific Unions, such as Antarctic, oceanic, space and water research, problems of the environment, genetic experimentation, solar - terrestrial physics and biotechnology, ICSU creates scientific committees or commissions. Activities in areas common to all the Unions, such as teaching of science, data and science and technology in developing countries, are also coordinated by ICSU Committees.

On the request of the IUGS President two proposals were prepared:

1. A proposal for an educational programme on environmental geology for developing countries, including both a programme for short term (1990-1991) and one for long term (1992-1996) activities. This proposal was prepared by ED DE MULDER and F. Chr. WOLFF and submitted to IUGS on the 18 July, 1990.

2. A proposal for the establishment of a data base of distribution and frequency of different types of environmental geological problems based on a circular to all national geological surveys and other appropriate national or regional contact, e.g. ESCAP etc. to register what types of environmental problems exist in each region of the world, for example, which need urgent study, which types and in which areas are such problems ignored or not appreciated by administrators and politicians, which local geological conditions could give rise to human problems due either to ill-constrained developments or by regional or global change, which are time bombs.

Objectives and aims

Some of the main objectives and aims of the Commission on Geology for Environmental Planning have briefly been mentioned in the minutes of the meeting of the Steering Committee in Washington DC. A further discussion will have to be carried out at the Santander meeting.

A draft for the topics to be discussed is given in the following:

1. How to promote both locally and internationally the utilization of geoscientific data and advice in environmental planning and ways to stimulate geoscientific activities in the service of environmental planning.

    This should be achieved by:

    -   Organizing international seminars and training courses. Some has already been successfully organized by the CGMW/SC-MEG, and others are planned within the framework of ICSU as proposed above. The objectives and contents of such training activities as well as the structure of its organization must be discussed in detail.

    -   Publishing proceedings and manuals. The steering Committee has nominated as one of its members the former editor of EPISODES, A.R. BERGER, Canada. We will try to make him take on the job to be the chief editor of our commission. I will further recommend that we discuss the election of a group of our members to work with Dr. BERGER to develop this field.

    -   We also will have to discuss how to promote the understanding of the necessity for the utilization of geoscientific information and expertise in environmental planning for national and local authorities. This calls for the set up of a special group of our members both to give general advice and to work on the local level. Here G.H. BRUNDTLAND'S words: *"think globally, work locally"* come into play.

2. To review the-state-of-art and suggest main lines and topics of research. This is a field that will need continuous updating, but a firm basis must be worked out at the Santander meeting. An interesting prerequisite in this discussion will be the results of programme for a data base of distribution and frequency of different types of environmental geological problems, proposed to ICSU through IUGS.

·3.   To establish agreements about types of environmental maps, scales, procedures, criteria and symbols for their preparation. This agreement has to be worked out in conjunction with the CGMW/SC-MEG (G. LÜTTIG) and other regional groups like the Working Club of Western European Geological Surveys (WEGS), Synthesis Project (E. STENESTAD) work on minimum standards.

Terms of reference

In a letter from the IUGS President (21 Aug 89) the steering committee was asked *"to go ahead with the preparation of suggestions for terms of reference for a new Commission on Environmental Geology and to formulate working activities and concrete projects. These suggestions should be submitted to the IUGS secretariat before mid november this year in order to be discussed at the next Executive Committee meeting in Moscow early next year".*

The IUGS standard form for annual reports of commissions has the following points which will be necessary to take into consideration when formulating the terms of reference:

1. Title of constituent body
2. Overall objectives
3. Fit within IUGS science policy
4. Organization
5. Extent of national/regional/global support
6. Interface with other international projects

A final completion of these points cannot be done at this stage, but to create a base for the discussion at the Santander meeting the following suggestions have been made:

1. Title of constituent body

COGEN Commission on Geology for Environmental Planning

2. Overall objectives

This point has been discussed to some extent in the chapter called "Objectives and aims" and can only be completed after a thorough discussion.

3. Fit within the IUGS policy

As pointed out by the IUGS President geoscientific aid to environmental planning must be considered to be the most important role of the geoscientist in the next decades. Moreover the aims of IUGS is first of all to encourage studies of world wide significance and facilitate international cooperation in geology and its related sciences.

4. Organization

The Commission will need a chairman, a vice-chairman and a secretary/treasurer (or both) to conduct the overall responsibility for planning and budgeting and contact with IUGS. We should also discuss the creation of working groups and regional representatives. Working groups could be established within for example for following fields:

- Training and education. This is important both in developing countries, where education in mineral exploitation has dominated, and in industrialized countries, where advice to established educational institutions is most demanded. The proposal to ICSU of 18 July will, if accepted, take care of educational programmes for the Third World. Suitable activities to aid educational systems in industrialized countries must also be developed within the framework of this working group.

- The need for written materials (textbooks etc.) in environmental geology has proved to be substancial and the writer has been approached by international publishers for compilation of such material. Editing of textbooks, proceedings etc. seems therefore to be one of the most important tasks for our Commission and a working group should therefore be established on this topic. As mentioned in an earlier paragraph the Commission has members who are experienced in this field.

- Hazard assessment is also of utmost importance and a special working group should be set up to establish the necessary basis for further objectives and actions. This group will also have to make an inventory of already ongoing activities within related organizations like IUGS Steering Committee on Global Change (KEN HSÜ) etc.

- Environmental management is also a topic worth while attention by a special group. Some of our members have experience in this field and should be nominated for this task.

- Geochemistry related to the environment embraces a wide field of great importance. Different projects are active in many parts of the world and our prime task must be to coordinate these activities and to study how an international structure of such activities may be organized.

- Coastal management. The strain on the nature in coastal areas increases every year as a result of long series of human activities (urbanization, tourism, transportation etc.). Events of catastrophic nature are reported almost every day and planning becomes increasingly necessary. These problems can only be tackled by a profound understanding of the geoscientific relations and processes. A working group is therefore most necessary to create activity in the field of coastal management.

- Land-use planning and urban geology. These two fields are both quite extensive and a number of our members have many years of experience in these fields (LÜTTIG; MATTIG, this volume). A thorough discussion will be necessary to see how these activities best can be organized, which colleagues should be elected to take responsibility and which already active, related organizations should be involved.

- Data processing in environmental geology. This activity is already organized as Working-Group 5 within COGEODATA (the IUGS Commission on Storage, Automatic Processing, and Retrieval of Geologic Data). Ties should be established with this group. Our group will have to explore the needs for data-technological support. To solve these problems we will have to consult the expert group under COGEODATA. Other possible activities may also be suggested by the members present at the Santander meeting.

5. Extent of local -- regional -- international support

One of the most important tasks of the first year for the Commission will be to explore what economical support can be found in addition to the seeding money from IUGS. This task will not rest only with the chairman and the treasurer but most certainly with all members of the Commission who will have to brush up all their contacts at all levels both locally and world-wide.

6. Interface with other internatinal organizations

In addition to the organization already mentioned:

CGMW Sub-Commission for Maps on Environmental Geology (SC-MEG)
The UNESCO, Earth Science Division
INQUA, Commission on Applied Quaternary Research (AQR)
AGID (Association of Geoscientists for International Development)
ESCAP (Economic and Social Committee Asian Planning)
IAEG (International Association of Engineering Geology)
IAH (International Association of Hydrogeologists).

The Commission will have to make an inventory of all international organizations that might be relevant for our activities or that may make use of our activities.

Conclusion

A steering Committee has been set up and a new IUGS Commission on Geology for Environmental Planning is likely to be established at the next IUGS Executive Committee meeting early 1990.

Terms of reference have been discussed, agreed and compiled during the SANTANDER meeting and sent to IUGS by mid November 1989.

Commission officers and working group chairmen should also be elected at the SANTANDER meeting.

Working groups and their objectives should also be discussed as well as possible sources for funding for the future work of our Commission.

References

ARNDT, P. & LÜTTIG, G.W. (1987): Mineral resources extraction, environmental protection and land-use planning in industrialized and developing countries.- 337 pp., 108 fig., 24 tabl., 2 folders, Stuttgart (Schweizerbart).

BRUNDTLAND, G.H. et al. (1987): Our common future, World Commission on Environment and Development.- Oxford (University Press).

LAFFERTY, V. (1987): What is IUGS?.- Washington (U.S.Government Printing Office).

MATTIG, U. (1991): Geoscientific Maps for Land-Use Planning.- This volume.

[WOLFF, F.C.] (1987): Geology For Environmental Planning.- Norg. geol. Unders. spec. Publ. **2**: 1-121, Trondheim.

Chapter 5:

Synthesis and Conclusions

Synthesis and Conclusions

ANTONIO CENDRERO [*]

The conclusions reached during the workshop have been summarized and grouped into the following categories: Research, Development, Communication, Training, Institutional (legal, management, organization). Naturally, it is difficult to make a strict separation between those categories, and some of the conclusions presented could be included under more than one heading. The main points are presented below.

Research

The following research topics were considered to be of interest:

A)   Resources

--   Fundamental research on Quaternary geology, mainly in connection with the possible influence of resource exploitation on present surficial processes.

--   Improved mineral resource extraction techniques, increasing efficiency and reducing side effects.

[*] Author's address: Prof. A. CENDRERO, DCITTYM, Division of Earth Sciences, University of Cantabria, Santander, Spain.

-- More efficient utilization of mineral resources, including substitution and recycling, in order to reduce the volumes extracted.

-- Effects of the different steps of the exploration-extraction-transport-transformation-use-disposal chain on surficial processes, biogeochemical cycles, flora and fauna. The study of the geochemical effects and the quantification of the influence on the erosion-sedimentation processes are considered to be of particular importance.

-- Improvement of restoration techniques in different climatic / geomorphological conditions and for different types of materials.

-- Quantitative indicators for the evaluation of geological and geomorphological resources of scientific and cultural interest should be developed.

B) Processes

-- Monitoring of river catchments and systematic geochemical mapping have proved to be extremely useful for understanding the factors influencing biogeochemical cycles and their relationship with human health. This, in turn, enables the establishment of treatments for the protection of the population affected by excesses or deficiences in some elements.

-- International interest on global change is too biased towards the climatological-biological aspects. However, the sedimentary and geomorphological record during the Quaternary, ice cores, etc. can furnish extremely useful information about past changes, their time framework and their environmental consequences.

-- Studies ought to be carried out comparing the effects of human influence on surficial processes of environmental significance (erosion, landslides, floods) in different parts of the world.

C)   Hazards

--   Research is needed to improve existing capabilities for identifying unstable slopes and for predicting landslide behaviour, especially with regards to large landslides. The use of SPOT images offers good prospects for monitoring the evolution of individual unstable masses. There is a need to develop methods for the prediction of volume, reach and type of movement with regards to potential landslides. Efforts should be made to date past movements in big landslides, in order to determine their period of recurrence and their relationships with climatic, seismic or anthropogenic events.

--   Research should be carried out on the identification and quantification of indicators for geological hazards.

--   Research is needed to develop means to express vulnerability of human components in relation to hazards using objective, quantitative indicators.

D)   Impacts

--   Better predictive models, either empirical or numerical, must be developed in order to improve the assessment of impacts and hazards and also to explain issues to non-specialists.

--   The improvement of impact-prediction capabilities can be approached through follow-up and monitoring studies of existing actions, and checking-up of predictions made in E.I.A. studies.

--   Cost-benefit analyses of applied earth science results in support of such activities must be made, through the study and evaluation of completed studies and the follow-up of later events and actions. The design of follow-up studies about the evolution/use of land resources in relation to given geoscientific advice would be of special interest.

-- It would be of interest to undertake studies for the comparison of land-use planning problems in different regions of the world and design of models of geoscientific maps to help in their mitigation/management.

E)   Mapping methods and map preparation

-- Existing maps/systems on the distribution, economic significance and reserves of the different mineral deposits should be transferred into systems better understandable to planners and decision makers.

-- Considerable efforts must be made to improve their layout, contents and usefulness for the "customer". Cost-benefit analyses of maps already made and of proposals contained in new maps for planning must be made.

-- Maps should be made showing the interactions between exploitations and the human environment, with special regard to natural and human-induced hazards.

-- Efforts should be concentrated on areas of specific interest (because of their present state or because of their production potential) and on materials with the highest incidence, like aggregates. The work ought to provide answers about the results to be expected from better management and about new possibilities that could be developed.

-- A "zoom" system of geoenvironmental mapping should be developed, so that it can be applied to the harmonious linking of plans at different levels of the planning-management process and at different scales.

-- It is necessary to develop new types of "derived maps" using computer techniques and new methods of evaluation, including economic aspects. Towards this end, it would be useful to select a zone well studied and mapped, and build from it adding economic aspects, etc., and using the maps to depict proposed solutions to existing problems.

-- Geomorphological studies for environmental impact assessment should focus on the identification of geomorphological resources and hazards and their representation by means of maps. The possible effects of a project on resources and hazards should be shown by means of impact and induced hazard maps. The combination of pre-existing and induced hazards with the vulnerability of human structures should be used to obtain risk maps.

Development

The following actions are suggested:

-- Geoscientific advisory work should be carried out step by step; first indicating problem areas; second, focusing on high impact/high risk areas and conducting detailes studies; next, developing predictive models and - if possible - adding the time and money factors to them.

-- Earth scientists should translate risks in terms of money as much as possible, despite the obvious limitations.

-- Inventorying -- in different regions/countries -- of possible sources of low unitary value resources, assessment of their quality and reserves and also of the possible impact of their exploitation.

-- Inventories of existing environmental problems related to active and abandoned exploitation and determination of possible corrective actions.

-- Inventories of general geoenvironmental problems, especially in developing countries, and of existing needs in order to solve them.

-- Many important geological sites have been irreversibly damaged in recent years and many others are in danger. This represents the loss of very valuable cultural, educational and scientific resources, as important information about past earth processes has disappeared. It is urgent to set-up national and international programmes - like the ones already under way in different countries - for the inventory, legal and physical protection and public utilization of important geological and geomorphological sites. Initiatives like the creation of the European Working Group on Earth Science Conservation should be promoted and supported by international agencies, such as UNEP.

-- Detailed urban geological mapping is a necessity for the sound development of cities. This task can be best carried out through the supervision, advice and control of boards to be created in the relevant municipal areas, so that interdisciplinary team-work can be achieved and actual important issues be addressed.

-- Mapping should be preferably follow a lithostratigraphic-geotechnical approach and it should be continuously updated. This can be carried out best through the use of computer data bases and GIS.

-- The development of standard criteria, procedures and legends for geoenvironmental maps is urgent. The establishment of common mapping standards should be carried out mainly in cooperation with the Sub-commission on Maps of Environmental Geology (CGMW).

-- Maps should be of the printed type for education and general information. Thematic computer maps (ad hoc) for planners.

-- New types of thematic maps including socio-economic aspects should be developed.

-- Standardization of data and procedures on environmental geology is highly desirable. Data should be digital, quality information.

-- A catalogue of data-sets and map sets (low-medium-high criteria) for common type problems should be developed, possibly by the SC-MEG and spread by COGEOEN-VIRONMENT. It should provide standard responses for standard problems.

Communication

-- Ignorance about essential geoscience aspects can cost dearly in environmental and economic terms. Communication with the public and authorities is essential to improve awareness about the utility of geoscientific advice.

-- Geoscientists must develop a dialogue with the public and with decision-makers and politicians. They should appeal to the most basic political/public issues: money and public safety.

-- Public discussion and explanation of geological findings in terms that the average intelligent citizen can understand should be pushed, so that the public becomes aware of the issues and can press decision-makers in the direction of better environmental management.

-- Efforts should be made to communicate with politicians, administrators, engineers, businessmen, and to persuade them that earth science advice can help them to save/make money or to solve problems of interest for their constituency.

-- The adaption of geological reports to the actual problems perceived by politicians and decision-makers, so that solutions can be proposed, is the means to convince them of the utility of geological advice.

-- In the preparation of any written materials, an executive summary should be prepared, concise, clear and simple, addressing the essential issues.

-- Communication efforts should be concentrated on topics that bring about strong public reactions, such as safe water supply, public safety, saving of public money, etc.

-- Case histories are perhaps the best way to illustrate principles. Clear and spectacular cases/problems should be selected for public presentation and discussion.

-- Although environmental legislation is important, its existence is certainly not sufficient to ensure environmental protection. Efforts must be made to increase public awareness.

-- Public reaction to the national plan for natural disasters prevention and mitigation in Colombia has shown that an adequate integration of social and geodynamic aspects, together with efforts in education and information to the public, are essential to obtain good results. Public awareness and acceptance of the plans can thus be increased very significantly.

-- Communication is extremely important for the application of earth science maps in the planning process. Earth scientists must address the right planning issues and provide answers for actual planners' problems. The general public must also be informed about the utility of earth science maps and expertise for solving environmental problems at the planning stage.

-- Geoscientists should become aware of the great political and economic impact of their reports - both in a positive and in a negative sense - on society.

-- Communication with the general public, authorities and other professionals can be improved, among other things, by means of adequate publications. Efforts should be devoted to the translation and transfer of geoscientific knowledge to non-specialists. The following actions are suggested:

* A newsletter like the one issued by AGID, with a similar method of funding.

* Development of a "minimum" list of items for land-use planning to include into any manual/leaflet/etc. They should be presented in a way that shows what can happen (in monetary value) if those things are not considered.

* A booklet or similar publication with advice on environmental earth science. It should be based on case studies and be clear and concise, to the point, in plain language, cheap and easily available. It is likely that industry sponsorship could be obtained for such a publication, and/or the interest of a commercial publisher.

* Organization of a Central Information Centre to gather literature, existing courses, information on actions/programmes, etc., setting-up a database and making the dissemination of information easier.

Training

-- Traditional geoscientists are not fully adapted to contemporary demands. Curriculum reform is necessary to provide an inroad of the geologist into planning.

-- Concrete proposals are needed for training programmes geared to the formation of a new type of geoscientists, so that they can play a greater role as environmental advisors or management experts.

-- Earth scientists working on environmental matters should have some training in economics and legislation, and they should learn to speak the common language of administrators and the public.

-- Planners, architects, engineers, and environmental scientists in general often do not have an adequate knowledge of geological concepts and facts of great relevance for planning and environmental management. Some training actions and publications should be designed specifically for these types of professionals.

-- The establishment of courses along the lines described is considered a necessary task. However, overlapping with somewhat similar courses, like the one organized in Yalta (USSR) by UNESCO, and perhaps others should be avoided. Courses should be accessible to participants from both the industrialized and developing countries.

-- Although focussing on the problems of developing countries is considered a priority, it is crucial not to forget parallel actions for industrialized countries. These could be the "locomotive" that will prepare further steps to the benefit of developing countries.

-- Training actions should be mainly at the "post-graduate" level. However, the problem of an appropriate definition of "post-graduate" level ought to be considered, taking into account the diversity of educational systems in the world.

-- To avoid the logistical problems of moving such courses from one place to another, the establishment of a permanent basis for "long" courses and the organization of short, "intensive" courses in other locations should be considered.

-- Financial support for these type of courses could come from international bodies, national agencies for international cooperation, and participants and their home institutions.

-- Manuals should be prepared as a guidance for the preparation of applied earth science studies for planning, development, environmental protection and conservation. These manuals should be of two types: one for earth scientists and a different one for planners.

-- Manuals should have a "core" of general, fundamental topics and an "appendix" to be added in each country about issues relevant there. The core should contain essential, common themes, not linked to national, and it should also have a chapter explaining how "unwise" actions lead to costs, but recognizing at the same time that most land-uses also produce benefits. It would be highly desirable to include in the mauals a series of "sample studies" showing the coupling of geologic and non-geologic analyses.

Institutional - Organizational

-- A commission should be created, within the framework of the international Union of Geological Sciences, to promote research, development, training and other actions in the field of earth science and environmental planning (see article of F.W. WOLFF, this volume).

  * The commission should concentrate its efforts, initially, on a few items, such as: increasing public awareness; providing appropriate training schemes; improving techniques of environmental geological representation and assessment.

* The establishment of common mapping standards should be promoted, mainly in cooperation with the Sub-Commission on Maps of Environmental Geology (CGMW) and other groups working on standardization, to avoid duplication of work. User contacts should be focussed on.

* Cooperation links with sister organizations such as IAEG, IAH and the like should be established, in order to avoid possible boundary problems.

* The commission should also contribute to internal communication among earth scientists, in matters related with environmental problems.

* The conduction of research, with emphasis on integrated analyses of problems resulting in outputs for more than one field should be promoted. Especially significant research items/cases should be selected.

* Planners should be incorporated into future activities, either in training, communication, research or development.

* Information should be collected on other organizations with similar interests and on their activities.

-- Although legislation on geoenvironmental matters exists in many countries, it is important that it should be extended to those countries where it has not been passed.

-- Actions like the ones contemplated within the U.N. Decade for the Reduction of Natural Disasters or the programmes on Global Change, should be undertaken, both at the national and international level. Especially important would be to undertake actions of a preventive nature, by correctly planning the use of the earth's surface.

-- Efforts should be made by international, national, regional and local administrations to undertake the restoration of older exploitations not affected by recent legislation.

-- International organizations (UNEP, UNESCO, UNDP, etc.) should participate in these tasks, especially concerning the support to developing countries, both with regards to financial and technological aspects.

Lecture Notes in Earth Sciences